Implications of Molecular and Materials Structure for New Technologies

NATO Science Series

A Series presenting the results of activities sponsored by the NATO Science Committee. The Series is published by IOS Press and Kluwer Academic Publishers, in conjunction with the NATO Scientific Affairs Division.

A. **Life Sciences**	IOS Press
B. **Physics**	Kluwer Academic Publishers
C. **Mathematical and Physical Sciences**	Kluwer Academic Publishers
D. **Behavioural and Social Sciences**	Kluwer Academic Publishers
E. **Applied Sciences**	Kluwer Academic Publishers
F. **Computer and Systems Sciences**	IOS Press

1. **Disarmament Technologies**	Kluwer Academic Publishers
2. **Environmental Security**	Kluwer Academic Publishers
3. **High Technology**	Kluwer Academic Publishers
4. **Science and Technology Policy**	IOS Press
5. **Computer Networking**	IOS Press

NATO-PCO-DATA BASE

The NATO Science Series continues the series of books published formerly in the NATO ASI Series. An electronic index to the NATO ASI Series provides full bibliographical references (with keywords and/or abstracts) to more than 50000 contributions from internatonal scientists published in all sections of the NATO ASI Series.
Access to the NATO-PCO-DATA BASE is possible via CD-ROM "NATO-PCO-DATA BASE" with user-friendly retrieval software in English, French and German (WTV GmbH and DATAWARE Technologies Inc. 1989).

The CD-ROM of the NATO ASI Series can be ordered from: PCO, Overijse, Belgium

Series E: Applied Sciences – Vol. 360

Implications of Molecular and Materials Structure for New Technologies

edited by

Judith A. K. Howard

Department of Chemistry,
University of Durham,
Durham, U.K.

Frank H. Allen

Cambridge Crystallographic Data Centre,
Cambridge, U.K.

Gregory P. Shields

Cambridge Crystallographic Data Centre,
Cambridge, U.K.

Kluwer Academic Publishers

Dordrecht / Boston / London

Published in cooperation with NATO Scientific Affairs Division

Proceedings of the NATO Advanced Study Institute on
Implications of Molecular and Materials Structure for New Technologies
Erice, Sicily
28 May – 7 June 1998

A C.I.P. Catalogue record for this book is available from the Library of Congress.

ISBN 0-7923-5816-3 (HB)
ISBN 0-7923-5817-1 (PB)

Published by Kluwer Academic Publishers,
P.O. Box 17, 3300 AA Dordrecht, The Netherlands.

Sold and distributed in North, Central and South America
by Kluwer Academic Publishers,
101 Philip Drive, Norwell, MA 02061, U S.A.

In all other countries, sold and distributed
by Kluwer Academic Publishers,
P.O. Box 322, 3300 AH Dordrecht, The Netherlands.

Printed on acid-free paper

Printed in the Netherlands.

TABLE OF CONTENTS

vi

PREFACE

This volume contains papers presented at the 27th Course of the International School of Crystallography held at the Ettore Majorana Centre for Scientific Culture, Erice, Sicily from 28th May to 7th June 1998. The Course on the *Implications of Molecular and Materials Structure for New Technologies* was designated a NATO Advanced Study Institute and the Directors of the Course acknowledge with thanks the considerable financial support provided by the Scientific Affairs Division of NATO.

The programme for the 27th Course developed from two earlier Courses on the *Static, Dynamic and Kinematic Aspects of Crystal Structures* held at Erice in 1985 and 1991. These courses recognised the wider value of the chemical and crystallographic knowledge that is obtained from each and every crystal structure determination. For the first time they dealt with the systematic analysis of the huge reservoir of such knowledge that was available, in order to address research problems in structural chemistry. Because of their broad scientific appeal, these Courses also forged links between crystallographers and other scientists who were interested in molecular and supramolecular structure: spectroscopists, theoreticians and molecular modellers.

New subject areas mature and develop over time, and this is reflected in the title and the programme for the 27th Course. The decade of the 1990s has seen a dramatic rise in the use of experimental structural knowledge to inform and improve the design of a very wide range of novel materials: pharmaceuticals, agrochemicals, molecular switches, catalysts and electro-optical materials. Today, materials design encompasses the complete chemical spectrum from proteins and viruses, through organic and metallo-organic compounds, to inorganics, minerals and metals. Crystallography is fully integrated into all of these areas, as the method of choice for experimental studies, and is used in concert with a wide variety of synthetic, computational and modelling methods. Further, materials design embraces and links academia and industry, and brings crystallography and crystallographers into close collaboration with scientists from many and varied disciplines.

It was not possible to address all of this diversity within a single Course, nor was it wise to arrange a programme that was too specialised. Instead, we tried to achieve scientific balance, a broad mutual understanding through lectures, computer tutorials and discussion, and an atmosphere that would catalyse future collaborations. Thus, the Course addressed, *inter alia*:

- Modern diffraction techniques, novel radiations and experimental conditions
- Databases and knowledge bases of experimental results
- Computational techniques, and their interplay with experimental information
- Hydrogen bonding and other intermolecular interactions
- Supramolecular assembly and crystal structure prediction
- Practical examples of materials design

To complement the expert lectures, the programme also comprised hands-on computer tutorials on five different topics, class discussions with lecturers on major scientific areas, and posters and short lectures contributed by the participants themselves.

Arrangement and delivery of a complex programme depends upon the efforts of many individuals. Firstly, and most of all, we thank the participants for their enthusiastic enjoyment of the Course, and for their excellent posters and oral contributions. The lecturers provided high quality notes, high quality talks, and have collaborated very effectively in the preparation of this volume. The Course Directors owe a special debt of gratitude to Greg Shields who has helped so much to bring this volume to press, and who is rightly included as one of its Editors.

We thank those who presented computer tutorials (P. Comba, S. Motherwell, G. Shields, F. Leusen, S. Wilke, H.-B. Buergi, S. Capelli, J. Hauser, C. Lecomte and V. Dahaoui), with special thanks to Silicon Graphics Inc., Zurich, for providing hardware and support, and to John Irwin of MRC, Cambridge, who looked after the computing infrastructure for the Course. We also thank Bruce Godfrey of the University of Cambridge Computing Service for his careful preparation of printed Course materials.

The Course also benefited from additional financial support and sponsorship from a number of organisations, institutions and companies: the Commission of the European Communities, the International Union of Crystallography, the British Council (Rome), the Italian National Research Council, the Cambridge Crystallographic Data Centre, and Molecular Simulations Inc.

The International Schools of Crystallography in Erice are made possible by the leadership and organisational skills of Lodovico Riva di Sanseverino of the University of Bologna, Paola Spadon of the University of Padua, and their local assistants. They were all magnificent in every respect. Finally, we thank Professor Sir Tom Blundell, Director of the International School, for the rewarding opportunity to organise the 27th Course.

Judith Howard

Frank Allen

LIST OF CONTRIBUTORS

1. Course Directors

Judith A.K. Howard
Department of Chemistry, University of Durham
South Road, Durham DH1 3LE, UK
j.a.k.howard@durham.ac.uk

Frank H. Allen,
Cambridge Crystallographic Data Centre
12 Union Road, Cambridge CB2 1EZ, UK
allen@ccdc.cam.ac.uk

2. Organisers

Paola Spadon
Dip. di Chimica Organica, Universita di Padova
Via Marzolo 1, 35131 Padova, ITALY
paola@pdchor.chor.unipd.it

Lodovico Riva di Sanseverino
Dip. di Scienze Mineralogiche,Universita di Bologna
Piazza di Porto San Donato 1, 40126 Bologna, ITALY
t54bom12@icineca.cineca.it

John Irwin
European Bioinformatics Institute
Hinxton Hall, Hinxton, Cambridge, UK
jji@ebi.ac.uk

3. Other Contributors

Y. G. Andreev
School of Chemistry, University of St. Andrews
St. Andrews, Fife KY16 9ST, SCOTLAND

Professor Joel Bernstein
Department of Chemistry, Ben-Gurion University of the Negev
P.O. Box 653, Beer Sheva, ISRAEL 84105
yoel@bgumail.bgu.ac.il

Dr Elena Boldyreva
Institute of Solid State Chemistry, Russian Academy of Sciences
Kutateladze 18, Novosibirsk 128, 630128 RUSSIA
elena@xray.nsk.su

Professor Dario Braga
Dipartimento di Chimica 'G. Ciamician', Università di Bologna
Via F. Selmi 2, 40126 BOLOGNA, Italy
dbxray@ciam.unibo.it

Professor Lee Brammer
Department of Chemistry, University of Missouri-St. Louis
8001 Natural Bridge Road, St. Louis, MO 63121-4499, USA
lee.brammer@umsl.edu

Professor Carolyn P. Brock
Department of Chemistry, University of Kentucky
Lexington, KY 40506-0055, USA
cpbrock@ukcc.uky.edu

Professor Peter Bruce
School of Chemistry, University of St. Andrews
St. Andrews, Fife KY16 9ST, SCOTLAND
pgbl@st-andrews.ac.uk

Professor H.-B. Buergi
Lab. fur Chemische und Mineralogische Kristallographie, Universitat Bern
Freiestrasse 3, CH-3012 Bern, SWITZERLAND
hbuergi@krist.unibe.ch

Dr Sylvia Capelli
Lab. fur Ghemische und Mineralogische Kristallographie, Universitat Bern
Freiestrasse 3, CH-3012 Bern, SWITZERLAND
capelli@krist.unibe.ch

Dr Hansong Cheng
Principal Research Scientist, Air Products and Chemicals, Inc.
7201 Hamilton Boulevard, Allentown, PA 18195-1501,USA
chengh@apci.com

Professor Peter Comba
Anorganisch-Chemisches Institut, Universitat Heidelberg
Im Neuenheimer Feld 270, 69120 Heidelberg, GERMANY
comba@akcomba.oci.uni-heidelberg.de

Dr J.K. Cockcroft
Department of Crystallography, Birkbeck College
Malet Street, London WC1 7HX, UK
cockcroft@gordon.cryst.bbk.ac.uk

Professor R.E. Davis
Department of Chemistry, University of Texas at Austin
Austin, TX 78712-1167, USA
redavis@mail.utexas.edu

Professor G.R. Desiraju
School of Chemistry, University of Hyderabad
Hyderabad 500 046, INDIA
grdch@uohyd.ernet.in

Professor J.D. Dunitz
Organic Chemistry Laboratory, Swiss Federal Institute of Technology
Universitatstrasse 16, CH-8092 Zurich, SWITZERLAND
dunitz@org.chem.ethz.ch

Dr Gerhard E. Engel
Molecular Simulations
240/250 The Quorum, Barnwell Road, Cambridge CB5 8RE, UK

G. MacGlashan
School of Chemistry, University of St. Andrews
St. Andrews, Fife KY16 9ST, SCOTLAND

Professor Carlo Maria Gramaccioli
Dip. Scienze della Terra - sez. Mineralogia, Universita' degli Studi di Milano
Via Botticelli, 23, 20133 Milano, ITALY
carlo@r10.terra.unimi.it

Dr Fabrizia Grepioni
Dipartimento di Chimica 'G. Ciamician', Università di Bologna
Via F. Selmi 2, 40126 Bologna, ITALY
grepioni@ciam.unibo.it

Dr Juerg Hauser
Lab. fur Chemische und Mineralogische Kristallographie, Universitat Bern
Freiestrasse 3, CH-3012 Bern, SWITZERLAND

Professor Kersti Hermansson
Inorganic Chemistry, The Angstrom Laboratory, Uppsala University
PO Box 538, S-75121 Uppsala, SWEDEN
kersti@kemi.uu.se

Professor Roald Hoffmann
Department of Chemistry, Cornell University
Ithaca, NY 14853-1301, USA
rh34@cornell.edu

Professor Claude Lecomte
Lab. de cristallographie et modelisation des materiaux mineraux et biologiques
(LCM3B), URA CNRS 809, Faculte des Sciences, Universite de Nancy I
BP 239, 54506 Vandoeuvre-les-Nancy, FRANCE
lecomte@lmpci.u-nancy.fr

Dr F.J.J. Leusen
Molecular Simulations
240/250 The Quorum, Barnwell Road, Cambridge CB5 8RE, UK
fleusen@msicam.co.uk

Dr W.D. Sam Motherwell
CCDC, 12 Union Road, Cambridge CB2 1EZ, UK
motherwell@ccdc.cam.ac.uk

Dr J.J. Novoa
Dpto. de Quimica Fisica, Universitat de Barcelona
Av. Diagonal 647, 08028-Barcelona, SPAIN
novoa@zas.qf.ub.es

Dr Sarah L. Price
Department of Chemistry, University College London
20 Gordon Street, London WC1H 0AJ, UK
s.l.price@ucl.ac.uk

Dr Greg P. Shields
CCDC, 12 Union Road, Cambridge CB2 1EZ, UK
shields@ccdc.cam.ac.uk

Dr Thomas Steiner
Institüt für Kristallographie, Freie Universität Berlin,
Takustrasse 6, Berlin,D-14159, GERMANY
steiner@çhemie.fu-berlin.de

Dr C.C. Wilson
ISIS Facility, CLRC Rutherford Appleton Laboratory
Chilton, Didcot, Oxon OX11 0QX, UK
c.c.wilson@rl.ac.uk

Dr Paul Verwer
CAOS/CAMM Centre, University of Nijmegen
PO Box 9010, 6500 GL Nijmegen, THE NETHERLANDS

Dr Lai-Sheng Wang,
Department of Physics, Washington State University
Richland, WA 99352, USA.

K. A. Wheeler
Department of Chemistry, Delaware State University
Dover, DE 19901, USA

J. K. Whitesell
Department of Chemistry, University of Texas at Austin
Austin, TX 78712-1167, USA

Dr S. Wilke
Molecular Simulations
240/250 The Quorum, Barnwell Road, Cambridge CB5 8RE, UK
swilke@msicam.co.uk

INTO THE NEW MILLENNIUM: THE PRESENT AND FUTURE OF CRYSTAL STRUCTURE ANALYSIS

JACK D. DUNITZ
Organic Chemistry Laboratory,
Swiss Federal Institute of Technology,
ETH-Zentrum,
CH-8092 Zurich, Switzerland.

The title of my introduction may well promise more than it can deliver because I am notoriously bad at prognostication. Indeed, the only reason I can imagine that I was asked to introduce this 27th Course of the International School of Crystallography is on account of my advanced age. As one of the older participants here, perhaps the oldest for all I know, I have lived through more than half a century of X-ray crystallography, have experienced its development from then until now: then, a difficult, long drawn and highly specialized method of studying crystal and molecular structure, requiring deep and intimate knowledge of the fundamentals as well as sometimes inspired guesswork to derive the correct crystal structure of a compound, even a moderately simple one; now, a set of more or less routine, highly automatized procedures that require more the ability to manipulate computer programs than deep thinking or soaring imagination. But the future rests on the present, and the present on the past. I am not old enough to have been present in the truly pioneering period, but when I started, Max von Laue, Paul Peter Ewald, Lawrence Bragg were still very much alive, and their brilliant followers, John Desmond Bernal, Dorothy Hodgkin, Kathleen Lonsdale, J. Monteath Robertson in the U.K., Linus Pauling, Ralph Wyckoff in the U.S.A., Johannes Martin Bijvoet in the Netherlands, were in their prime. Max Perutz was busy with problems that most of his contemporaries regarded as insoluble; Francis Crick and Jim Watson had not yet been heard of. Who could have guessed that things would progress so far, that by the end of the century the structure analysis of medium to large organic molecules would have become more or less routine? Or even that structure analyses of many classes of proteins would become almost commonplace — one or two in each weekly issue of Nature or Science? Quite likely there were a few optimists who could look forward to such fantastic possibilities — as I recall, Bernal was one — but I, certainly, was not among them. For me, it has been a marvellous experience to follow these developments, and I am glad to have the opportunity to share some of my memories and impressions with you.

1

J.A.K. Howard et al. (eds.), Implications of Molecular and Materials Structure for New Technologies, 1–9.
© 1999 *Kluwer Academic Publishers. Printed in the Netherlands.*

1. How it was.

When I started my apprenticeship in Glasgow with J. Monteath Robertson, crystal structure analysis was based mainly on the interpretation of visually estimated intensities of a few hundred reflections recorded on sets of photographic films. The days of arguing from cell dimensions alone were past. For simple inorganic structures, powder diffraction patterns sometimes had to suffice. For organic structures, with a few notable exceptions (such as penicillin) a successful analysis was possible only when fairly reliable information was available about the approximate arrangement of the atoms in the molecule. This was before direct methods had been developed, and most structures were solved by a trial and error procedure: you postulated a model, an arrangement consistent with the available chemical and crystallographic information, you calculated structure factors for this model — by hand — and checked whether they were in qualitative agreement with the observed X-ray intensities. If they were, then you calculated a density map by Fourier synthesis, usually a two-dimensional projection down the shortest unit-cell direction — by hand again, with the help of Beevers-Lipson strips or Robertson templates (does anyone still remember what they were?) — adjusted the parameters of the trial model accordingly, recalculated the structure factors, checked whether any signs of Fourier terms had changed (we were more or less limited to centrosymmetric projections), and repeated the process. If the agreement was bad, and this was a matter of judgement, then you started again with a new trial model. Occasionally, if one was lucky, the structure to be solved contained a heavy atom, which considerably simplified the task of guessing a suitable trial model. It was all hard work, it took a long time, but you were only expected to solve one or two structures in the course of a normal doctoral research project. Those of us who survived look back on it as a heroic age.

As the years passed, direct methods were first envisaged and then developed. Their general applicability had to wait for advances in the power and availability of electronic computers to carry out the protracted calculations needed. Try to imagine the time required just to make a preliminary list of reflection triplets or quartets without computer assistance. It would be a matter of months; I know from experience; I tried it for triples in 1951, and gave up after a month. Similarly, it was only with improvement in computer power that least-squares algorithms for refining atomic positions and "thermal" parameters gradually became standard procedures. Again, it was computer power that enabled automated three- and four-circle diffractometers to collect hundreds, if not thousands, of reflection intensities per day. Think of calculating all the angles required by hand and cranking the circles into the correct positions! I read recently that computer speed doubles about every eighteen months. In 50 years, that gives an improvement of about 2^{33}; what now takes a second would then have needed more than the age of the universe.

And that is roughly where we are from the technical point of view. We can look forward to new developments, but most of the ones I can think of are essentially improvements in existing methods rather than anything radically new: on the experimental side, more powerful radiation sources, making it possible to obtain diffraction patterns from very tiny crystals; the routine use of area detectors to image large parts of reciprocal space at once, rather than to observe it point by point; on the computational side, we can expect further applications of maximum entropy methods and more routine structure determination of medium-sized organic molecules from analysis of powder diffraction patterns. With the exception of the last, the main thrust of these will be to overcome some of the present limitations in the area of biomolecular structure analysis, but I do not expect them to change small-molecule crystallography in any radical way. As compared to serial diffractometry, the use of area detectors should make it easier to detect and study incommensurate and disordered structures, but the principal effect will be to produce normal crystal structures still more rapidly and more routinely than at present, in other words, they will lead to a still more rapid accumulation of information about crystal and molecular structure. Whether this will lead to a corresponding increase in knowledge is another matter, which I shall attempt to discuss later.

2. What have we achieved?

2.1. MOLECULAR STRUCTURE

There is no need for me to emphasize here that the preoccupation with molecular structure is at the heart of chemistry. By the 1950s a change was becoming apparent; X-ray analysis was being introduced with great success to determine the molecular structures of complex natural products. This task had been regarded as one of the principal undertakings of organic chemistry, in the fulfillment of which much basic knowledge had been accumulated over the years. The intrusion of X-ray analysis into natural product chemistry may even have been regarded by some chemists at the time as a kind of threat to one of their traditional activities, but from the present viewpoint there can be no question that the liberation from the burden of proof of structure by chemical degradation and synthesis has had the effect of unleashing tremendous new energies in chemistry. Today, X-ray analysis is called in not only as a big gun, to solve the difficult problems, but almost routinely, and most major chemistry departments now have their own X-ray analysis service facilities. NMR spectroscopy may be of comparable importance, but where the evidence is equivocal crystal structure analysis still provides the most clear-cut decision. In one step it answers questions of constitution, configuration, and conformation, besides providing metrical information about interatomic distances and angles.

A unique contribution has been the determination of absolute configuration. By the early years of our century, it was possible to relate the configurations of hundreds of optically active compounds among each another — i.e., to establish their configurations relative to some reference compound. Emil Fischer took this as (+)-glyceraldehyde, which was arbitrarily assigned configuration I and represented by projection formula II. Within this convention, (+)-tartaric acid was known to be represented by III and the naturally occurring amino acids by IV. However, there was no way to decide whether (+)-glyceraldehyde actually corresponded to structure I or to the mirror image. Once this could be settled, stereochemistry could be placed on an absolute footing, but until mid-century there seemed no way to answer the question. Indeed, when I was a student, we were told that it was *impossible* to answer this question, even by X-ray diffraction, because of Friedel's law. As we all now know, Bijvoet used anomalous scattering of X-rays to show that the absolute structure assigned by Fischer was indeed correct. It was not necessary to rewrite all the formulas in the textbooks! In the meantime the absolute structure of thousands of chiral and polar crystals have been determined by the anomalous scattering method and we use the Prelog-Ingold-Cahn (CIP) system to specify the sense of chirality at tetrahedral centres. What seemed an insoluble problem has become routine.

One might also mention here that it was the metrical information about molecular conformation that provided an essential basis for the development of what has come to be known as molecular mechanics. One could require of a proposed force field at least that it should reproduce the observed conformations of a series of related molecules, say cyclic hydrocarbons, where different types of force can be assumed to act in oppostion to one another and where a factorization of the strain energy among various types of non-bonded interactions was therefore possible. Crystallography, together with gas-phase electron diffraction and molecular spectroscopy, provided the required information. For all I know, it may be possible nowadays to obtain the required structural information by high grade quantum mechanical calculations, but this was not possible when the information was actually needed.

2.2. ATOMIC MOTION IN SOLIDS

Once least-squares methods came into general use it became standard practice to refine not only atomic positional parameters but also the anisotropic "thermal parameters" or displacement parameters (ADPs), as they are now called. These quantities are calculated routinely for thousands of crystal structures each year, but they do not always get the attention they merit. It is true that much of the ADP information is of poor quality, but it is also true that ADPs from reasonably careful routine analyses based on modern diffractometer measurements can often yield physically significant information about atomic motions in solids. Cruickshank taught us in 1956 how analysis of ADPs can yield information about molecular rigid-body motion, and many improvements and modifications have been introduced since then. In particular, various programs are available to estimate the amplitudes of simple postulated types of internal molecular motion (e.g., torsional motions of atomic groupings about specified axes), besides the overall rigid-body motion, from analysis of ADPs. Caution may be required in interpreting results of such analyses because of extensive coupling among the effects produced by the postulated motions on the various atoms. As the amplitudes of motion are temperature dependent, multi-temperature measurements can be very useful in assessing the physical significance of results derived from such analyses. In particular, since different kinds of motion show different kinds of temperature dependence, some of the ambiguities inherent in the analysis of single-temperature data may be resolved. As the technical possibilities for carrying out accurate diffraction measurements at high and low temperatures come into general use, more attention should be given to the interpretation of ADPs and the physical significance of the results. There is much to be learned in this area.

2.3. EXPERIMENTAL CHARGE DENSITY DISTRIBUTIONS

Electron density maps have been used for decades to give images of molecules in crystals, and it has always been realized that such maps might also tell us something about the "nature of the chemical bond". It is fortunate that the electron density in a molecular or ionic crystal is closely similar to the superposition of the densities of the separated atoms, placed at the positions they occupy in the crystal, for it is this similarity that made it possible in the first place to use standard, spherically symmetrical scattering factors in solving crystal structures and in refining them by least-squares methods. In fact, when crystallographers take pride in their low R factors they pay tribute to the goodness of the pro-crystal approximation as well as to the accuracy of their measurements.

The difference $\Delta\rho(\mathbf{X}) = \rho(\mathbf{X}) - \rho_M(\mathbf{X})$ between the actual density and the pro-molecule density is known as the deformation density and can be interpreted as the reorganization in the electron density that occurs when a collection of independent, isolated, spherically symmetric atoms is combined to form a molecule in a crystal. Since $\Delta\rho$ is only a very small fraction of ρ in the region of the atoms, it is very susceptible to experimental error in the X-ray measurements and to inadequacies in the model, namely errors in the assumed atomic positions and ADPs. In one approximation, a deformation density map is obtained by direct subtraction of the two densities. The density map obtained in this way is smeared by vibrational motion of the atoms, but its peaks and troughs can often be interpreted in terms of some model of chemical bonding — e.g., peaks between bonded atoms being identified with "bonding density" and so on. Alternatively, $\Delta\rho$ can be expressed in parametric form as the sum of a number of suitably designed functions, e.g., a set of multipoles, each multiplied by a radial function and centred at an atomic position. The Fourier coefficients of the various functions are then added to the free-atom form factors with variable coefficients, which are refined, together with the atomic positional coordinates and ADPs in one giant least-squares analysis. The density map obtained in this way is sometimes known as a static deformation map. In contrast to the difference map, it represents the charge density reorganization on going from the vibrationless pro-molecule to the vibrationless molecule in the crystal.

It has been recognized for some time that dependable deformation densities require an expenditure in experimental time and effort far in excess of that involved in routine X-ray analysis. The experimental charge density can be used to calculate values of molecular properties such as electrostatic potential, but with increase in computational power it has also become possible to obtain such properties by quantum mechanical calculations. Of course, one can use one method to check the other, but if and when the reliability of the theoretical methods can be established, will there be any point in trying to extract the information from experiment?

2.4. OTHER ACHIEVEMENTS

Although we are supposed to limit ourselves to the subject of this school — implications of molecular and materials structure for new technologies — I can hardly avoid mentioning here the prodigious achievements in the area of biomolecular crystal structure analysis. Protein crystallography was regarded as a wildly visionary target when I was young but it has become a more or less standard method with hundreds of structures being produced annually. These have provided unparalleled insights (the word is here used literally) into enzymic active sites and catalytic mechanisms, antibody structure and specificity, DNA-protein recognition phenomena, light-harvesting assembly systems and other targets that were once believed to be far beyond the range of experimental structural study. Each year stretches the limits of X-ray analysis still further; there seems to be no end.

I should also mention the discovery of quasi-crystals. Apart from any possible technological importance, this has radically changed our views on the definition of the crystalline state. Once we admit that perfect periodicity is not the only kind of regularity that can characterize a crystal, quasi-crystals are not a "new form of matter" but simply another variation on the theme of non-periodic regularity.

3. Chemical Crystallography

3.1. CRYSTAL PACKING

The focus of interest for many crystallographers has been shifting more and more from the molecular to the intermolecular (supramolecular) level of organization. When I began my work, the structures of ionic crystals were reasonably well understood in terms of a few simple rules (Pauling's Rules). For organic structures there were no obvious regularities, apart from the hydrogen bond, whose importance as a structure directing element had been recognized at an early stage. One problem was that there were not many structures from which to draw general conclusions. It was Kitaigorodskii with his theory of close packing of molecules in crystals who paved the way for future developments. Today there is much interest in the study of weak (non-covalent) interactions, a topic I shall discuss in my second lecture. For the present, I merely remark that we are still unable to predict with confidence the crystal structure of a compound, given its molecular formula. This is one of the great challenges for the future.

3.2. POLYMORPHISM, CRYSTAL NUCLEATION AND GROWTH

A polymorph has been defined as "a solid crystalline phase of a given compound resulting from the possibility of at least two different arrangements of the molecules of that compound in the solid state". Computed lattice energies based on various types of atom-atom potentials suggest that even for quite simple molecules there may be several arrangements within an energy range of say 10 kJ mol^{-1}. One can conclude from this that polymorphism must be ubiquitous, or, alternatively, that crystallization does not depend only on thermodynamic factors. In fact, one of the great gaps in our knowledge concerns the initial steps of crystal nucleation. There are many hints that the formation of viable nuclei is the rate-determining step in crystallization. Especially for molecules with enough conformational freedom to adopt several shapes in solution or in the liquid phase, there is no reason why the conformer present in the thermodynamically stable crystal form should be the most stable conformer in solution. Thus, formation of the more stable crystal modification may be hampered by a low concentration of the particular conformer required.

Likewise, there is good evidence for the importance of a rate-limiting nucleation step in solid-solid phase transitions. Here nucleation may depend on the presence of suitable defects, such as micro-cavities and other surface irregularities between different crystal domains. Depending on the nature of such defects, nuclei of the new phase may be formed at slightly different temperatures and grow at different rates. In a sense, the defects act as catalysts for the structural transformation. Here again, we encounter limits to our present-day knowledge. A better understanding of crystal nucleation seems essential for further progress, but it is not clear how this is to be attained.

As far as crystal growth is concerned, studies of the influence of "tailor-made" impurity additives on differential face development has led to insights (and, as an unexpected bonus, to an independent confirmation of the correctness of absolute configurations determined by anomalous dispersion methods!).

3.3. SOLID STATE CHEMISTRY

When I was young, solid-state phase changes and chemical reactions were regarded as a nuisance rather than as an area worthy of serious study and attention. It was the topochemical approach to solid-state chemical reactions, pioneered by Gerhard Schmidt, that transformed the subject for me and for many others. The textbook example is the photochemical dimerization of *trans*-cinnamic acids; in solution such compounds yield mixtures of the various possible stereoisomeric products but irradiation of a crystal leads to a single stereo-specific product, or to no reaction, depending on the crystal structure. Thus one could determine the relative positions of the atoms before the reaction and after it, and hence deduce the metrical relationships that needed to be satisfied for reaction to proceed. In the meantime we have learned that not all chemical reactions in solids are topochemical. Some proceed not in the ordered bulk of the crystal but at defects, on the surface, or at other irregularities. The subject is now of very general interest and will undoubtedly continue to be in the forefront of chemical crystallography.

3.4. STRUCTURE CORRELATION.

With the establishment of "standard" bond lengths and angles functional groups used to be a characterized as having a more or less fixed structure. Gradually, it was recognized that this is not the case. For many groupings, structural changes occur in different crystal and molecular environments, and, moreover, the changes in individual structural parameters are often correlated in ways characteristic of the grouping itself. The connection with chemistry comes with the assumption that observed structures tend to concentrate in low lying regions of the potential energy surface, leading to what has been called the Principle of Structure Correlation:

If a correlation is found between two or more independent parameters
describing the structure of a given fragment in a variety of environments,
then the correlation functions maps a minimum energy path in the
corresponding parameter space.

This approach thus provides a link between the "statics" of crystals and the "dynamics" of reacting chemical systems and has been applied to map reaction paths for several types of prototypal chemical reactions.

4. Where are we going?

The influence of crystallography on 20th century chemistry has been profound — in fact, many areas of present day chemistry are unthinkable without the contributions of X-ray analysis. It may seem ironic that this progress, almost unimaginable 50 years ago, has been accompanied by the virtual disappearance of crystallographic research and teaching in many University chemistry departments. Crystal structures are now often, perhaps even mostly, determined as part of a service, and, while this is generally done with great speed and efficiency, the service crystallographer almost always has too much to do, she has no time to think about the broader implications of her results — or even to check them for possible mistakes and misinterpretations. The world production of single-crystal X-ray analyses now runs at something more than 10,000 structures a year, and with the advent of area detectors and still more efficient computers, this mass production of essentially unchecked crystallographic data is going to increase still further. In any case, results of many current crystallographic studies remain unpublished or receive only scant mention in chemical journals — a computer drawn picture of a molecules and a brief footnote are often all the information provided. A vast amount of metrical information about molecular structure and about intermolecular interactions is thus being accumulated, but not all of it is published. Much of this information — the published part at least — is being collected in computer readable form in the Cambridge Structural Database (CSD). Whatever the original intentions may have been, this has now developed into a scientific instrument for studying the systematics of molecular and supramolecular structure.

Forty years ago, one could read Acta Crystallographica with interest and even with excitement to broaden one's general education. One can still benefit from a perusal of the early issues — the 1956 volume is my personal favourite. Nowadays, information about crystal structure analyses is mostly confined to footnotes of papers in various chemistry journals and to the brief standardized accounts in Acta Crystallographica Section C and other specialist crysallographic journals. Many of these reports are doomed never to be read by anybody, once they appear in print. Today, more than ever, the only practicable way to find details of particular structures or classes of structures is through the CSD for organic and organometallic structures and through the Brookhaven Protein Data Bank (PDB) for biomolecular structures. It is essential that these two compilations continue to be kept running smoothly and efficiently for the foreseeable future, otherwise the only people acquainted with any particular crystal structure will be the people who solved it — or the people for whom it was solved — and their friends

In the preparation of this article, I am indebted to hundreds of colleagues for their friendly help and encouragement over the last fifty years.

MODERN NEUTRON DIFFRACTION METHODS

C.C. WILSON
ISIS Facility
CLRC Rutherford Appleton Laboratory
Chilton, Didcot, Oxon OX11 0QX, UK

1. Introduction

This paper will briefly review the facilities, instruments and methods available for neutron diffraction experiments now and also look ahead to future possibilities. It will not attempt to cover other than in an illustrative way the science performed using these techniques, examples of which will be found elsewhere in this volume.

2. The use of neutron diffraction

Neutron diffraction is the method of choice for many crystallographic experiments. Neutrons are scattered by the nucleus rather than the electrons in an atom, and many of its most useful properties in studying structure stem from this.

- The scattering power does not have the strong dependence on Z found for many other scattering techniques such as X-ray or electron diffraction. Thus (i) it is easier to sense light atoms, such as hydrogen, in the presence of heavier ones, (ii) neighbouring elements in the periodic table generally have substantially different scattering cross sections; for light elements in particular this is the only direct method of distinguishing neighbouring elements and (iii) isotopes of the same element can have substantially different neutron scattering lengths, thus allowing the technique of isotopic substitution to be used to yield structural and dynamical details. The use of contrast variation where the scattering density of an H_2O-D_2O mixture is chosen to highlight part of the system under study is particularly powerful and has been a key to many successful applications of the technique of neutron scattering in chemistry and biology.

- The scattering power from atoms for neutrons does not fall off as a function of $\sin\theta/\lambda$, which makes neutron scattering ideal for high resolution studies. Of course, the scattered intensity in a diffraction pattern will still decrease at higher angles due to the Debye-Waller factor, but the effect is considerably less severe for neutron experiments. The consequent ability to access very high resolution data (short d-spacings) makes neutron diffraction a highly precise method of studying structure. Hence it is ideal for the study of disordered structures, where the presence of a wide range of data including that at very high $\sin\theta/\lambda$ helps to decouple the strong

11

J.A.K. Howard et al. (eds.),
Implications of Molecular and Materials Structure for New Technologies, 11–22.
© 1999 *Kluwer Academic Publishers. Printed in the Netherlands.*

correlations normally found between temperature factors and site occupancy values. This is particularly true of data collected at pulsed sources.

- Neutrons interact weakly with matter and are therefore non-destructive, even to complex or delicate materials. Correspondingly, neutrons are a bulk probe, allowing us to probe the interior of materials, not merely the surface layers probed by techniques such as X-rays, electron microscopy or optical methods. This also helps to remove systematics and makes the data less susceptible to error than other techniques. A further consequence is that neutrons are ideal for studying samples contained inside complex sample environments, since it is possible to get the incident and diffracted beam through a significant thickness of containment material to the sample and thence to the detectors. The use of complicated sample environment facilities means that neutron diffraction has always lent itself naturally to the study of structural trends as a function of external variables, with a consequent high impact in the areas of systematic structural studies and structure-property relationships;

- Neutrons have a magnetic moment, allowing magnetic structure (the distribution of magnetic moments within a material) to be studied in a way not possible with other forms of radiation.

The applications of neutron diffraction methods are widespread. In powder diffraction, Rietveld analysis of neutron powder diffraction patterns gives accurate and precise results for a wide range of materials in physics, solid state and materials chemistry and in materials science including engineering. On the single crystal side, neutrons are used in chemistry, physics and biomolecular sciences, typically for providing accurate anisotropic displacement parameters and accurate hydrogen atom positions and thermal motions.

3. Neutron Sources

The characteristics of the two types of neutron sources (nuclear reactors and accelerator-driven spallation sources) used for condensed matter studies are very different. In a spallation source all neutrons are effectively produced at the same time (t_0, when the proton beam hits the target). The tight pulsed structure inherent in the production process must be retained for use of the time-of-flight technique, which is essential at such a source . The high energy neutrons are thus slowed down to thermal energies (and to energies just above the thermal region) by using small hydrogenous moderators which effectively "under" moderate the neutrons and retain the pulse structure. For sources where neutron production is continuous, it is more important to slow down as many neutrons as possible, regardless of how long this takes, and so "full" moderation using large moderators tends to be used at reactor sources. The methods for using the neutrons so produced for diffraction experiments are discussed later.

3.1. TWO EUROPEAN SOURCES: THE ILL AND ISIS

The high flux reactor at the ILL, Grenoble, is the highest flux reactor in the world dedicated to neutron scattering and the study of condensed matter. It has a comprehensive suite of high quality instruments, currently numbering 25, offering the full range of possibilities for neutron scattering experiments including both powder and single crystal neutron diffraction. Details can be found on the web page *http://www.ill.fr/*.

The use of pulsed spallation sources for neutron scattering science has largely been established in the past decade or so. The highest flux pulsed neutron source is the ISIS facility at Rutherford Appleton Laboratory in the UK. The instrument suite is again comprehensive, and again offers wide scope for powder and single crystal diffraction experiments. The novelty of pulsed white beam techniques means that there is still much scope for further development of both the instrumentation and data collection techniques. Details can be found on the web page *http://www.isis.rl.ac.uk/*.

4. Methods For Powder Diffraction

Developments in powder diffraction over the past 30 years have progressed on two fronts. First, the instruments have been improved, with an increased emphasis on high resolution and also the exploitation of increasingly high flux sources. Secondly, the introduction of the Rietveld method for structural refinement has been exploited and extended. We have now reached the stage where it is routine to perform Rietveld refinement and extract detailed atomic structural information from most high resolution powder diffraction patterns. This is especially true of neutron powder diffraction, where only a few categories of experiments are excluded from the possibilities of Rietveld refinement - the most rapid time-resolved measurements, highly multi-phase samples, samples showing texture or preferred orientation, weak and complex magnetic structures - and even these areas are being tackled.

4.1. POWDER DIFFRACTION ON A STEADY STATE SOURCE

The scanning of the diffraction pattern for a steady state source is almost always carried out using a monochromatic beam and scanning the 2θ angle:

$$\lambda = 2d_{hkl}\sin\theta_{hkl} \Rightarrow d_{hkl} = \lambda/2\sin\theta_{hkl}$$

The thermalised white beam is incident on a monochromator which selects the desired wavelength. The monochromatic beam is incident on the sample and the diffracted beam is then detected over the required range of scattering angles. This can be achieved either by moving a single detector round in very small 2θ steps ('step-scanning'), or by using a curved one-dimensional position-sensitive detector (PSD) subtending the scattering angles of interest.

Constant wavelength diffraction geometry can be advantageous in the following ways:

- The time-averaged flux available on a steady state source is higher than that available at even the best pulsed neutron sources. This gives such instruments the potential for extremely high count rate;

- The simplicity of the peak shape (which is symmetric) can make the refinement of the powder diffraction profiles more straightforward and also allow for more simple deconvolution of sample-related effects. The peak shape in a pulsed source is asymmetric, which can have benefits in peak separation.

4.2. POWDER DIFFRACTION ON A PULSED SOURCE

When a wavelength-sorted white beam is available from a pulsed neutron source, the diffraction geometry is significantly different. The assignment of a unique wavelength to each neutron by the time-of-flight (tof) technique removes the need to make the incident beam monochromatic. The tof technique exploits the fact that the neutrons produced at an accelerator-based pulsed neutron source are created at a known time, t_0, the time at which the high energy particle beam hits the neutron-producing target. Subsequently, as a consequence of the massive nature of the neutron, the velocity of each neutron is related to its wavelength. By recording the arrival time, t, of each neutron at the detector, a known distance from the source, it is therefore possible to calculate its time-of-flight, $t-t_0$, its velocity and hence its wavelength.

The variable wavelength incident beam then means that the whole diffraction pattern can be measured, if desired, at a single scattering angle. This is achieved by using the variation of λ_{hkl} to perform the scan, keeping θ_{hkl} constant:

$$\lambda_{hkl}=2d_{hkl}\sin\theta_{hkl} \Rightarrow d_{hkl}=\lambda_{hkl}/2\sin\theta_{hkl}$$

Time-of-flight diffraction geometry has significant advantages in several ways:

- The resolution of a tof powder diffractometer can be enhanced by increasing the overall flight path of the neutrons, improving the $\Delta\lambda/\lambda$ term in the resolution. The resolution term due to the scattering angle is a function of $\cot\theta$, which is thus minimised as higher scattering angles. The ability of a pulsed source instrument to measure at a fixed angle, often close to backscattering, allows for improved resolution. Equally importantly, the resolution is constant over the whole diffraction pattern, often with significant benefits;
- The fixed geometry arrangement means that incoming and outgoing beams can be well collimated. This allows the sample to be enclosed in complex sample environment apparatus, without the scattering from that apparatus interfering with the detected diffraction pattern.

5. Methods For Single Crystal Diffraction

On a steady state source, traditional 4-circle diffractometer techniques are normally used, with a monochromatic beam and a single step-scanning detector. The region of reciprocal space accessed in a single measurement can be increased by using an area detector. In this case it is still necessary to scan the crystal or the detector to observe the diffracted intensity. Alternatively it can be combined with a broad band (white) beam, and used for Laue or quasi-Laue diffraction, with a stationary crystal and detector.

Structure factor data collected on a monochromatic steady state source presently yields the ultimate in accuracy for neutron single crystal structure determination. The benefits of such instrumentation can be summarised as follows:

- Well established 4-circle diffractometry techniques can be utilised;
- All reflections are observed with the same neutron wavelength, eliminating the need for wavelength dependent corrections. Large area detectors are also less important, removing the systematic deviations caused by fluctuations of detector response;
- The time averaged flux at current high flux steady state sources is substantially higher than at present-day pulsed sources, allowing better counting statistics to be obtained in a unit time, and allowing the study of smaller crystals or larger unit cells on reactor-based instruments.

These factors tend to lead to more accurate structure factors, better internal agreement and ultimately to lower crystallographic R factors and somewhat more precise atomic parameters. Constant wavelength single crystal diffraction is the method of choice if ultimate precision is required in an individual structure determination and also for larger unit cells or smaller crystals.

The time-of-flight Laue diffraction (tofLD) technique used at pulsed sources exploits the capability of a single crystal diffractometer on such a source to access large volumes of reciprocal space in a single measurement. This is due to the combination of the wavelength-sorting inherent in the time-of-flight (tof) technique with large area position-sensitive detectors. tofLD gives a genuine sampling of a large three dimensional volume of reciprocal space in a single measurement with a stationary crystal and detector.

Structure factor data collected on an instrument on a pulsed source can have certain advantages in structural refinement:

- The collection of many Bragg reflections simultaneously in the detector allows the accurate determination of crystal cell and orientation from a single data frame (collected in one fixed crystal/detector geometry). It is also worthy of note that for some applications this single frame may be the only data required;
- The white nature of the incident beam enables the straightforward measurement of reflections at different wavelengths, which can be useful in the precise study of wavelength dependent effects such as extinction and absorption;
- The collection of data to very high $\sin\theta/\lambda$ values (exploiting the high flux of useful epithermal neutrons from the undermoderated beams) can allow more precise parameters to be obtained, enabling subtle structural features to be examined;
- The Laue method allows greater flexibility for the rapid collection of data sets. This flexibility, long appreciated in synchrotron Laue methods for studying protein structures, has recently become recognised as a great strength of time-of-flight neutron Laue diffraction methods.

Time-of-flight single crystal diffraction is the method of choice for rapid surveying of reciprocal space, for rapid structure determinations and for the following of structural changes using a subset of reflections. It also provides good accuracy and precision in standard structural refinements.

6. Neutron applications in materials chemistry

This section will not concentrate on specific examples, instead considering the general areas tackled by neutron diffraction and pointing out the particular benefit gained from using the technique.

6.1. POWDERS

Ab initio structure determinations [1-3]
Until the advent of very high resolution diffractometers, it was felt unlikely that genuinely unknown structures could be solved using standard techniques such as direct methods. However, with improved instrumentation offering the opportunity to measure more peaks in a powder pattern with better accuracy, this situation has changed. The solution of structures from powder data is still not routine (apart from the traditional areas of structural development and quasi-isomorphous structures) but there have been some significant triumphs. The trend is expected to continue, both with improved instrumentation and also with improved computational methods.

High resolution - subtle structural features [4]
Resolution in this context can be taken in three ways:
- Accessing short d-spacing data means that the scattering density maps produced in the subsequent Fourier reconstruction (which uses the measured structure factors as coefficients) are a more reliable representation of the structure. Thus more real space information can be obtained from high reciprocal space resolution;
- Separating peaks which are close together lends high resolution powder diffraction the ability to examine phase transitions, for example if the symmetry of the system changes and certain peaks in the pattern split;
- Determining peak positions to an accuracy of 1 part in 10^6, manifest in the highest resolution powder diffractometers such as HRPD at ISIS, allows lattice parameters to be determined with unprecedented accuracy. This is important in surveying parameter space when looking for physical effects such as subtle phase transitions.

Temperature scanning and time-resolved studies [5-7]
The availability of high flux sources and large detector systems have made it possible to collect complete diffraction patterns in ever shorter periods of time, opening up the possibility of carrying out systematic temperature-dependent or time-resolved studies. Here, the evolution of a diffraction pattern as a function of time can be followed and the structural changes occurring in a material during a chemical or physical process can be observed and investigated. The information obtained can be extremely important in studies of the thermodynamics of a system, and in following chemical reactions, crystallisations and phase transformations. Of particular interest is the ability to detect transient species, which may remain unobserved during static measurements.

High T_c superconductors - location of oxygen atoms and stoichiometry [8]
Bringing together the neutron strengths of resolving disordered structures and locating light atoms, the first complete and definitive structure of the $YBa_2Cu_3O_{7-x}$ structure was carried out using high resolution neutron powder diffraction.

Electrolyte and battery materials [9,10]
Work in this technologically important area exploits many aspects of neutron powder diffraction, including the location of light atoms (e.g. lithium), rapid structural characterisation of many materials of varying composition, examination of defect or non-stoichiometric structures and the ability to follow structural changes in "real-time", for example during battery discharge.

Ionic and superionic conductors [11,12]
Of particular relevance for these systems are the determination of light atoms, for example protons or lithium ions, studies of defect or disordered structures and the ability to correlate structure with physical properties such as conductivity.

High pressure studies [13]
These exploit the ability of neutrons to penetrate containment materials (the anvils and gaskets used in the application of high pressures). Powder diffractometers at pulsed sources are particularly useful in that they can measure the entire diffraction pattern at a single scattering angle, allowing the incoming and diffracted beam to be tightly collimated, reducing still farther the extraneous scattering from the pressure cell.

6.1.2. Specific examples in solid state and materials chemistry

Development of lithium battery electrodes [9]. Powder neutron diffraction structural work has been used in programmes to develop new cathode materials to form the basis of a new generation of high energy density, rechargeable lithium batteries. Neutron diffraction work on just such a new material, layered $LiMnO_2$, has been geared towards understanding the structural features underlying the retention or loss of Li storage capacity. Detailed information on lithium content and location can only be achieved using neutron diffraction, and the materials have been examined at various stages of their charge/discharge cycles and with varying degrees of lithium depletion.

Negative thermal expansion materials [14,15]. Recent reports of negative thermal expansion in ZrW_2O_8 stimulated both technological and academic interest in this unusual phenomenon. Structural work is aimed at understanding mechanistic features and hence allowing the development of enhanced, application-specific materials. Recent neutron diffraction experiments have allowed the structure to be characterised at no less than 260 temperatures between 2 and 600 K. Full Rietveld refinements of the high resolution powder diffraction patterns revealed detailed structural information on the unit cell parameters, atomic positions and atomic vibrations over the whole of this range.

Chemical basis of the magnetoresistance in GMR materials [16,17]. To understand the recently discovered phenomenon of greatly enhanced magnetoresistance in perovskite phases of the type $Ln_{1-x}M_xMnO_3$, it is vital to gain knowledge of the detailed positions and occupations of all atoms in the structure. Oxygen displacements and vacancy orderings, as well as disorder in the cation distribution, are extremely important in controlling the electronic structures and hence properties of this class of materials, and neutron diffraction is the only appropriate structural method of defining these. Much current work is aimed at understanding the chemical basis for generating the required structural characteristics for GMR behaviour, and also understanding the magnetism of these phases at low temperature. Neutron diffraction facilities both at ISIS and the ILL have been used in these studies, utilising the combination of high resolution to study small oxygen displacements and subtle phase separation phenomena, and high flux to allow the observation of weak diffraction features and magnetic scattering.

6.2. SINGLE CRYSTALS

Accurate location of hydrogen atoms [18,19]
The ability of neutron to locate hydrogen atoms and refine their positions and thermal parameters is unlikely to be challenged in the foreseeable future. Neutrons can locate the hydrogens in metal clusters, for example hydride ligands, far more reliably than with any other method. Neutrons have a long and successful history in hydrogen bonded systems, such as the study of amino acids, nucleic acid components, carbohydrates, cyclodextrins and in the examination of short O...O hydrogen bonds where the ability of the method to determined hydrogen anisotropic displacement parameters allows deduction of the shape of the potential well in which the hydrogen atom sits.

Charge density studies / X-N [20]
The ability of neutrons to provide complementary information to that available from X-ray diffraction is well illustrated by X-N studies, where the neutron parameters fix the nuclear positions and the X-ray data determine the electron density, specifically the electron density involved in the bonding and non-bonding interactions. The success of such methods requires careful monitoring of the consistency between the two data sets.

Thermal parameter studies
The accurate determination of anisotropic displacement parameters can allow assessment of the effect of thermal motions on apparent bond lengths, especially for bonds involving hydrogen atoms, detailed comparisons of X-ray and neutron determined parameters, and using analyses of the thermal motion by rigid body (TLS analysis) or harmonic oscillator models to understand the forces operating inside the molecule.

Quantitative diffuse scattering: defect structures, local clusters [21,22]
Single crystal neutron diffraction has been widely used in the study and modelling of diffuse scattering (which can occur anywhere in reciprocal space, not just at the Bragg peak positions), looking for example at correlated atomic positions in disordered materials and local structural distortions in non-stoichiometric systems.

Variable temperature measurements [23,24]

The information available from structures determined at a range of temperatures can be valuable in understanding the energetics and dynamics of a system. Used both as a method for removing systematics in the thermal parameter behaviour of molecular systems and as a probe of phase transitions, this technique can yield powerful information from structural studies under varying conditions.

7. Technique Advances

Many advances have been achieved at neutron sources to date, in support of the development of strong user programmes with modern instrumentation and techniques.

7.1. NEW DETECTORS

Traditional gas detectors (linear tubes and position-sensitive area detectors) have been partly superseded by developments in scintillator technology including ^{6}Li-glass and ZnS plastic scintillator materials. Both pixellated detectors for powder diffraction and position-sensitive area detectors for single crystal work have been successfully constructed. At the ILL, microstrip detectors are being increasingly exploited with a large linear array of such strips installed in the new D20 instrument [25]. Developments towards 2D microstrip detectors are also well advanced. In addition on the steady state source the potential of image plates is being exploited, using a gadolinium coated layer to convert the image plate to a neutron detector. Because of the necessity for off-line processing, this technique is not applicable to time-of-flight methods at a pulsed source.

7.2. INSTRUMENTATION IMPROVEMENTS

These have developed both in terms of resolution, using better monochromators at steady state sources and longer flight paths at pulsed sources, and in flux, principally by the provision of more and larger detectors, but also by the use of neutron guides and improved monochromators and collimators. Among the prime concerns has been the reduction of background levels, allowing smaller signals to be measured more reliably, and the pursuit of the twin goals of enhancing flux and resolution together.

7.3. SAMPLE ENVIRONMENT EQUIPMENT

The routine availability of complex sample environment kit is one of the enduring strengths of neutron diffraction. Variable temperature experiments can be carried out inside helium flow cryostats to $T=1.2$ K and even lower, in closed cycle helium refrigerators to 10 K with very low background levels, and in furnaces to 2000 K and beyond. High pressure experiments are possible up to ~1 GPa inside gas cells, up to 3 GPa in traditional clamped cells such as the McWhan cell and now to beyond 30 GPa in modern clamped cells [13]. Facilities also exist to apply stress *in situ* on the beamline, allowing the measurement of the induced strains inside materials of interest to engineering, such as metals or composites.

7.4. SOFTWARE

The software for processing the raw data, to reduce the measured pattern to a normalised profile or to a set of structure factors, tends to be customised locally to the requirements of the particular instrument. For structural refinement, however, many of the widely available programs are suitable for the refinement of neutron diffraction data. The choice of program system to use is often problem-dependent, with different programs having particular benefits, such as for multiple phase samples, or in allowing the joint refinement of X-ray and neutron data, or for performing magnetic structure refinements.

8. Future Applications and Developments

It is of course impossible to predict where the future applications of neutron diffraction will lie, though one can summarise the general areas where expected trends in solid state and materials chemistry might benefit from the technique. Possible future areas include:

- New materials. Recent examples include studies of high T_c superconductors, fullerenes, GMR compounds, battery materials, etc;
- Heterogeneous systems. Structural information from more complex materials or mixtures of materials;
- Disordered systems. Improved and more quantitative treatment of diffuse scattering (as well as the Bragg intensities traditionally associated with structural refinement) will allow the study of materials exhibiting disorder, such as ionic conductors;
- Phase transitions and transformations. Likely applications include *in situ* studies of reaction kinetics, transformations under increasingly high pressure, the precise derivation of order parameters and critical exponents;
- *In situ* diffraction methods. Including sol-gel processing of ceramics, hydrothermal synthesis, in-situ film growth by techniques such as chemical vapour deposition, monitoring chemical reactions as they take place in the neutron beam;
- "Diffraction plus" measurements. The novel aspect of making complementary non-structural measurements simultaneously with diffraction will also expand. The complementary techniques being developed include conductivity and calorimetry;
- Supramolecular chemistry and molecular engineering [26]. Neutron single crystal diffraction can define the crucial hydrogen atoms involved in patterns of "weak" intermolecular interactions in complex molecular and supramolecular structures;
- Drug materials. In the study of the small structural and conformational changes which govern polymorphism and control morphology. If the preferred form is stabilised from alternatives by very small energy differences, an exact description of the entire structure, including the hydrogens, is required to understand these;
- Advanced materials. Including superalloys, metal-matrix composites, zeolites, catalysts [6,27], battery charge and discharge processes [9,10], etc.

8.1. NEW TECHNIQUES

There will be many new techniques made available to users of neutron diffraction in the next few years to meet the continuing challenges of those studying novel materials for applications. The ability to study structures in combined environments is already established and will extend. For example in combined high p and high T the current capability is around 1 GPa at 1000 K [28], while high p and low T reaches ~10 GPa at 100 K. Ultra-high pressure facilities will extend beyond the 30 GPa currently available using Paris-Edinburgh cell technology [13]. There will be demand for more sophisticated reaction vessels, gas flow arrangements for *in situ* diffraction. There are also possibilities to study magnetic structure under high (and pulsed) magnetic fields and studying materials under the influence of an externally applied electric field, including applications in *in situ* electrochemistry and in real-time studies of battery discharge. Pulsed sources also offer the possibility of applying an external field stroboscopically in phase with the pulse of neutrons, yielding potentially very high time resolutions [29].

8.2. THE FUTURE: NEW INSTRUMENTS AND NEW SOURCES

The facilities provided for neutron diffraction can also be enhanced by improving the instruments and increasing the available neutron flux to meet the challenges being thrown up by novel materials development. The new D20 instrument at the ILL has a very large detector to obtain the highest possible count-rate [25] while a complementary high flux instrument, GEM, at ISIS will combine high count-rate with high resolution to allow the rapid collection of fully refineable diffraction data [30].

In terms of sources, there is unlikely to be a new steady state research reactor built in the foreseeable future. The inherent limits of reactor technology mean that a factor of five in flux is the most one could realistically hope to achieve, and this will not overcome the environmental and political concerns regarding reactor construction. ISIS will remain the brightest pulsed neutron source for several more years, but if plans for construction of 1 MW pulsed sources in Japan and the US come to fruition, these will be 6 times more powerful than ISIS. Looking further forward, the proposed European Spallation Source would provide a 5 MW pulsed neutron source, 30 times the power of ISIS, which would combine the time-averaged flux of the best reactor source with the enormous benefits of the pulsed structure from a spallation source - a true next generation source [*http://www.isis.rl.ac.uk/ISISPublic/ESS.htm*].

The current sources and planned new sources will take neutron scattering and neutron diffraction into the next millenium in a healthy position to meet the emerging challenges posed by physicists, solid state chemists and materials scientists in their continuing quest to develop new and useful materials.

9. References

1. Christensen, A.N., Lehmann, M.S., Nielsen, M. (1985) Solving crystal structures from powder diffraction data, *Aust. J. Phys.* **38**, 497-505.

2. Cheetham, A.K., David, W.I.F., Eddy, M.M., Jakeman, R.J.B., Johnson, M.W. and Torardi, C.C. (1986) Crystal structure determination by powder neutron diffraction at the Spallation Neutron Source, ISIS, *Nature* **320** 46-48.

3. Ibberson, R.M. and Prager, M. (1995) The *ab initio* structure determination of dimethyacetylene using high resolution neutron powder diffraction, *Acta Cryst. B* **51** 71-76.

4. David, W.I.F., Ibberson, R.M. and Matsuo, T. (1993) High resolution neutron powder diffraction: a case study of the structure of C_{60}, *Proc. Roy. Soc. Lond. A* **442**:1914, 129-146.

5. Pannetier, J. (1986) Time-resolved neutron powder diffraction, *Chemica Scripta* **26A**, 131-139.
6. Epple, M. (1994) Applications of temperature-resolved diffraction methods in thermal-analysis , *J. Thermal Analysis* **42**, 559-593.
7. Wilson, C.C. and Smith, R.I. (1997) Pulsed neutron diffraction: New opportunities in time-resolved crystallography, in J.R. Helliwell, P. Rentzepis (eds), *Time-resolved Diffraction*, Oxford University Press, pp. 401-435.
8. David, W.I.F., Harrison, W.T.A., Gunn, J.M.F., Moze, O., Soper, A.K., Day, P., Jorgensen, J.D., Hinks, D.G., Beno, M.A., Soderholm, L., Capone, D.W., Schuller, I.K., Segre, C.U., Zhang, K. and Grace, J.D. (1987) Structure and crystal chemistry of the high-T_c superconductor $YBa_2Cu_3O_{7-x}$, *Nature* **327**, 310-312.
9. Bruce, P.G. (1997) Solid-state chemistry of lithium power sources, *Chem. Commun.* **1997**, 1817-1824.
10. Chabre, Y. and Pannetier, J. (1995) Structural and electrochemical properties of the proton γ-MnO_2 system, *Prog. in Solid State Chem.* **23**, 1-130.
11. Skakle, J.M.S., Mather, G.C., Morales, M., Smith, R.I., West, A.R. (1995) Crystal structure of the Li^+ ion-conducting phases, $Li_{0.5-3x}RE_{0.5+x}TiO_3$ - RE=Pr, Nd, x~0.05, *J. Mater. Chem.* **5**, 1807-1808.
12. Hull, S. and Keen, D.A. (1996) Superionic behaviour in copper(I) chloride at high pressures and high temperatures, *J. Phys.: Condensed Matter* **8**, 6191-6198.
13. Besson, J.M. and Nelmes, R.J. (1995) New developments in neutron-scattering methods under high pressure with the Paris-Edinburgh cells, *Physica B* **213**, 31-36.
14. Mary, T.A., Evans, J.S.O., Vigt, T. and Sleight, A.W. (1996) Negative thermal-expansion from 0.3 to 1050 K in ZrW_2O_8, *Science* **272**, 90-92.
15. David, W.I.F., Evans, J.S.O. and Sleight, A.W. (1997) Zirconium tungstate: the incredible shrinking material, *ISIS 1997 Annual Report*, Rutherford Appleton Laboratory TR-97-050, pp. 38-39.
16. Battle, P.D., Green, M.A., Laskey, N.S., Millburn, J.E., Radaelli, P.G., Rosseinsky, M.J., Sullivan, S.P. and Vente, J.F. (1996) Crystal and magnetic-structures of the colossal magnetoresistance manganates $Sr_{2-x}Nd_{1+x}Mn_2O_7$ (x=0.0, 0.1), *Phys. Rev. B* **54**, 15967-15977.
17. Rodriguez-Martinez, L.M. and Attfield, J.P. (1996) Cation disorder and size effects in magnetoresistive manganese oxide perovskites, *Phys. Rev. B* **54**, 15622-15625.
18. Braga, D., Grepioni, F. and Desiraju, G.R. (1997) Hydrogen bonding in organometallic crystals - a survey, *J. Oranomet. Chem.* **548**, 33-43.
19. Ceccarelli, C., Jeffrey, G.A. and Taylor, R. (1981) A survey of O-H...O hydrogen bond geometries determined by neutron diffraction, *J. Mol. Struct.* **70**, 255-271.
20. Coppens, P. (1992) Electron-density from x-ray-diffraction, *Ann. Rev. Phys. Chem.* **43**, 663-692.
21. Hull, S. and Wilson, C.C. (1992) The defect structure of anion-excess $(Ca,Y)F_{2+x}$ *J. Sol. State Chem.* **100**, 101-114.
22. Li, J.-C., Nield, V.M, Ross, D.K., Whitworth, R.W., Wilson, C.C. and Keen, D.A. (1994) Diffuse neutron scattering study of deuterated ice Ih *Phil. Mag. B* **69**, 1173-1181.
23. Kampermann, S.P., Sabine, T.M., Craven, B.M. and McMullan, R.K. (1995) Hexamethylenetetramine - extinction and thermal vibrations from neutron-diffraction at 6 temperatures, *Acta Cryst.* A**51**, 489-497.
24. Wilson, C.C., Shankland, N. and Florence, A.J. (1996) A single crystal neutron diffraction study of the temperature dependence of hydrogen atom disorder in benzoic acid dimers, *J. Chem. Soc. Faraday Trans.* **92**, 5051-5057
25. Convert, P., Berneron, M., Gandelli, R., Hansen, T., Oed, A., Rambaud, A., Ratel, J. and Torregrossa, J. (1997) A large high counting rate one-dimensional position sensitive detector: the D20 banana, *Physica B* **234**, 1082-1083.
26. Allen, F.H., Howard, J.A.K., Hoy, V.J., Desiraju, G.R., Reddy, D.S. and Wilson, C.C. (1996) First neutron diffraction analysis of an O-H...π hydrogen bond: 2-ethynyl-2-adamantanol, *J. Amer. Chem. Soc.* **118**, 4081-4084.
27. Thomas, J.M. (1996) Catalysis and surface science at high resolution, *Faraday Disc.* **105**, 1-31
28. Hull, S., Keen, D.A., Done, R., Pike, T., Gardner, N.J.G. (1997) A high temperature, high pressure cell for time-of-flight neutron scattering, *Nucl. Inst. Meth. A* **385**, 354-360.
29. Steigenberger, U., Eckold, G. and Hagen, M. (1995) Time-resolved studies on a millisecond time-scale by elastic neutron-scattering - transient properties of the ferroelectric phase-transition in Rb_2ZnCl_4, *Physica B* **213**, 1012-1016.
30. Day, P., Ibberson, R.M., Soper, A.K. and Williams, W.G. (1998) GEM, the General Materials Diffractometer at ISIS, *Physica B*, in press.

ELECTRON DENSITIES AND ELECTROSTATIC PROPERTIES OF MATERIALS FROM HIGH RESOLUTION X-RAY DIFFRACTION

C. LECOMTE
Laboratoire de Cristallographie et Modélisation des Matériaux Minéraux et Biologiques, Université Henri Poincaré,
Nancy 1, Faculté des Sciences,
BP 239, 54506 Vandœuvre-lès-Nancy Cédex

1. Introduction

Since the pionering work of P. Coppens [1] thirty years ago, experimental electron research has become a mature field: in chemistry, physics and molecular biology, the precision of the density obtained either permits us to calibrate theory or gives accurate values for properties which are not, nowadays, accessible by theoretical calculations at the same level of accuracy due to inadequacy of the theories or to the limited capability of the computers. On the experimental side, an important step was the use of analytical models of the electron density (Stewart [2], Hirshfeld [3], Coppens, Hansen *et al.* [4]), which permit the calculation of electrostatic properties [5] like electrostatic potential, electric field and multipole moments, allowing applications in various fields from biology to material science. Furthermore, the results of the topology of the experimental charge density also permits the classification of interactions and open new avenues of exploration.

The aim of this paper is to review some applications of charge density research with a special emphasis on the results which *are not easily attainable by other experimental or theoretical techniques*. The first part will describe some results strictly concerning the electron density model: comparison of experiment and theory, success and pitfalls, derived properties (electrostatic potential, multipole moments, topology) from the X-ray experiment; then, some applications to material science (Non-Linear Optical materials, zeolites, hydrogen bonds, coordination chemistry *etc.*) will be described. This review focuses only on few applications but the reader can find all the background and the most important results in the excellent book "X-ray charge densities and chemical bonding" by Coppens [7].

J.A.K. Howard et al. (eds.),
Implications of Molecular and Materials Structure for New Technologies, 23–44.
© 1999 *Kluwer Academic Publishers. Printed in the Netherlands.*

2. Electron Density Models and Calculation of Electrostatic Properties

2.1. THE MULTIPOLE MODEL

Experimental charge density analysis by crystallographic methods requires accurate low temperature X-ray diffraction measurements on single crystals in order that thermal vibrational smearing of the scattering electron density distribution is small; the experiments, the techniques and the processing for accurate data are given in specific references [8-12] and will not be discussed here. These data give a set of accurate moduli of the Fourier components of the thermally smeared electron density in the unit cell $\rho_{av}(\underset{\sim}{r})$ which are called structure factors:

$$F(\underset{\sim}{H}) = \int \rho_{av}(\underset{\sim}{r}) \, e^{2\pi i \underset{\sim}{H} \cdot \underset{\sim}{r}} \, d^3\underset{\sim}{r} \tag{1}$$

$$\rho_{av}(\underset{\sim}{r}) = \int \rho_{static_j} (\underset{\sim}{r} - \underset{\sim}{u}) \, P(\underset{\sim}{u}) \, d^3\underset{\sim}{u} \tag{2}$$

where $P(\underset{\sim}{u})$ and $\rho_{static_j}(\underset{\sim}{r})$ are respectively the probability distribution function and the static electron density of the atom. This continuous static density which may be compared to the density calculated from *ab initio* methods (HF or DFT) is divided into pseudo-atomic charge densities:

$$\rho_{static}(\underset{\sim}{r}) = \sum_{j=1}^{Na} \rho_{static_j} (\underset{\sim}{r} - \underset{\sim}{R}_j) \tag{3}$$

where Na is the number of atoms in the asymmetric unit or in the molecule.

In conventional X-ray least squares refinements $r_{static_j}(\underset{\sim}{r} - \underset{\sim}{R}_j)$ is the electron density of the free neutral atom j which has a spherically averaged shape; when this free atom density is summed over all the atoms of the molecule (formula (3)), it is called the promolecule density ($\rho^{pro}(\underset{\sim}{r})$).

However, due to chemical bonding and to the molecule-molecule interactions, the electron density is not spherical and the deformation density is obtained from X-rays by the inverse Fourier transform:

$$\delta\rho_{dyn}(\underset{\sim}{r}) = \rho_{dyn}^{obs}(\underset{\sim}{r}) - \rho_{dyn}^{pro}(\underset{\sim}{r}) = V^{-1} \sum_{\underset{\sim}{H}} \left[\left|F_o(\underset{\sim}{H})\right| e^{i\varphi m} - \left|F_c(\underset{\sim}{H})\right| e^{i\varphi c} \right] e^{-2i\pi \underset{\sim}{H} \cdot \underset{\sim}{r}} \tag{4}$$

where $|F_o|$ and $|F_c|$ are respectively the moduli of the observed structure factor and of the dynamic structure factor of the promolecule with their respective phases (m = multipolar).

However, these dynamic density maps do not readily lead to numbers describing charges and electrostatic properties; alternative and much more elegant methods are those using aspherical pseudo-atom least squares refinements. These refinements permit access to the positional and thermal variables of the atoms as well as to the electron density parameters.

Several pseudo-atom models of similar quality exist [2-4,7] and are compared in reference [13]. In general, these models describe the continuous electron density of the unit cell as a sum over pseudo-atom densities centered at the nuclear sites:

$$\rho_{stat}(\underset{\sim}{r}) = \sum_{j=1}^{Na} \left(\rho_j(\underset{\sim}{r} - \underset{\sim}{R_j}) + \delta\rho_j(\underset{\sim}{r} - \underset{\sim}{R_j}) \right)$$

where ρ_j and $\delta\rho_j$ are either the core density and the perturbed refinable non-spherical valence pseudo-atom density or the free atom total density and the deviations from this density. In any case, the ρ and $\delta\rho$ functions are centered at the nuclei, $\delta\rho$ being the product of radial functions (usually Slater type $R_n(r') = N\, r^n\, e^{-\alpha r'}$, sometimes a Laguerre function) with a set of orientation dependant functions $A_n(\theta,\varphi)$ defined on a local axis centered on the atoms (figure 1).

$$\delta\rho_j(\underset{\sim}{r}') = \delta\rho_j(\underset{\sim}{r} - \underset{\sim}{R_j}) = \sum_n C_n R_n(r') A_n(\theta',\varphi') \qquad (5)$$

where the C_n coefficients are obtained from least squares refinement against the X-Ray structure factors.

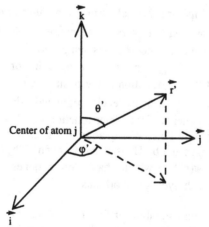

Figure 1. Coordinate system on the atoms

The models used in this chapter are due to Coppens, Hansen and coworkers [4]. First the Kappa formalism [4a] permits an estimation of the net charge of the atom and allows for the expansion or contraction of the perturbed valence density (spherical average): for each atom, the density is described as:

$$\rho_{stat}^{at}(r') = \rho_{core}^{at}(r') + P_v \, \kappa^3 \, \rho_{val}^{at}(\kappa r')$$

where ρ_{core}^{at} and ρ_{val}^{at} are the spherically averaged core and valence electron density of the free atom, calculated from the best available wave functions. P_v is the valence shell population and Kappa is the expansion or contraction coefficient of the perturbed density. If Kappa is larger than one, the observed valence density of the atom at distance r' corresponds to the valence density of the free atom at a larger distance; it means that the real observed density is contracted compared to the free valence distribution. Coppens justified the existence of the Kappa parameter by the variation in electron-electron repulsion with electron population and showed that there is a linear correlation between Kappa and the net charge obtained from P_v as predicted by Slater rules [4a].

To take into account the non-spherical shape of the valence electron distribution, the Kappa model has been improved by the addition of multipole parameters [4b]. Then, the pseudo-atomic density is written (Molly program):

$$\rho^{at}(\underset{\sim}{r}\,') = \rho_{core}^{at}(r') + P_v \, \kappa^3 \, \rho_v(\kappa r') + \sum_{l=0}^{l\,max} \kappa'^3 \, R_l(\kappa' r') \sum_{m=-l}^{l} P_{lm} \, y_{lm}(\theta',\varphi') \qquad (6)$$

$$\int |y_{lm}| \, d\Omega = 2 \text{ if } l \neq 0 \text{ and } 1 \text{ when } l = 0$$

where the y_{lm} are the multipolar spherical harmonic angular functions in real form, the $R_l = N_l \, r^n \exp - (\kappa' \, \zeta \, r')$ are Slater type radial functions in which N_l is a normalization factor. The P_{lm} are the multipole coefficients which are refined in the least squares process. The normalization of y_{lm} implies that a P_{lm} value of 1 transfers one electron from the negative lobe of the y_{lm} function to the positive lobe. The zeta parameters are chosen to be consistent with atom or molecule optimized orbital exponents α ($\zeta = 2\alpha$ since $\rho(r) \propto \psi^2$). The n exponents of the Slater function are chosen with $n \geq 1$ for proper Coulombic behavior satisfying Poisson's equation as r goes to zero. Values of n for the multipoles were at first suggested by Hansen and Coppens [4b] based on the product of Slater orbitals $\psi(n'l')\,\psi(n''l'')$ which have pre-exponential radial dependence $r^{n'-1}$ and $r^{n''-1}$ by analogy with hydrogenic orbitals.

These values have been found suitable for first row atoms. For second row atoms, the optimal value of n must be found by inspection of the residual density maps:

$$\Delta\rho_{res}(\underset{\sim}{r}) = V^{-1} \sum_{\underset{\sim}{r}} [(|F_o| - |F_m|) e^{i\varphi_m}] \exp(-2\pi i H \cdot \underset{\sim}{r}) \quad (7)$$

where the m suffix designates the multipole atom model in the structure factor calculation; φ_m is the phase of the structure factor calculated with the multipole model.

In the multipole model, the refinable parameters are P_v, P_{lm}, κ and κ'. The limit $l_{max} = 4$ is used for the description of second row atoms and first row transition metals because of the d orbitals (l = 2 for the wave functions) whereas l_{max} is usually taken equal to 3 for C, O, N atoms and 1 for hydrogen.

The local axis on each atom is defined by the program's user (figure 1); this flexibility is very interesting for large molecules possessing non-crystallographic local symmetry and/or containing chemically equivalent atoms. These symmetry and chemical constraints permit us to reduce the number of the κ, P_v, P_{lm} electron density parameters in the least squares process [14]. For example, all atoms of a benzene ring may be constrained to have the same density parameters and a local symmetry mm2 can be applied to each atom.

Modeling the electron density by spherical harmonic functions is equivalent to modifying the form factor of the atom by adding [13]:

$$\Delta f(H) = i^l P_{lm} \times f_l(H) \times y_{lm}(u,v)$$

One application of the multipole refinement is the calibration of theory: for example, Souhassou *et al.* have shown [15] that almost quantitative agreement on the bond peaks in the deformation density maps between theory and experiment can be obtained only if using extended triple ξ basis sets with polarization functions in the theoretical calculations. Furthermore, the difficulty of experimental electron density increases only moderately with the size of the molecule compared to the fourth power dependence of theoretical calculations. *This is a significant advantage for experimental studies compared to theory*, especially when one has to study supramolecules or biological molecules like polypeptides or small proteins. Stevens and Klein have experimentally studied chemical carcinogens and opiate molecules with a good precision [16] and the calculation of the experimental electrostatic potential permits us to understand their reactivity. The experimental electron density of leu-Enkephalin, $3H_2O$ (C_{28} N_5 O_7 H_{37} . $3H_2O$; tyr^1-gly^2-gly^3-phe-leu) in its folded conformation has been published by Pichon-Pesme *et al.* [17]. The maps obtained are very accurate: figures 2a,b show for example the static deformation maps:

$$\delta\rho_{stat}(\underset{\sim}{r}) = (P_v K^3) \rho_v(Kr) - N\rho(\underset{\sim}{r}) + \Sigma \Sigma K'^3 R_{nl}(Kr') P_{lm} Y_{lm} \quad (8)$$

of the tyrosine group and of the phenylalanine residue: the bonding density in the C-C bonds agrees quantitatively with that obtained in smaller molecules. *Theoretical studies of such big molecules by HF SCF methods are at present beyond the computational possibilities and experimental determination of electron density is therefore the only tool to get accurate electrostatic parameters.*

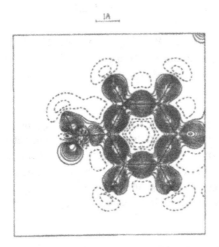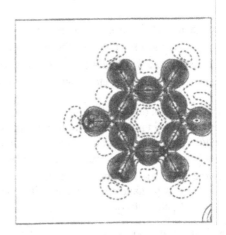

Figure 2. Experimental static deformation maps in the plane of the tyrosine (a) and of the phenylalanine residue, (b) in leu-enkephalin [17b]

In conclusion, the multipole model has proved its value in modeling valence electron densities; however, some work has still to be done to access its accuracy in the intermolecular region. Also, a comparison with other imaging methods like maximum entropy has to be done [18].

2.2. ELECTROSTATIC POTENTIAL DERIVED FROM THE X-RAY EXPERIMENT

2.2.1. *Electrostatic potential*
Electron density mapping permits a direct comparison with theory but does not provide much information about chemical reactivity or intermolecular interactions. This information can be obtained by inspection of the electrostatic potential generated by molecules in their outer part. Bertaut (1952) then Stewart (1979) have used Fourier components of the electron density to evaluate a variety of electrostatic properties [19] of molecules in the crystal: for example, the electronic part of the electrostatic potential at a point $\underset{\sim}{r}$ inside the crystal:

$$V_e(\underset{\sim}{r}) = \int \frac{\rho(\underset{\sim}{r'})}{|\underset{\sim}{r}-\underset{\sim}{r'}|} \, d^3\underset{\sim}{r}\,'$$

can be calculated by expanding $\frac{1}{|\underset{\sim}{r}-\underset{\sim}{r'}|}$ in reciprocal space, one gets:

$$V_e(\underset{\sim}{r}) = \frac{1}{\pi V} \sum_{\underset{\sim}{H}} \frac{F(\underset{\sim}{H})}{H^2} \exp{-2i\pi(\underset{\sim}{H}.\underset{\sim}{r})} \tag{9}$$

The electrostatic potential V_e is the inverse Fourier transform of $H^{-2} F(\underset{\sim}{H})$. However, there is a singularity for $\underset{\sim}{H} = \underset{\sim}{0}$ [20]. In order to avoid this problem, one can calculate the deformation electrostatic potential at $\underset{\sim}{r}$:

$$\Delta V(\underset{\sim}{r}) = \frac{1}{\pi V} \sum_{\underset{\sim}{H}} \frac{1}{H^2} (|F_m|e^{i\phi_m} - |F_s|e^{i\phi_s}) \exp{(-2\pi i\, \underset{\sim}{H}.\underset{\sim}{r})}$$

where $|F_m|$, ϕ_m, $|F_s|$, ϕ_s are the moduli and phases of the static structure factors ($U_{ij} = 0$) calculated respectively from the multipole model and from the promolecule.

Another method is to calculate the molecular electronic electrostatic potential by replacing $\rho(\underset{\sim}{r}')$ by its multipole formulation: the quantity obtained represents the electrostatic potential of a molecule removed from the crystal lattice. The first calculations have been performed by the Pittsburgh group (Stewart, Craven, He and coworkers) [21]; electrostatic potential calculations were also derived from the Hansen Coppens electron density model [22]: the total electrostatic potential including the nuclear contribution may be calculated as:

$$V(\underset{\sim}{r}) = V_{core}(r) + V_{val}(r) + \Delta V(\underset{\sim}{r}) \tag{10}$$

with (figure 3)
$$V_{core}(r) = \frac{Z}{|\underset{\sim}{r}-\underset{\sim}{R}|} - \int_0^\infty \frac{\rho_{core}(\underset{\sim}{r}')d^3\underset{\sim}{r}\,'}{|\underset{\sim}{r}-\underset{\sim}{R}-\underset{\sim}{r}'|}$$

$$V_{val}(r) = - \int_0^\infty \frac{\rho_{val}(\underset{\sim}{r}')d^3\underset{\sim}{r}\,'}{|\underset{\sim}{r}-\underset{\sim}{R}-\underset{\sim}{r}'|}$$

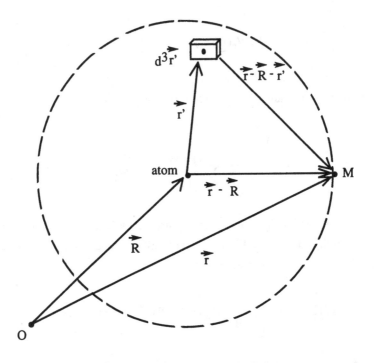

Figure 3. Calculation of electrostatic potential

where ρ_{val} is the spherical term of the total valence electron density. $\Delta V(\underline{r})$ is the deformation potential due to the asphericity of the electronic cloud (for explicit formulation, see reference [22b]). This aspherical term calculated from dipolar, quadripolar, octopolar and hexadecapolar terms of the electron density vanishes very quickly as soon as the distance between the molecule and the observation point increases.

2.2.2. *Fit of the electrostatic potential in terms of charge and moments [23]*

The electrostatic potential calculated for a molecule extracted from the crystal can be fitted by atomic charges and multipole moments; given a good choice of the sampling points where $V(\bar{r})$ is calculated, the spherical part of the electron density (from a kappa refinement for example) is perfectly fitted by point charges centered on atoms. The R factor, defined by:

$$R = \left[\sum_{}^{M} |V_0 - V_{cal}|^2 / \sum_{}^{M} V_0^2 \right]^{1/2}$$

was found always less than 1%. Furthermore, the obtained net charges are independent of the molecular conformation and are as expected, equal within standard deviations to the charge determined from the kappa refinement [23a]. Fitting the total potential including the aspherical part of the electron density, necessitates however higher atomic moments [23b]: the electrostatic potential may be fitted against multipole moments using a Buckingham expansion:

$$V(\underline{r}) = \sum_{j} \sum_{lm} (q_{jlm} / |\underline{r} - \underline{R}_j|^{l+1} y_{lm\pm}(\theta,\varphi) \tag{11}$$

where q_{jlm} represents the m moment of order l of atom j and $|\underline{r} - \underline{R}_j|$ the distance between the observation at $\underline{r}$ and the atom j ($\underline{R}_j$). Figure 4 gives the experimental molecular electrostatic potential (MEP) calculated from formula (9) and from reference [22b], compared to the potential calculated from the multipole moments fit. The agreement is excellent, 1 Å outside the molecule. Then, these moments should be used for further modeling.

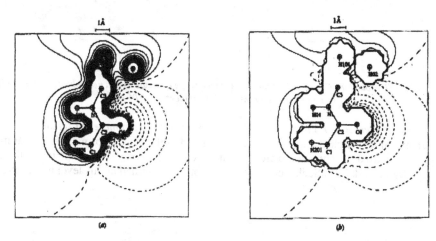

Figure 4. Comparison between observed MEP and the potential calculated from the multipole fit; the potential is calculated aroud a peptide plane [23b] (contours 0.05 e / Å)

2.2.3. Examples of Molecular Electrostatic Potential (MEP)

The electrostatic potential has been calculated on small organic and biological molecules [24,25,26]. The potential, which is very conformation dependent, reveals precisely the nucleophilic sites of the molecule. It is an excellent tool to predict information about complexation and host-guest interactions. For example, the charge density for 2,2'-dimethyl 6,6'-diphenyl 4,4'-bipyrimidine in an s-*trans* conformation was used to predict the potential for the s-*cis* conformation of the molecule and hence to rationalize Cu+ complexation [27]; this prediction of the Cu+ conformation was based on the transferability concept [28] *i.e.* the charge density at first order is very nearly invariant under a conformation change. Another application of this calculation is the characterization of hydrogen bonds as a valley of constant and close to zero potential

32

along the H...Y bond [24,26,29]. As an example, figure 5 shows the modification of $V(\vec{r})$ (formula (10)) due to H bonds in the tripeptide tyrosyl-glycyl-glycine [29]. The negative region created by O_{31} and O_{32} (minimum -0.5 e Å) (figure 5a) is almost totally neutralized by four hydrogen bonds ($O_{32}...HN_{11}$, $O_{31}...HW_1$, $O_{31}...HN_2$ and $O_{32}...HO_5$).

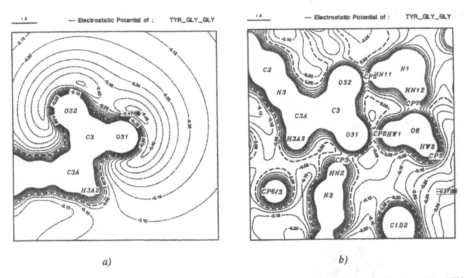

a)   b)

Figure 5. MEP around the COO group of the isolated tyr-gly-gly molecule a) and neutralization of the MEP with intermolecular interactions b) (contours 0.05 e Å⁻¹)

2.2.4. Experimental $V(\vec{r})$ in the crystal provides information in material science

As an example, the electrostatic potential generated in the cavities of natrolite, a natural zeolite, was used to estimate the electrostatic interaction energy of the Na⁺ ion with the zeolite framework [30]. Such a calculation cannot be performed at this level by theory.

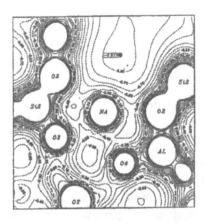 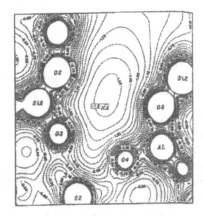

Figure 6. Electrostatic potential in natrolite with (a) and without (b) the sodium contribution in natrolite [30] (contours 0.1 e Å⁻¹)

Figure 6 gives the total electrostatic potential (calculated from (8)) in the oxygen cavity where the sodium ion lies; this potential is calculated (a) with and (b) without the sodium ion. The Na^+ cation lies exactly at the minimum of (b) and then, the electrostatic energy of Na^+ is -21.6 eV.

2.2.5. *Comparison between experimental and ab initio calculated electrostatic potential [31]*

Using extended basis sets of good quality (triple zeta + polarization functions), the theoretical evaluation by HF calculation of non bonded interactions has been performed on L-arginine phosphate monohydrate [31] where hydrogen bonding occurs in the asymmetric unit of the crystal; $V(\vec{r})$ was compared to the experimental potential derived from an accurate X-(X+N) diffraction experiment: when arginine and phosphate groups interact via the N_1-H_4...O_2 hydrogen bond, the region around the phosphate group splits into two negatives zones separated by the H_4...O_2 line. In both experimental (figure 7a) and theoretical (figure 7b) models, a region of $\sim$ 0.10, 0.15 e $Å^{-1}$ characterized by a saddle point (at 0.8 Å from H_4) appears; as discussed by Craven and coworkers [24] and as shown in figure 5 above, this is the signature of an H bond. The $\Delta V = V_{exp} - V_{theo}$ map shows that the experimental potential is slightly higher (0.05 e $Å^{-1}$) than the theoretical one. As the difference is of the order of magnitude of the experimental error, this is not significant, therefore a quantitative agreement is reached only when high level calculations are used.

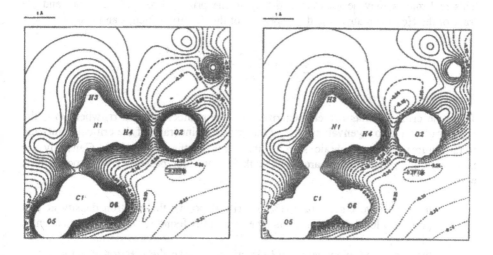

Figure 7. Electrostatic potential in the HB region: a) experimental, b) HF calculation (contours 0.05 e $Å^{-1}$)

2.3. TOPOLOGY OF THE ELECTRON DENSITY: APPLICATION TO INTERMOLECULAR INTERACTIONS

Bader discovered a method [32] based on the topology of the total electron density, which leads to an atomic classification of the properties of matter: "the form of the total electron distribution in a molecule is the physical manifestation of the forces acting within the system". The characteristics of the total electron density topology may be analyzed by a search of the critical points (minima, maxima, or saddle points) located at given points r for which the gradient of the density is zero.

$$\nabla\rho(\underset{\sim}{r}) = \underset{\sim}{0}$$

Whether a function is a minimum or a maximum at an extremum is determined by the sign of its second derivative or curvature at this point; in three dimensional space for a given set of coordinates axis, the curvature is determined by the Hessian matrix whose elements are:

$$H(i,j) = \frac{\partial^2\rho(\underset{\sim}{r})}{\partial x_j \partial x_i}$$

This real matrix may be diagonalized to give the principal axis of curvature and the trace of the Hessian matrix, *i.e.* the Laplacian of the density, which is an invariant.

$$\nabla^2\rho = \underset{\sim}{V} \cdot \underset{\sim}{V}\rho = \frac{\partial^2\rho}{\partial x^2} + \frac{\partial^2\rho}{\partial y^2} + \frac{\partial^2\rho}{\partial z^2}$$

Then, the critical points are characterized by two numbers, ω and σ, where ω is the number of non-zero eigenvalues of H at the critical point (rank of the critical point) and σ (signature) is the algebraic sum of the signs of the eigenvalues. Generally for molecules, the critical points are all of rank 3; then, four possible critical points may exist:

- three positive eigenvalues ((3,+3) critical point): the electron density at that point is a local minimum; this type of point is found for example in the center of a cage;
- three negative eigenvalues ((3,-3) critical point): the electron density at that point is a local maximum; they are usually found at the positions of the nuclei;
- two positive eigenvalues ((3,+1) critical point). The density is a minimum in the plane containing the two positive curvatures and maximum in the perpendicular direction; this type of critical point is found at the center of a ring formed by n atoms covalently bonded; a hydrogen bond may replace a covalent bond;

- two negatives curvatures ((3,-1) critical point). The density is maximum in the plane containing the two negative curvatures and minimum along the perpendicular line (bond path); these points are found in covalent bonds, when associated with a high value of ρ.

The Laplacian of the electron density is also related to the total energy by the virial theorem. The sign of the Laplacian determines which of the kinetic energy or potential energy is in excess in the total energy. In regions of space where the Laplacian is negative and electronic charge is concentrated, the potential energy dominates (this is the case for covalent bonds and lone pairs). Conversely, in regions where the Laplacian is positive, the total energy is dominated by the kinetic energy (hydrogen bonds, ionic bonds *etc.*) and there is a local electron density depletion.

Then, analysing the electron density topology requires the calculation of $\underline{\nabla}\rho$ and of the Hessian matrix; after diagonalization one can find the critical points; in a covalent bond characterized by a (3,-1) critical point, the positive curvature λ_3 is associated with the direction joining the two atoms covalently bonded *i.e.* the bond path and the λ_2, λ_1 curvatures characterize the ellipticity of the bond by:

$$\varepsilon = \left| \frac{\lambda_1 - \lambda_2}{\lambda_2} \right|$$

For example, ε would increase with the π character of a double bond, but would be close to zero for a triple bond.

I will only focus on some experimental results. The experimental topological analysis [34] can be performed for molecules removed from the crystal lattice in the same way as for the electrostatic potential calculations (see above) or in the crystal. The crystal field effects are not absent, even in the first case.

Few topological analyses of X-X experimental densities have been performed to date. Due to the finite resolution of the experiments, they require a combination of experimental results for the valence electron distribution, more diffuse in real space – *i.e.* more contracted in reciprocal space - with theoretical core electron density usually calculated from good quality atomic wave functions [34]. As an example, De Titta and N. Li [33] collected high resolution, very high quality, 100K, X-Ray data on two forms of glycouryl (Cmcm and Pnma) and on biotin and chainless biotin. Glycouryl is a bicyclic, *cis* fused ring compound, each ring of which resembles chemically the ureido ring of biotin. Souhassou [34] has performed a topological analysis of the multipolar electron density resulting from a Hansen-Coppens refinement against De Titta data. It is very interesting to see that an excellent transferability between the experimental properties of the fragments exists, confirming Bader's work [32] on theoretical densities of peptide bonds as well as the results of Pichon *et al.* [28] (transferability of experimental electron density parameters, see below).

36

The same type of calculations have been performed using experimental X-Ray structure factors on crystalline phosphoric acid, N-acetyl-α,ß-dehydrophenylalamine methylamide and N-acetyl-l-tryptophan methylamide by Souhassou [34], on urea, 9-methyladenosine, on imidazole by Stewart [35], on l-alanine [36] and on annulene derivatives [37] by Destro and coworkers. These last authors collected their X-Ray data at 16K [38]. Stewart [35] showed that the positions of the (3,-1) critical points from the promolecule are very close to those of the multipole electron density, but that large differences appear when comparing the density, the Laplacian and the ellipticities at the critical points. Destro showed that the results obtained may be slightly dependent on the refinement model.

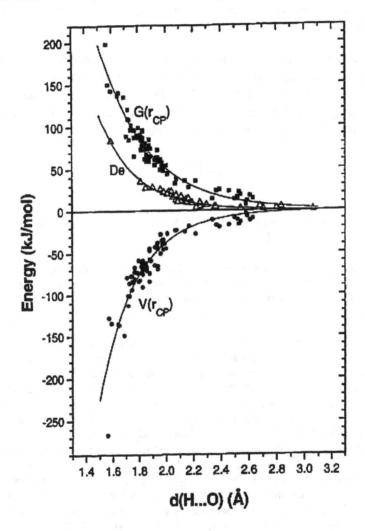

Figure 8. Kinetic energy density, $G(r_{cp})$, (K/Mole per atomic unit volume), potential energy density, $V(r_{cp})$, and dissociation energy, De, (K/Mole) versus H...O (Å) [39b]

An interesting and new application of the topological analysis of the charge density is the characterization of hydrogen bonds [39]. This analysis permits us to correlate the properties of the electron density at the hydrogen bond critical point with the interatomic distance. Furthermore, the analysis of the Laplacian of ρ shows without any doubt the covalent character of strong hydrogen bond as in ferroelectric (KH_2PO_4) or antiferroelectric ($NH_4H_2PO_4$) material (O...H...O with O...O = 2.5 Å). Using the Abramov equation [40] which estimates the kinetic energy density at the critical point, it also was shown that, for closed shell interactions, the kinetic energy density and the potential energy density at the critical point depend exponentially on the H...O distance (figure 8). Furthermore, a correlation with hydrogen bond dissociation energy De was found (also shown on figure 8).

3. Some Other Applications

3.1. NON LINEAR OPTICAL MATERIALS

Intense research work and many crystallographic studies are devoted to non linear optical materials, due to the applications in telecommunication, lasers, *etc*. Inorganic or organic non centrosymmetric materials with highly delocalized electrons may be excellent candidates for second harmonic generation. The most promising inorganic material is KTP (potassium titanium oxide phosphate, $KTiOPO_4$) [41]. KTP crystallizes in the polar space group $Pna2_1$; X-ray high resolution studies were performed at 10K, 100K and at room temperature [41,42] in order to determine precisely the π electron density of the titanium oxygen shortest bond (Ti = 0 = 1.73 Å compared to $\approx$ 2.10 Å for the other bonds) and also the spontaneous polarization:

$$\left|\vec{Ps}\right| = \frac{\left|\vec{\mu}_{unitcell}\right|}{V}$$

has been calculated after a judicious choice of the origin *i.e.* the center of symmetry of the high temperature phase (Pnam). The author [42] finds 0.16 e C m^{-2} compared to 0.20 from optical macroscopic measurements.

Baert and Zyss [43] have been working on organic non-linear materials; as an example, they have recently performed a careful electron density study of N-(4-nitrophenyl)-L-prolinol, which possesses donor and acceptor groups at *para* positions of a phenyl $\bullet$ system. From this experiment, they were able to calculate the dipole moment by several methods: Molly refinement, discrete Boundary, Stockholder principle ($6.9 < \left|\vec{\mu}\right| < 9.6$

Debyes, $\vec{\mu}$ almost aligned along the donor-acceptor axis of the molecule). Furthermore, given some restrictions (no local field), they are able to estimate the hyperpolarizability [44] tensor which is a function of the octopolar moment of the molecule.

3.2. d-ORBITAL POPULATIONS

Experimental charge density is also a powerful tool to visualize d electrons in first row transition metals [7]. Then, a test of wave functions may also be applied to first row transition metal complexes for which X-ray can give accurate d electrons populations of the metal ground state. Hence, the experiment allows distinction between contenders for the leading contributing configuration to the ground state, as the different configurations usually correspond to very different spatial distribution (which are seen by X-rays), even if these configurations are very close in energy. Therefore, *two configurations very close in energy cannot be distinguished by theory whereas they can by X-rays*. As an example, the reader is referred to the electron densities of metal porphyrins [14].

However, the multipole model in this case may fail due to inadequate radial functions: Larsen, Iversen and Figgis [45] show that more than one radial function is necessary to describe accurately the metal-ligand interaction after a careful analysis of 10K data. In fact, as soon as experiment becomes more and more reliable (very low temperature, two dimensional detectors, synchrotron data, multi-wavelengths, *etc.*), more sophisticated models have to be used and the need of more and better radial functions becomes evident. Another problem related to radial functions is the estimate of electron density in the intermolecular region: the multipole refinement deals with atom centered electron density and is maybe not adequate for extracting electron density in the intermolecular region. Spackman is presently working on this problem.

3.3. TRANSFERABILITY OF ELECTRON DENSITY: APPLICATION TO MACROMOLECULAR SYSTEMS, SUPRAMOLECULES OR PROTEINS

A future direction of research is the use of the electron density parameters, determined for small molecules, in modeling macromolecular properties. Pichon *et al.* have reported [28] *the transferability of such parameters and their potential use in biocrystallography*: transfer of multipole parameters as fixed contributors in the least squares refinement gives much more precise thermal motion parameters. A prospective paper on the application to protein refinement and related charge density is given by Jelsch *et al.* [46]. This transferability concept has also been used in materials science to study the evolution of U^{ij} thermal parameters in ferroelectric and antiferroelectric materials [47] ($NH_4H_2PO_4$, $K_{(x)}NH_{4(1-x)}H_2PO_4$).

3.4. INTERMOLECULAR INTERACTIONS AND ONE DIMENSIONAL MATERIALS [48]

The topology of the experimental electron density of the one-dimensional organic metal *bis* (thiodimethylene)-tetrathiafulvalene tetracyanoquinodimethane (BTDMTTF-TCNQ) (figure 5) has been performed after an X-ray multipolar refinement against 100K data; at this temperature, the disorder induced by the charge density wave, is very small and permits accurate charge density measurements. In this material, segregation between anionic and cationic stacks along $\bar{c}$ is observed.

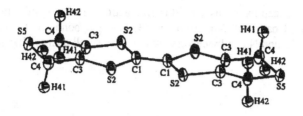

BTDMTTF

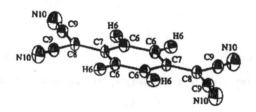

TCNQ

Figure 9a. ORTEP view of BTMTFF-TCNQ: cation and anion

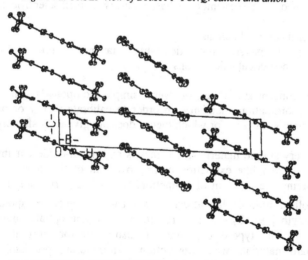

Figure 9b. ORTEP view of BTMTTF-TCNQ: cation and anion stacks in the crystal

The gradient of ρ map (figure 10) shows that the charge transfer (0.7 e⁻ measured from Kappa refinement of the X-X experiment) occurs via an interaction between S_s (cation) and the C≡N triple bonds of the anion. It leads to a (3,-1) critical point between S5 and the C≡N triple bond. This (3,-1) critical point behaves as a two dimensional attractor

40

which is a very exceptional case; as this interaction involving 2D attractors is very unstable, the bond bath will flip from the (3,-1) triple bond CP to the stable (3,-3) critical points of C_9 and N_{10} and vice versa with a very small energy change.

a) $\vec{\nabla}\rho$ BTDMTTF   b) TTF ...S$_5$

Figure 10. $\nabla\rho(\vec{r}\ )$ *functions of BTDMTTF in the cation plane a) and of the whole BTDMTTF-TCNQ in the*

$$C_9\text{-}S_5\text{-}C_9.$$

4. Conclusions and Future Directions

The results obtained from experimental charge density in most scientific areas provide:

- tests for theoretical calculations,
- estimation of physical quantities which may not be presently obtained from accurate theoretical calculations.

Therefore, we concentrate either on very simple compounds, on intermolecular interactions or coordination chemistry, in order to calibrate theory or on complicated systems with potential interest for which theory does not give a satisfactory answer.

In the case of very simple materials, like metals and metal-oxides, many of problems have still to be solved like extinction or improvement of the charge density model. Rather than solving the problem of extinction, which is a very complicated task, one should try to collect multiple data at various short wavelenghts, extrapolating the I($\vec{H}$) to zero wavelength [49]. Synchrotron radiation is the appropriate source to perform these experiments. This type of experiment is also useful for coordination complexes, minerals and hard materials, where absorption is an important problem. Moreover, the high flux of the SR beam, combined with the low divergence permits us to measure weak reflections with excellent statistics. This leads to a simple question: is it possible nowadays to collect synchrotron diffraction data with an accuracy better than what we obtain in our laboratories? Problems like non stability of the beam, *i.e.* monitoring the X-ray beam "seen" by the crystal, data reduction on two dimensional detectors, CCD and Imaging Plates, are not totally solved. Recently, however, extinction-free synchrotron data on Al_2O_3 ($\lambda = 0.214$ Å) have produced high quality deformation electron density maps which compare very well with theory [50].

Most crystallographic laboratories will possess in the very near future diffractometers equipped with CCD detectors. Even if all problems in acquiring very accurate data are not completely solved now, one can guess that numerous, good or excellent quality [51], charge density studies will be published soon due to the considerable time gains (less than a week, whatever the unit cell studied for CCD compared to more than one month for medium to big systems on a CAD4 diffractometer). This huge amount of data will totally change the experimental charge density field; then, if we suppose now that we have accurate F's questions about the electron density modeling arise. Is the multipolar model robust enough, is it refinement strategy dependent, user dependent? Do we need better radial functions? To answer part of these questions, a new International Union of Crystallography (IUCr) project concerning the electron density modeling exists: both theoretical and experimental F's of Al_2O_3 are available to any researcher who wants to participate in the project. Any participant will get these F's and refine them with his appropriate model in order to recover the best $\rho(\vec{r})$; then, anonymous comparisons will be made.

Multipole refinement is not the only way to describe the electron density: all the methods deriving from maximum entropy, MEM, (Collins [52], Bricogne [18,53], Sakata [54], Feil, Briels and de Vries [55], Papoular [56]) can be used and programs like BUSTER (Bricogne) SMEEDT (Sakata) begin to be available. This new method must be calibrated and tested because it is more and more evident that MEM results highly depend on the prior probability used in these calculations [18]: a very stimulating discussion concerning this problem was held at the XVIIth IUCr meeting in Seattle during one of the Charge, Spin and Momentum densities sessions which concluded that the results coming from MEM calculations must be interpreted with very great care especially when one looks to very tiny details of charge density like non-nuclear attractors (large dynamic range of the features observed by MEM). Therefore this mathematical technique still needs basic research before being applied routinely by non-specialists.

One can guess that all these developments both experimental (CCD, synchrotron) or theoretical (new models, MEM) will lead to more and more interesting results, especially for molecular material, biomolecules and will help the development of solid state electron density theory.

Finally, diffraction experiments on excited states [57], or under electric field [58,59], redeveloped by Coppens' group will certainly appear in the coming years with application to theoretical chemistry and material science: for example, Coppens *et al.* [59] were able to determine the crystal structure at 138K of sodium nitroprusside $[Fe(CN)_5NO]^{2-}$ ion in an extremely long-lived electronic excited state produced by long term Ar^+ laser illumination ($\lambda = 485$ nm) of a single crystal, with $\underset{\sim}{K} // \underset{\sim}{a}$ and polarization $// \underset{\sim}{c}$. A metastable population of about one third excited molecules was obtained. Their X-ray analysis shows a 0.060(9) Å lengthening of the Fe-N bond and C-

42

Fe-N, C-Fe-C angular variations. Furthermore, they were also able to analyse their data with a multipolar model and to determine the deformation density of the excited state of the molecule, knowing the ground state from another X-Ray experiment. These types of experiments will certainly enable us to get information on second order properties of matter like non linear susceptibilities, very important for example for the understanding and the design of non linear optical materials.

5. Acknowledgments

Some of the work described here has been made in the Laboratoire de Cristallographie et Modélisation des Matériaux Minéraux et Biologiques (LCM³B - UPRESA 7036). I am very grateful to my collaborators: Pr Y. Dusausoy, Drs N. Bouhmaida, S. Dahaoui, E. Espinosa, N.E. Ghermani, C. Jelsch, N. Hansen, H. Lachekar, V. Pichon-Pesme, M. Souhassou and Mrs. A. Clausse who I thank very much. This work was performed under grants of the University Henri Poincaré, Nancy 1 and of the CNRS.

6. References

1. Coppens, P. (1967) *Science* **158**, 1577.
2. a) Stewart, R.F. (1969) *J. Chem. Phys.* **51**, 4569-4577.
    b) Stewart, R.F. (1973) *J. Chem. Phys.* **58**, 1668-1676.
    c) Stewart, R.F. (1976) *Acta Cryst.* **A32**, 565-574.
3. a) Hirshfeld, F.L. (1971) *Acta Cryst.* **B27**, 769-781.
    b) Hirshfeld, F.L. (1977) *Isr. J. Chem.* **16**, 198-201.
4. a) Coppens, P., Guru Row, T.N., Leung, P., Stevens, E.D., Becker, P. and Yang, Y.W. (1979) *Acta Cryst.* **A35**, 63-72.
    b) Hansen, N.K. and Coppens, P. (1978) *Acta Cryst.* **A34**, 909-921.
5. a) Stewart, R.F. (1979) *Chem. Phys. Lett.* **65**, 335-338.
    b) Stewart, R.F. (1982) *God. Jugosl. Cent. Kristallogr.* **17**, 1.
    c) Ghermani, N., Lecomte, C. and Bouhmaida, N. (1993) *Z. Naturforsch.* **48a**, 91-98.
    d) Spackman, M.A. and Weber, H.P. (1988) *J. Phys. Chem.* **92**, 794-796.
    e) He, X.M. (1984) Ph. D Thesis, University of Pittsburgh, PA, U.S.A.
    f) Stewart, R.F. and Craven, B.M. (1993) *Biophys. J.*, 000.
    g) Su, Z. and Coppens, P. (1992) *Acta Cryst.* **A48**, 188-197.
6. Stewart, R.F. (1969) *J. Chem. Phys.* **51**, 4569-4577.
7. Coppens, P. (1997) *X-ray charge densities and chemical bonding*, IUCr text on Crystallography **4**, IUCr, Oxford Science Publications.
8. Lehman, M.S. (1980) Electron and Magnetization Densities in Molecules and Crystals, edited by P.J. Becker, *NATO Advanced Studies Institute*, New York Plenum **B48**, pp. 287-322 and pp. 355-372.
9. a) Seiler, P. (1985) Static and Dynamic Implications of Precise Structural Information, edited by A. Domenicano in Lectures Notes, *International School of Crystallography*, Hargittaï I. and Murray-Riest P., CNR, Rome, Italy, pp. 79-94.
    b) Seiler, P. (1987) *Chimia* **41**, 104-116.
10. Blessing, R.H. (1990) Lectures Notes, tutorial on Accurate Single Crystal Diffractometry, edited by Blessing R.H., Am. Cryst. Assoc. Meeting, New Orleans, LA, Dayton, Ohio, Polycrystal Book Service, USA.
11. a) Larsen, F.K. (1991) The Application of Charge Density Research to Chemistry and Drug Design, edited by G.A. Jeffrey and J.F. Piniella, *NATO Advanced Studies Institute*, New York, Plenum **B250**, pp. 187-208.

b) Blessing, R.H. and Lecomte, C. (1991) The Application of Charge Density Research to Chemistry and Drug Design, edited by G.A. Jeffrey and J.F. Piniella, *NATO Advanced Studies Institute*, New York, Plenum **B250**, pp. 155-185.

12. Blessing, R.H. (1986) *Cryst. Rev.* **1**, 3-58.

13. Lecomte, C. (1991) The Application of Charge Density Research to Chemistry and Drug Design, edited by Jeffrey G.A. and Pinella J.F., *NATO Advance Studies Institute* **B250**, pp. 121-153.

14. Lecomte, C., Blessing, R.H., Coppens, P. and Tabard, A. (1986) *J. Amer. Chem. Soc.* **108**, 6942-6950 and publications cited in this paper.

15. Souhassou, M., Lecomte, C., Ghermani, N.E., Rohmer, M.M., Wiest, R., Bénard, M. and Blessing R.H. (1992) *J. Amer. Chem. Soc.* **114**, 2371-2382.

16. Stevens, E.D. and Klein, C.L. (1991) The Application of Charge Density Research to Chemistry and Drug Design, edited by Jeffrey G.H. and Piniella J.F., *NATO Advanced Studies Institute* **B250**, pp. 319-336.

17. a) Pichon-Pesme, V., Lecomte, C., Wiest, R. and Bénard, M. (1992) *J. Amer. Chem. Soc.* **114**, 2713-2715.

b) Wiest, R., Pichon-Pesme, V., Bénard, M. and Lecomte, C. (1994) *J. Phys. Chem.* **98**, 1351-1362.

18. Roversi, P., Irwin, J.J. and Bricogne, G. (1998) *Acta Cryst.* **A54**, 971-996.

19. a) Bertaut (1952) *J. Phys. Radium* **13**, 499.

b) Stewart, R.F. (1979) *Chem. Phys. Lett.* **65**, 335-338.

c) Stewart, R.F. (1982) *God. Jugosl. Cent. Kristallogr.* **17**, 1.

20. Becker, P. and Coppens, P. (1990) *Acta Cryst.* **A46**, 254.

21. Stewart, R.F. and Craven, B.M. (1993) *Biophys. J.* **65**, 998.

22. a) Su, Z. and Coppens, P. (1992) *Acta Cryst.* **A48**, 188-197.

b) Ghermani, N., Lecomte, C. and Bouhmaida, N. (1993) *Z. Naturforsch.* **48A**, 91-98.

23. a) Ghermani, N.E., Bouhmaida, N. and Lecomte, C. (1993) *Acta Cryst.* **A49**, 781-789.

b) Bouhmaida, N., Ghermani, N.E. and Lecomte, C. (1997) *Acta Cryst.* **A53**, 564-575.

24. a) Swaminathan, S. and Craven, B.M. (1984) *Acta Cryst.* **B40**, 511-518.

b) Weber, H.P. and Craven, B.M. (1987) *Acta Cryst.* **B43**, 202-209.

c) Kloosten, W.T., Swaminathan, S., Naumi, R. and Craven, B.M. (1992) *Acta Cryst.* **B48**, 217-227.

d) Weber, H.P. and Craven, B.M. (1990) *Acta Cryst.* **B46**, 532-538.

e) Spackman, M.A., Weber, H.P. and Craven, B.M. (1988) *J. Amer. Chem. Soc.* **110**, 775-782.

f) Stewart, R.F. (1991) The Application of Charge Density Research to Chemistry and Drug Design, edited by Jeffrey G. and Piniella J.F., *NATO Advanced Studies Institute* **B250**, pp. 63-102

g) Swaminathan, S., Craven, B.M., Spackman, M.A. and Stewart, R.F. (1984) *Acta Cryst.* **B40**, 398.

h) Epstein, J., Ruble, J.R. and Craven, B.M. (1982) *Acta Cryst.* **B38**, 140.

25. a) Lecomte, C., Souhassou, M., Ghermani, N., Pichon-Pesme, V. and Bouhmaida, N. (1990) Studies of Electron Distributions in Molecules and Crystals, edited by Blessing R.H., *Trans. Amer. Cryst. Ass.* **26**, pp. 91-103.

b) Lecomte, C., Ghermani, N.E., Pichon-Pesme, V. and Souhassou, M. (1992) *J. Mol. Struct. (Theochem)* **255**, 241-260.

26. Moss, G. and Feil, D. (1981) *Acta Cryst.* **A37**, 414-421.

27. Ghermani, N.E., Bouhmaida, N., Lecomte, C., Papey, A.L. and Marsura, A. (1994) *J. Phys. Chem.* **98**, 10202-10211.

28. Pichon-Pesme, V., Lecomte, C. and Lachekar, H. (1995) *J. Phys. Chem.* **99**, 6242.

29. a) Lachekar, H., Pichon-Pesme, V., Souhassou, M. and Lecomte C. (1998) *Acta Cryst.* **B**, in press.

b) Lachekar, H. (1997) Thèse de l'Université Henri Poincaré, Nancy 1.

30. a) Ghermani, N.E., Lecomte, C. and Dusausoy, Y. (1996) *Phys. Rev.* **B9**, 5231-5239.

31. a) Espinosa, E., Lecomte, C., Ghermani, N.E., Devemy, S., Rohmer, M.M., Bénard, M. and Molins, E. (1996) *J. Am. Chem. Soc.* **118**, 2501-2502.

44

b) Espinosa, E., Lecomte, C., Molins, E, Veintemillas, S., Cousson, A. and Paulus, W. (1996) *Acta Cryst.* **B52**, 519-531.

32. a) Bader, R.F.W. (1990) Atoms in molecules. A quantum theory, *Oxford University Press*, Oxford, U.K.
    b) Bader, R.F.W. and Laidig, K.E. (1990) in ref [55], pp. 1-21.
    c) Bader, R.F.W. and Essen, H. (1984) *J. Chem. Phys.* **80**, 1943.

33. Li, N., De Titta, G.D., Blessing, R.H. and Moss, G. (1990) 40th A.C.A. Meeting, New Orleans, Abst. PD05, p. 79.

34. a) Souhassou, M. (1993) *Personal communication.*
    b) Souhassou, M. and Blessing, R.H. (1998) *J. Applied Cryst.*, in press.

35. Stewart, R.F. (1991) The Application of Charge Density Research to Chemistry and Drug Design, edited by Jeffrey G. and Piniella J.F., *NATO Advanced Studies Institute* **B250**, pp. 63-102

36. a) Destro, R., Bianchi, R., Gatti, C. and Merati, F. (1991) *Chem. Phys. Lett.* **186**, 47-52.
    b) Gatti, C., Bianchi, R., Destro, R. and Merati, F. (1992) *J. Mol. Struc. (Theochem)* **255**, 409-433.

37. Bianchi, R., Destro, R. and Merati, F. in [13], p. 340.

38. Destro, R., Marsh, R.E. and Bianchi, R. (1988) *J. Phys. Chem.* **92**, 966-974.

39. a) Espinosa, E., Souhassou, M., Lachekar, H. and Lecomte C. (1998) *J. Amer. Chem. Soc.*, submitted.
    b) Espinosa, E., Molins, E. and Lecomte, C. (1998) *Chem. Phys. Lett.*, in press.

40. Abramov, Yu A. (1997) *Acta Cryst.* **A53**, 264-272.

41. Hansen, N., Protas, J. and Marnier, G. (1991) *Acta Cryst.* **B47**, 660-672.

42. Dahaoui, S. (1996) Thèse Université Henri Poincaré, Nancy 1, France.

43. Fkyerat, A., Guelzim, A., Baert, F., Paulus, W., Heger, G., Zyss, J. and Perigaud, A. (1995) *Acta Cryst.* **B51**, 197-209.

44. Baert, F. *Private communication.*

45. Figgis, B.N., Inversen, B.B., Larsen, F.K. and Reynolds, P.A. (1993) *Acta Cryst.* **B49**, 794-806.

46. Jelsch, C., Pichon-Pesme, V., Lecomte, C. and Aubry, A. (1998) *Acta Cryst.* **D54**, 1306-1318.

47. Boukhris, A., Souhassou, M., Lecomte, C., Wyncke, B. and Thalal, A. (1998) *J. Phys. Cond. Matt.* **10**, 1621-1641.

48. Espinosa, E., Molins, E. and Lecomte, C. (1997) *Phys. Rev.* **B56**, 1820-1833.

49. a) Palmer, A. and Jauch, W. (1995) *Acta Cryst.* **A51**, 662-667.
    b) Hester, J.R. and Okamura, F.P. (1996) *Acta Cryst.* **A52**, 700-704.

50. Graafsma, H., Souhassou, M., Puig-Molina, A., Harkema, S., Kvick, A. and Lecomte, C. (1998) *Acta Cryst.* **B54**, 193-195.

51. Dahaoui, S., Jelsch, C., Lecomte, C. and Howard, J.A.K. (1998) BCA, Spring Meeting, April 1998.

52. Collins, D.M. (1982) *Nature* **298**, 49.

53. a) Bricogne, G. (1988) *Acta Cryst.* **A44**, 517-545.
    b) Bricogne, G. (1993) *Acta Cryst.* **D49**, 37-60.

54. Sakata, M. and Sato, M. (1990) *Acta Cryst.* **A46**, 469.

55. De Vries, R., Briels W. and Feil, D. (1994) *Acta Cryst.* **A50**, 383.

56. Papoular, R., Vekhter, Y. and Coppens, P. (1996) *Acta Cryst.* **A52**, 397.

57. Pressprich, M.R., White, M.A. and Coppens, P. (1993) *J. Amer. Chem. Soc.* **115**, 6444-6445.

58. Paturle, A., Graafsma, H., Shen, H.S. and Coppens, P. (1991) *Phys. Rev.* **B43**, 14683-14691.

59. Graafsma, H., Heunen, G., Dahaoui, S., El Haouzi, A., Hansen, N. and Marnier, G. (1997) *Acta Cryst.* **B53**, 565-567.

DYNAMIC PROCESSES AND DISORDER IN MATERIALS AS SEEN BY TEMPERATURE-DEPENDENT DIFFRACTION EXPERIMENTS

H.B. BUERGI, S.C. CAPELLI
Labor für Kristallographie
Freiestr. 3, Universität
CH-3012 BERN (Switzerland)

1. Introduction

The primary results of crystal structure determinations are long lists of uninformative numbers: tables of atomic coordinates and anisotropic displacement parameters (ADP's). To be useful the numbers need to be transformed. Atomic coordinates are easily converted into interatomic distances, angles, conformational descriptors and other quantities of interest in structural chemistry. Atomic displacement parameters which encode information on motion and disorder, are more difficult to decipher. This is because *interatomic* or *correlation* ADP's which describe the coupling of atomic displacements, are lost in Bragg diffraction. Here it is shown how information on correlated atomic displacements can be retrieved from the temperature dependence of ADP's.

Models of atomic and molecular motion derived from ADP's are complementary to those obtainable from solid state NMR and vibrational spectroscopies. They are a basis for estimating specific heat, entropy and other thermodynamic functions. They provide information on the intra- and intermolecular forces and thus on some of the factors determining crystal packing, occurrence of polymorphs and phase transitions. Many material properties such as conductivity and superconductivity, are intimately related to the interactions between nuclear and electronic motions. Since the procedures described below allow to distinguish motion from disorder, they also provide a more detailed description of the static aspects of structure, a point which is relevant for the determination of accurate electron density distributions. Quite generally, a physically consistent interpretation of ADP's over a range of temperatures adds a dynamic dimension to crystal structure analysis.

2. Background

Atomic positions and displacement parameters derived from X-ray or neutron Bragg scattering experiments represent averages over time and space, over countless vibration periods and billions of unit cells. In a perfectly ordered crystal ADP's represent the extent of vibrational displacements of atoms from their equilibrium positions. They

J.A.K. Howard et al. (eds.),
Implications of Molecular and Materials Structure for New Technologies, 45–58.
© 1999 *Kluwer Academic Publishers. Printed in the Netherlands.*

increase with increasing temperature. As early as 1913 Debye showed that this effect can be studied experimentally, because it decreases Bragg intensities, especially at high scattering angles [1]. Real crystals deviate to a smaller or larger degree from exact space group symmetry. An atom may occupy one equilibrium position at a given time and in a given unit cell, but different ones at other times or in other unit cells. If the distance between possible positions in the averaged unit cell are small compared to the experimental resolution, only a mean atomic position can be obtained and the effects of disorder are found in the ADP's [2].

2.1. PROBABILITY DENSITY FUNCTION

The distribution of an atom in a crystal is specified by its *probability density function* $P(\Delta r)$. $P(\Delta r)dV$ denotes the probability of finding an atom in the volume element dV displaced by the vector Δr from the mean atomic position r. In X-ray and neutron diffraction studies, atomic pdf's are usually taken to be Gaussian

$$P(\Delta r) = [(\det U^{-1})^{1/2}/2\pi)^{3/2}] \cdot \exp(-\Delta r^T U^{-1} \Delta r/2) \tag{2.1}$$

U^{-1} is the inverse of the second moment matrix $U = \langle \Delta r \, \Delta r^T \rangle$ of the distribution. The displacements can be dynamic or static or both. Provided that sufficient high angle diffraction data are available, the higher cumulants of non-Gaussian distributions can also be determined [3].

If all eigenvalues of U are positive, the surfaces of constant probability defined by the quadratic form

$$\Delta r^T U^{-1} \Delta r = \text{const} \tag{2.2}$$

are ellipsoids enclosing some definite probability for atomic displacement. This is the basis for the „vibration ellipsoids" shown in many illustrations of crystal structures [4] (for an example see *Figure 3*). The mean square displacement amplitude in a direction defined by a unit vector n is

$$\langle u^2 \rangle_n = n^T U n \tag{2.3}$$

with n referred to the unit vectors a^*/a^*, b^*/b^*, c^*/c^*. By varying the direction of n, surfaces can be constructed, even for tensors with positive and negative eigenvalues. Such surfaces are useful for inspecting difference tensors ΔU between experimental U tensors and those calculated from kinematic or dynamic models of motion [5] (for examples see *Figure 4* and *5*).

2.2. CORRELATION OF MOLECULAR MOTION

Atomic motions are correlated to a smaller or larger degree. For example, two bonded atoms tend to move in concert in the direction of the bond; similarly, the displacements of the three hydrogen atoms in a rotating methyl group are highly correlated even though they take place in different directions. In contrast atoms at opposite ends of a long flexible molecule may move in a practically uncorrelated fashion. ADP's pertaining to a single temperature contain none of this information [6]. It can only be guessed at on the basis of general knowledge about the flexibility of bond distances, angles and torsion angles. From such guesses one can attempt to parametrize atomic ADP's in terms of a limited number of motions of the molecule as a whole, or of one part of a molecule relative to another.

2.3. RIGID BOND AND RIGID MOLECULE TESTS

For a perfectly rigid molecule the interatomic distances do not change, no matter how the molecule moves. Therefore all atoms are displaced in phase. For every pair of atoms, the mean square amplitudes of displacement must be equal along the interatomic unit vector $\mathbf{n}_{kl} = (\mathbf{r}_k - \mathbf{r}_l)/|\mathbf{r}_k - \mathbf{r}_l|$. It follows that

$$\Delta(kl) = \mathbf{n}_{kl}^T (U(k) - U(l)) \, \mathbf{n}_{kl} = 0 \tag{2.4}$$

Although no molecule is perfectly rigid, the atomic displacements due to bond length deformation in typical organic molecules are much smaller than those of other vibrations and tend to be the same along the bond vector. As a consequence $\Delta(kl)$ along bonds is small, often on the order of $\sigma(\Delta)$ and generally less than 10^{-3}Å^2. This is Hirshfeld's rigid bond criterion; it has been found to hold for well-refined organic structures based on good diffraction data [7]. The rigid bond test has been generalized to all interatomic vectors in a molecule to detect the presence or absence of significant molecular non-rigidity [8].

2.4. RIGID BODY MODEL

If a molecule can be considered rigid, its motion is described in terms of six *molecular* displacement coordinates, three translations and three librations. The corresponding second moments are the mean square translation amplitudes **T**, the mean square libration amplitudes **L**, and the mean square screw couplings **S** between rotation and translation [9]. Expressed in terms of **T**, **L** and **S**, the ADP's of an atom at position $\mathbf{r}_k = \{x_k \; y_k \; z_k\}^T$ become

$$U(k) = A_k \begin{bmatrix} T & S \\ S^T & L \end{bmatrix} A_k^T, \quad A_k = \begin{bmatrix} 1 & 0 & 0 & 0 & z_k & -y_k \\ 0 & 1 & 0 & -z_k & 0 & x_k \\ 0 & 0 & 1 & y_k & -x_k & 0 \end{bmatrix} \tag{2.5}$$

The matrix A_k performs the transformation from molecular to atomic mean square displacements. It is the link between the uninformative list of ADP's and a more intelligible description of molecular motion. $U(k)$ is a small part of the more general, molecular mean square amplitude matrix Σ^x. For a molecule with N atoms the dimensions of Σ^x are 3N by 3N and the $U(k)$'s are the diagonal 3 by 3 blocks of Σ^x

$$\Sigma^x = A \begin{bmatrix} T & S \\ S^T & L \end{bmatrix} A^T \tag{2.6}$$

The full matrix $\mathbf{A}$ is built from the atomic blocks $\mathbf{A_k}$ according to $\mathbf{A}^T = [\mathbf{A_1}^T$ $\mathbf{A_2}^T...\mathbf{A_k}^T...\mathbf{A_N}^T]$.

The off-diagonal blocks of Σ^x are the *interatomic* or *correlation* ADP's. They contain the information on the couplings of atomic displacements. Unfortunately they cannot be obtained from Bragg diffraction. They could be calculated if $\mathbf{T}$, $\mathbf{L}$ and $\mathbf{S}$ were known. However, in a rigid-body analysis the reverse route is followed: the elements of $\mathbf{T}$, $\mathbf{L}$ and $\mathbf{S}$ are extracted from the U(k). This raises the question whether the parameters of rigid body motion can be determined uniquely and completely from the incompletely known Σ^x. They cannot! The lack of correlation ADP's implies, that only differences between the diagonal elements of $\mathbf{S}$ can be determined. Their sum, the trace of $\mathbf{S}$, is indeterminate [9]. For practical calculations the arbitrary assumption is made that the trace of $\mathbf{S}$ is zero (unless symmetry requires it to be zero). This implies that the numbers obtained for the diagonal elements of $\mathbf{S}$, the screw couplings, differ from their true values by an unknown additive constant.

For nonrigid molecules additional molecular coordinates φ can be introduced, *e.g.* an internal rotation coordinate. The relationship between atomic and molecular displacement becomes

$$\Sigma^x = \mathbf{A} \begin{bmatrix} \mathbf{T} & \mathbf{S} & \langle \varphi t \rangle \\ \mathbf{S}^T & \mathbf{L} & \langle \varphi l \rangle \\ \langle \varphi t \rangle^T & \langle \varphi l \rangle^T & \langle \Phi \rangle \end{bmatrix} \mathbf{A}^T = \mathbf{A} \, \Sigma^I \, \mathbf{A}^T \qquad (2.7)$$

$\langle \Phi \rangle$ is the mean square internal displacement, $\langle \varphi t \rangle$ and $\langle \varphi l \rangle$ represent the couplings between internal motion and translation or libration. The entire molecular mean square amplitude matrix is abbreviated as Σ^I. The elements $\mathbf{A_k}$ of $\mathbf{A}$ now show an extra column $\{a_{k1} \ a_{k2} \ a_{k3}\}^T$ which transforms the displacements along the molecular coordinate φ into the ADP's. Adding more internal motions increases the dimensions of Σ^I and $\mathbf{A}$ accordingly. Elegant as this approach may seem, it has its problems. The number of indeterminate parameters or linear combinations of parameters of Σ^I increases rapidly with the number of internal displacement coordinates [10, 11]. As in the case of the trace of $\mathbf{S}$, this is due to the fact that no correlation ADP's are available from Bragg diffraction. Again, the problems are circumvented by making arbitrary assumptions. The results of such analyses obviously depend on the validity of the assumptions.

These difficulties raise the question whether the information on collective atomic displacements is lost irretrievably in Bragg scattering. Fortunately, it is not. Correlation ADP's, *i.e.* the off-diagonal blocks of Σ^x, can be recovered from the temperature dependence of the ADP's, *i.e.* the diagonal blocks of Σ^x [12].

3. Temperature Dependence of Mean Square Amplitudes

The motions of a molecular crystal can be viewed in two alternative ways. In one model, the vibrations of the crystal as a whole are considered. They are the result of the interactions between all atoms in the crystal, especially between those in the same molecule and those in neighbouring ones. The couplings within a molecule tend to be stronger than those between molecules. This is the phonon or lattice dynamical point of view [13]. In the second model the strongly coupled atoms in a molecule are assumed to move under the average influence of all neighbouring molecules, *i.e.* of the crystal lattice. This is called the „molecular mean field model"; and is used throughout the rest of the discussion. A well known and simple example is the harmonic oscillator.

3.1. THE HARMONIC OSCILLATOR

The harmonic oscillator describes the motion of an atom of mass m (or of a molecule of reduced mass μ) subjected to a restoring force proportional to the displacement Δx. Such a force gives rise to a parabolic potential *(Figure 1)*. The corresponding Schrödinger equation is

$$\left[-(\hbar^2/2m)\,\partial^2/\partial\Delta x^2 + f\Delta x^2/2 \right]\psi(\Delta x) = E\,\psi(\Delta x) \tag{3.1}$$

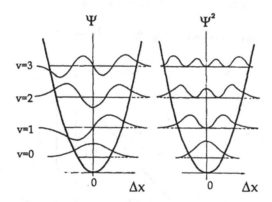

Figure. 1. Harmonic oscillator: Potential energy surface V(Δx), eigenstates E_v and wavefunction $\psi_v(\Delta x)$ (left), probability density functions $|\psi_v(\Delta x)|^2$ (right).

The quantizied energy states of the harmonic oscillator are:

$$E_v = hv\,(v+1/2) = \hbar\,\sqrt{f/m}\,(v+1/2) \tag{3.2}$$

The wavefunctions expressed in terms of the variable Δx are:

$$\psi_v(\Delta x) = \left(2^v\,v!\,\sqrt{\pi}/\alpha\right)^{-1/2} H_v(\alpha\Delta x)\,e^{-(\alpha\Delta x)^2/2} \tag{3.3}$$

50

H_v are Hermite polynomials and $\alpha = \sqrt{mf/\hbar}$. The square of the wavefunctions for a given state v represent the probability density function for that state:

$$P_v (\Delta x) = |\psi_v(\Delta x)|^2 \qquad (3.4)$$

The total probability density function is the Boltzmann weighted sum over all states v and depends on temperature

$$P(\Delta x,T) = N \Sigma |\psi_v(\Delta x)|^2 \exp(hv(v + \tfrac{1}{2})/k_BT) \qquad (3.5)$$

where N is a normalization factor. The second moment $<\Delta x^2>$ of $P(\Delta x,T)$ is the mean square amplitude measured in diffraction experiments.

$$<\Delta x^2>(T) = \int \Delta x^2\, P(\Delta x,T)d\Delta x = h/(8\pi^2 mv)\, \coth(hv/2k_BT) \qquad (3.6)$$

In the simple case of a harmonic oscillator, v and $<\Delta x^2>$ are alternative and equivalent quantities for characterizing motion.

At very low temperatures, $<\Delta x^2>$ is temperature independent and inversely proportional to v, while at very high temperature it is proportional to T and inversely proportional to v^2 (*Figure 2*)

$$<\Delta x^2>(0) \qquad = \hbar / (4\pi mv) \qquad = \delta_o \qquad (3.7)$$

$$<\Delta x^2>(T\to\infty) = k_BT/(4\pi^2 mv^2) = sT \qquad (3.8)$$

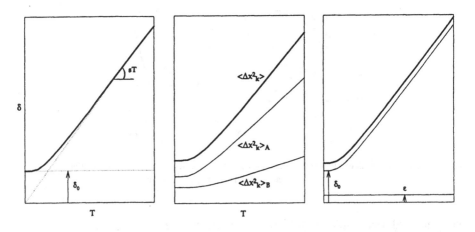

Figure. 2. Mean square displacements as a function of temperature. Left: displacement of a single harmonic oscillator; the zero point limiting value δ_o and the slope s reached assymptotically at very high temperatures are indicated. Center: displacement $<\Delta x^2_k>$ composed of contributions $<\Delta x^2_k>_A$ and $<\Delta x^2_k>_B$ from two independent oscillators A and B. Right: displacement composed of a temperature dependent contribution and a temperature independent one. The latter may be due to either high frequency motion or disorder.

There are two simple tests to check experimental values of $<\Delta x^2>$ for harmonic behaviour. The first requires that the high temperature values of $<\Delta x^2>$ extrapolate to zero for $T = 0$ (eq. 3.8). The second one requires that

$$s/\delta_o^2 = 4mk_B/\hbar^2 \tag{3.9}$$

i.e. the ratio of the quantities characteristic of high and low temperature behaviour must be proportional to the mass of the oscillator, but independent of its frequency (eq. 3.7). For experimentally determined ADP's it is often found that the first condition is met, whereas the second one is not. The reason for this will become clear after discussing the slightly more complicated, but also more interesting case of a diatomic molecule.

3.2. THE HOMONUCLEAR DIATOMIC MOLECULE

The motions along the axis of this diatomic can be described in terms of molecular translation and bond length deformation. The motions perpendicular to the axis are translation and libration. In both cases the mean square amplitudes can arise from one, the other or from any mixture of these motions (*Figure 3*). This indeterminacy is sometimes called the second phase problem in crystal structure determination: ADP's measure the extent of atomic displacements but not their relative phases.

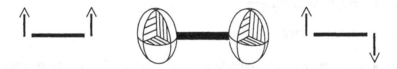

Figure. 3. Homonuclear diatomic. Two possible interpretations of the vertical components of the ADP's in terms of molecular translation (left) or libration (right; ADP's given as equal probability surfaces, eq. 2.2, [4]).

Assuming quadratic potentials for the two motions A and B, the total mean square amplitude of the two atoms k and l in a given direction, *e.g.* along the molecular axis x, is a sum of two terms

$$\begin{aligned}
<\Delta x_k^2> (T) &= <\Delta x_l^2> (T) \\
&= <\Delta x_k^2>_A (T) + <\Delta x_k^2>_B (T) \\
&= \hbar/ (4\pi \mu_A \nu_A) \coth (h\nu_A/2k_BT) + \hbar/ (4\pi \mu_B \nu_B) \coth (h\nu_B /2k_BT)
\end{aligned} \tag{3.10}$$

For the high and low temperature regimes, the expressions become

$$<\Delta x_k^2> (0) = \hbar (\mu_A^{-1} \nu_A^{-1} + \mu_B^{-1} \nu_B^{-1})/(4\pi) \tag{3.11}$$

$$<\Delta x_k^2> (T \to \infty) = k_BT (\mu_A^{-1} \nu_A^{-2} + \mu_B^{-1} \nu_B^{-2})/(4\pi^2)$$

where μ_A and μ_B are the reduced masses for motions of type A and B, respectively, and can be expressed in terms of atomic masses and positions.

From measurements at a single temperature it is impossible to distinguish between the separate contributions $<\Delta x_k^2>_A$ and $<\Delta x_k^2>_B$ (*Figure 2*). If $<\Delta x_k^2>$ can be measured in both the zero point and classical regimes, however, v_A and v_B and thus $<\Delta x_k^2>_A$ and $<\Delta x_k^2>_B$ can be determined, because eqs. 3.11. are linearly independent in these v's. This solves the second phase problem mentioned above. As a consequence it becomes possible to calculate the interatomic or correlation mean square amplitudes. For the homonuclear diatomic $\Delta x_k = \Delta x_l$ for motion A, $\Delta x_k = -\Delta x_l$ for motion B and the result is simply

$$\begin{aligned}<\Delta x_k \Delta x_l> (T) &= <\Delta x_k \Delta x_l>_A (T) + <\Delta x_k \Delta x_l>_B (T) \qquad (3.12)\\ &= <\Delta x_k^2>_A (T) - <\Delta x_k^2>_B (T)\end{aligned}$$

In terms of the notation used in section 2 (eqs. 2.6 and 2.7), we write

$$\Sigma^x = \begin{bmatrix} \left\langle \Delta x_k^2 \right\rangle & \left\langle \Delta x_k \Delta x_l \right\rangle \\ \left\langle \Delta x_k \Delta x_l \right\rangle & \left\langle \Delta x_l^2 \right\rangle \end{bmatrix} \qquad (3.13)$$

Note that the off-diagonal elements of Σ^x which are not available from Bragg diffraction experiments, have been determined indirectly from the observed temperature dependence of the diagonal elements.

3.3. A SPECIAL SITUATION

Suppose the temperature range available for diffraction ($0<T<T_{melt}$) is too small for normal mode B to be excited significantly. In this case the limiting behaviours are

$$<\Delta x_k^2> (0) = \hbar (\mu_A^{-1} v_A^{-1} + \mu_B^{-1} v_B^{-1})/(4\pi) \qquad (3.14)$$

$$<\Delta x_k^2> (T\to\infty) = k_B T/(4\pi^2 \mu_A v_A^2) + \hbar/(4\pi\mu_B v_B)$$

Extrapolation of the high temperature expression to 0 K leads to the temperature independent contribution $\hbar/(4\pi\mu_B v_B)$, different from zero. The latter is not necessarily due to high frequency motion only, it can also have contributions from positional disorder of the atoms (*Figure 2, right*). In the special case discussed here, as well as in the more general situation discussed above, the ratio s/δ_0^2 obtainable from $<\Delta x_k^2>(T)$ no longer follows the simple relationship given in eq. 3.9.

The *important message* in this section is quite simple. Although ADP's measured at a single temperature say nothing about the correlation of atomic displacements, the temperature dependence of ADP's does contain such information.

4. Molecular Mean Field Theory of Atomic Displacement Parameters [12]

To describe the ADP's of a three dimensional molecule, many more than two harmonic oscillators are needed. For an N-atomic, nonrigid molecule there are 3N normal modes. The problem is to relate atomic displacements to molecular displacements, and then to temperature dependent normal mode displacements [14]. The first transformation has been described in terms of the matrix A in eq. 2.7. The second transformation is

$$\Sigma^l = L \delta L^T \tag{4.1}$$

The elements δ_i of δ are the temperature dependent mean square amplitudes of the normal modes

$$\delta_i = \hbar/(4\pi\nu_i) \coth (h\nu_i/2k_BT) \tag{4.2}$$

as given in section 3. Note that in the general theory the mass is included in L according to

$$L L^T = G \tag{4.3}$$

G is the matrix of reduced masses which depends only on atomic masses, bond distances, angles and torsion angles [15]. L can be written as the product of the orthonormal eigenvector matrix V and the lower triangular matrix g.

$$L = g V \tag{4.4}$$

where $g g^T = G$. Introducing this into eq. 4.1 and combining with eq. 2.7, gives the desired relationship

$$\Sigma^x(T) = A \Sigma^l(T) A^T = A g V \delta(T) V^T g^T A^T \tag{4.5}$$

It expresses the temperature dependence of the ADP's, including the diagonal and off-diagonal blocks of Σ^x (T), in terms of the temperature dependent mean square amplitudes of the normal modes.

In practice some of the 3N normal modes are of high frequency. Their contribution to the ADP's is generally small and does not change significantly in the temperature range $0<T<T_{melt}$ available for single crystal diffraction. This requires that eq. 4.5 is split into a temperature dependent first part accounting for low frequency vibrations and a temperature independent correction ε accounting for high frequency vibrations.

$$\Sigma^x (T) = A g V \delta(T) V^T g^T A^T + \varepsilon \tag{4.6}$$

Note that the relevant dimensions of A, g, V and δ are now smaller (and most of the time much smaller) than 3N. The known quantities in eq. 4.6 are: the diagonal 3 by 3 blocks of Σ^x (T), i.e. the ADP's U(k;T) observed at several suitably chosen temperatures

T; the matrix A which only depends on the type of internal coordinates chosen for the analysis; the matrix g which only contains the known atomic masses and coordinates. The unknown quantities, to be determined from the $U(k;T)$, are: (1) The effective frequencies v_i occurring in the elements δ_i of the diagonal matrix δ; (2) the elements of V relating normal modes to mass adjusted molecular coordinates; (3) the elements of the matrix ε which contains 3 by 3 diagonal blocks $\varepsilon(k)$, one for each atom.

The observations $U(k;T)$ are nonlinear functions of the v_i and the elements of V and linear functions of the elements of ε. Expansion into a Taylor series to first order leads to the following expression for each of the six elements of $U(k)$:

$$U_{ij}(k;T) = [A\,g\,V_0\,\delta(v_0,T)V_0^T\,g^T\,A^T\,]_{ij} + \tag{4.7}$$

$$[A\,g\,\{\Delta V\,\delta(v_0,T)\,V_0^T + V_0^T\,\delta(v_0,T)\Delta V^T +$$

$$V_0\,d\delta(v_0,T)/dv\,\Delta v\,V_0^T\}g^T A^T]_{ij} + [C\varepsilon(k)C^T]_{ij}$$

$$= U_{ij,0}(k\;;T) + \Delta U_{ij}(k, \Delta V, \Delta v, \varepsilon\,(k)\;;T)$$

with V_0, v_0 representing trial values of the unknowns. The $\varepsilon(k)$ are defined in a local atomic coordinate system; C is a transformation matrix to the working coordinate system. This allows $\varepsilon(k)$ to be the same for groups of related atoms in the molecule. The orthonormality condition $V\,V^T = I$ is expanded analogously, leading to

$$0 = V_0\Delta V^T + \Delta V\,V_0^T \tag{4.8}$$

where 0 is the appropriate zero matrix. A computer program based on these expressions has been written to determine the elements of V and $\varepsilon(k)$ and the frequencies v by an iterative least-squares procedure from the observed $U(k;T)$. The program which is still under development, is available from the authors.

The above approach is based on a harmonic model of motion. Diffraction data collected at different temperatures show effects of anharmonicity: cell constants increase with increasing temperature and consequently the dependence of ADP's on temperature may differ from that shown in *Figure 2*. We are presently expanding the model of eq. 4.5 to include anharmonic effects. Systematic errors in the observed ADP's may also be a problem. They must be minimized by appropriate corrections of the Bragg diffraction data for extinction, absorption and crystal decay. In our experience systematic errors can sometimes be absorbed into an overall ε-tensor affecting all atoms equally or they tend to accumulate in the normal modes with large translational contributions and affect the other results much less.

5. Examples

The general theory will be illustrated with two examples, hexadeuterobenzene and a complex between benzene and silver perchlorate. The examples allow to test the adequacy of the theory for explaining the observed temperature dependence of ADP's, to assess the influence of an organic and an ionic environment on the motions of the benzene molecule, to compare results from different isotopic species of the same molecule, and to study the motion of the spherical ClO_4^- ion which is often disordered.

Neutron diffraction data of C_6D_6 have been measured at 15 and 123 K [16]. The ADP's have been analyzed in terms of six temperature dependent contributions, accounting for molecular librations and translations, and two temperature independent ϵ-tensors accounting for the intramolecular vibrations. The ϵ-tensors, one for carbon and one for deuterium, are defined in local atomic coordinate systems (Footnote to Table 2). The rms deviation between observed and calculated ADP's is 0.0003 $Å^2$, the rms value of $\sigma(U)$ is the same. The goodness of fit defined as $[\Sigma w(U_{obs} - U_{calc})^2/(n_{obs} - n_{par})]^{1/2}$ is 1.6 for 72 observations and 24 parameters. The wR-value defined as $[\Sigma w(U_{obs} - U_{calc})^2/\Sigma w\, U^2_{obs}]^{1/2}$ is 0.019; w is a weighting factor equal to $1/\sigma^2(U_{obs})$ [12].

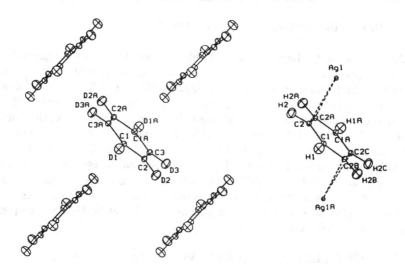

Figure 4. Packing of C_6D_6 in the crystallographic *a, c*-plane (15K, left) and coordination of C_6H_6 to Ag^+ in the ClO_4^- complex (18K, right); ADP's given as root mean square displacement surfaces, scale factor 1.54, eq. 2.3, [5]).

Neutron diffraction data of $AgClO_4.C_6H_6$ have been measured at 18, 78 and 158 K [17]. The ADP's of C_6H_6, ClO_4 and Ag have been analyzed separately. ADP's of the two molecules were modeled in terms of temperature dependent librations and translations, separate ϵ-tensors for C and H as for C_6D_6, and an overall diagonal ϵ-tensor for all atoms to account for systematic error of unknown origin (orthorhombic crystal coordinate system, diagonal elements 12, 12, -15(1).10^{-4} $Å^2$). For C_6H_6 the number of independent observations is 60, the number of refined parameters is 23, the goodness of fit is 0.98, the wR-factor is 0.019, for ClO_4^- the corresponding quantities are 33, 15, 2.4 and 0.026.

TABLE 1. Effective vibration frequencies and force constants for molecular libration and translation of C_6D_6 and of C_6H_6 in the $C_6H_6.AgClO_4$ complex.

Libration	$\parallel$ C1 – C1A	$\perp$ C1 – C1A	$\perp C_6$-ring
ν (C_6D_6)	70(1)	84(2)	60(1) cm^{-1}
ν (C_6H_6)	155(6)	103(2)	70(1)
f(C_6D_6)[a]	0.31(1)	0.45(2)	0.46(1) mdynÅ/rad^2
f(C_6H_6)[a]	1.2(1)	0.56(2)	0.51(1)
Translation			
ν (C_6D_6)	43(1)	45(1)	51(1) cm^{-1}
ν (C_6H_6)	57(1)	65(1)	81(2)^c
f(C_6D_6)[b]	0.10(1)	0.10(1)	0.13(1) mdynÅ^{-1}
f(C_6H_6)[b]	0.15(1)	0.19(1)	0.30(1)^c

[a]) $f = I_{ii} \nu^2 \cdot 5.891 \cdot 10^{-7}$, I_{ii} in amu Å^2 ν in cm^{-1}.
[b]) $f = M \nu^2 \cdot 5.891 \cdot 10^{-7}$, M in amu, ν in cm^{-1}
[c]) Eigenvector inclined to ring normal by ~ 30 deg and pointing in Ag1...Ag1A direction (*Figure 4*).

5.1. BENZENE IN DIFFERENT ENVIRONMENTS

Table 1 shows that the force constants for libration and translation are generally higher for the Ag-complex than for pristine benzene in agreement with the stronger interactions in the ionic compound. The biggest difference is found for libration about the C1-C1A axis (*Figure 4*). In $C_6H_6.AgClO_4$ this corresponds to changing the Ag...C bonds, whereas in C_6D_6 this motion affects only the weaker van der Waals contacts. An analogous, but smaller difference is found for the translations normal to the ring planes. The force constant for libration about the normal can be translated into a barrier B for molecular reorientation if a sixfold periodic potential is assumed (B = 288 f/n^2 (kcal/mol) for V = B(1-cosnϕ)/2). The two barriers are very similar, 3.7 and 4.1 kcal/mol. A recent reinterpretation of NMR measurements on benzene gives a value of 3.8 kcal/mol [18]; barriers reported for various transition metal-benzene complexes are in the range 3.7 to 6.8 kcal/mol [19].

5.2. ISOTOPE EFFECTS ON INTERNAL VIBRATIONS

The mean square amplitudes due to all internal vibrations are given by the ϵ-tensors compiled in table 2. The values extracted from the ADP's are compared with those obtained independently from an *ab-initio* force field [20]. Diffraction and spectroscopic values agree with respect to order of magnitude. The primary isotope effects for H and D are well reproduced. Note that ϵ_{11}(neutron) is systematically larger than ϵ_{11}(spectr), ϵ_{33}(neutron) is systematically smaller than ϵ_{33} (spectr). The discrepancy in ϵ_{11} points to inadequate treatment of anharmonicity in C-H/C-D stretching vibrations and that in ϵ_{33} to deficiencies in the out-of-plane force field. However, a detailed explanation of these differences requires further study.

TABLE 2. Mean square amplitudes of C, D and H from all internal vibrations (ε-tensors times $10^4Å^2$)[a]

	$\varepsilon_{11}(C)$	$\varepsilon_{22}(C)$	$\varepsilon_{33}(C)$	$\varepsilon_{11}(D/H)$	$\varepsilon_{22}(D/H)$	$\varepsilon_{33}(D/H)$
C_6D_6						
neutron diffraction	13(2)	7(2)	15(3)	53(2)	83(4)	110(5)
spectroscopy	13	8	16	44	89	133
C_6H_6						
neutron diffraction	18(1)	15(1)	15(-)	73(2)	128(2)	175(3)
spectroscopy	9	12	14	61	130	202

[a]) ε_{11} along C-D/C-H bonds, ε_{22} in plane perpendicular to C-D/C-H bonds, ε_{33} perpendicular to benzene plane.

5.3. THE PERCHLORATE ION

The vibration frequencies are in the range from 40 to 80 cm^{-1}. The contributions from internal motion are 3 to $9.10^{-4}Å^2$ for the chlorine atom, and 3 to $17.10^{-4}Å$ for the oxygen atoms. However, the discrepancy between observed and calculated ADP's is significantly worse (*Figure 5*). Preliminary calculations have shown that this could be due to a breakdown of the harmonic approximation. The vibrations which are affected most are (1) the libration about the axis perpendicular to the O2-Cl-O2A plane and passing through Cl and (2) the translation in the O2...O2A direction. Assuming a Grüneisen type dependence of these frequencies on temperature improves the fit significantly ($v(T) = v_0 (1 - \alpha T)$, α is an anharmonicity constant). The anharmonicity seems to be real since it coincides with anharmonic behaviour of the mean square amplitude of Ag in the direction of O2 and O2A and because the benzene shows no noticeable deviation from harmonic behaviour in this direction (corresponding to C1-C1A).

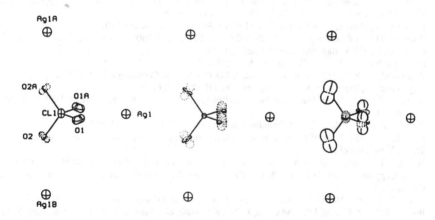

Figure 5. Difference displacement parameters U_{obs}-U_{calc} at 18, 78 and 158 K (given as root mean square displacement surfaces, solid lines indicate positive differences, dots indicate negative differences, scale factor 15.4, [5]).

58

6. Outlook

The examples discussed above show that consistent and physically plausible results can be obtained from the temperature dependence of ADP's. It has been shown elsewhere that the problems of indeterminacy encountered in rigid and non-rigid body analyses of single temperature ADP's, disappear with multitemperature data [12]. The effects of motion can be distinguished from those of disorder and to some extent, of systematic error in the ADP's [21]. Other examples studied (but mostly not published) include internal rotation in 2,2'-dimethylstilbene, N-methyluracil [12] and benzamide, out-of-plane deformations of urea [12] and anthraquinone. In some cases there is evidence for significant anharmonic contributions to the ADP's (Ag $ClO_4.C_6H_6$, hexamethylene-tetramine). The potential of the method to analyse motion and disorder in larger molecules, *e.g.* oligo- and polypeptides or nucleotides, remains to be explored. Since ADP's show directly which atoms move and by how much, they are complementary to spectroscopic frequencies which provide this information only very indirectly. The ADP's required for a comprehensive analysis are obtained relatively easily by the standard methods of structure determination. With the availability of fast CCD technology the measurement of reasonably accurate diffraction data over a range of temperatures is no longer a limiting factor for such studies.

7. References

1. Debye, P. (1913) *Verh. Dtsch. Phys. Ges.* **15**, 738-752.
2. Trueblood, K.N., Bürgi, H.B., Burzlaff, H., Dunitz, J.D., Gramaccioli, C.M., Schulz, H.H., Shmueli, U., and Abrahams, S.C. (1996) *Acta Crystallogr.* **A52**, 770-781.
3. Coppens, P. (1993) *International Tables for Crystallography*, Vol. B, *Reciprocal Space*, U. Shmueli, ed., Section 1.2.11. Kluwer Academic Publishers, Dordrecht.
4. Burnett, M.N., Johnson, C.K. (1996) *ORTEP-III, Oak Ridge Thermal Ellipsoid Plotting Program for Crystal Structure Illustrations*, ORNL-6895, Oak Ridge National Laboratory.
5. Hummel, W., Hauser, J., and Bürgi, H.B. (1990) *J. Mol. Graphics* **8**, 214-220.
6. Born, M. (1942) *Rep. Prog. Phys.* **9**, 294-333; von Laue, M. (1960) *Röntgenstrahl-Interferenzen*, 3rd ed., Akademische Verlagsgesellschaft, Frankfurt/Main.
7. Hirshfeld, F.L. (1976), *Acta Crystallogr.* **A32**, 239-244.
8. Rosenfield, Jr., R.E., Trueblood, K.N., Dunitz, J.D. (1978) *Acta Crystallogr.* **A34**, 828-829.
9. Schomaker, V., Trueblood, K.N. (1968) *Acta Crystallogr.* **B24**, 63-76.
10. Johnson, C.K. (1970) In *Thermal Neutron Diffraction*, Willis; B.T.M., ed.; Oxford University Press, London p. 132-160.
11. Dunitz, J.D., Schomaker, V., and Trueblood, K.N. (1988) *J. Phys. Chem.* **92**, 856-867.
12. Förtsch, M. (1997) *Normal Mode Analysis from Atomic Mean Square Displacement Amplitudes*, Ph.D. Thesis, University of Bern.
13. Willis, B.T.M., and Pryor, A.W. (1975) *Thermal Vibrations in Crystallography*, Cambridge University Press. Coppens, P. (1997) *X-ray Charge Densities and Chemical Bonding*, Oxford University Press, Chapter 2.
14. Cyvin, S.C. (1968) *Molecular Vibrations and Mean Square Amplitudes*, Universitets Forlaget, Oslo.
15. Wilson, E.B., Jr., Decius, J.C., and Cross, P.C. (1955) *Molecular Vibrations*, McGraw Hill, New York.
16. Jeffrey, G.A., Ruble, J.R., McMullan, R.K., and Pople, J.A. (1987) *Proc. Roy. Soc. London* **A414**, 47-57.
17. McMullan, R.K., Koetzle, T.F., Fritchie, Jr., C.J. (1997) *Acta Crystallogr.* **B53**, 645-653.
18. Goc, R. (1997) *Z. Naturforsch.* **52a**, 477-484.
19. For a review see: Braga, D. (1992) *Chem. Rev.*, **92**, 633-665.
20. Goodman, L., Ozkabak, A.G., and Thakur, S.N. (1991) *J. Phys. Chem.* **95**, 9044-9058.
21. Capelli, S.C., Bürgi, H.B. (1997) ECM-17, meeting abstract MS2.7-3, Lisboa, Portugal,).

ATOMIC DISPLACEMENT PARAMETERS, VIBRATIONAL SPECTRA AND THERMODYNAMIC FUNCTIONS FOR CRYSTALS: A STRONG CONNECTION

C. M. GRAMACCIOLI
Department of Earth Sciences, University of Milan
Via Botticelli 23, I-20133 Milan, Italy.

In crystallography, atomic motion is usually accounted for by considering the Fourier transform of the probability density function (pdf) of an atom around its mean position, which is often called a "temperature factor". For "anisotropic" displacement, such "temperature factors" $T_i(hkl)$ can be written as:

$$T_i(h) = \exp(-2\pi^2 \mathbf{H}^T \mathbf{U} \mathbf{H}) \tag{1}$$

where the $\mathbf{U}$'s ($=<u_i u_j>$) are second-rank tensors and are called *anisotropic atomic displacement parameters* or ADP's and $\mathbf{H}$ is the reciprocal-lattice vector referred to a set of Cartesian axes, or (more frequently) to a set of reference axes of unit length coinciding in direction with the reciprocal axes. There are also more complex formulations, involving higher terms in the cumulant expansion, leading to higher-rank tensors (see, for instance, Johnson [22-24]).

The importance of such parameters has long been underestimated by most crystallographers: even at present too many of them consider ADP's only as a relevant number of additional unknowns to be determined in the final refinement of crystal structures, with the only purpose of lowering the final R index notably. However, if accurate data are to be derived from X-ray crystallography for comparison with theoretical results, a physically reliable model accounting for thermal motion of the atoms involved and all the available experimental information in this respect becomes necessary.

Therefore, it might be quite important to find out what possible relationships there are with other disciplines. A possibility is given by *vibrational spectroscopy*. For an isolated molecule, in the harmonic approximation a "classic" procedure leads to the following expression for the equations of motion (for each atom, there are three of them, referring to the axes x, y and z):

$$m(\kappa)\partial^2 v(\kappa,t)/\partial t^2 = -\Sigma_\kappa \cdot \Phi(\kappa\kappa')v(\kappa',t) \tag{2}$$

where $m(\kappa)$ is the mass of the atom κ, $v(\kappa,t)$ is the displacement of the κ atom at the time t and $\Phi(\kappa\kappa')$ is the force constant relative to the interaction between the atoms κ and κ'.

J.A.K. Howard et al. (eds.),
Implications of Molecular and Materials Structure for New Technologies, 59–70.
© 1999 *Kluwer Academic Publishers. Printed in the Netherlands.*

After some rearrangement, the above expression can be written in matrix form as:

$$\omega^2\xi = D\xi \tag{3}$$

where ω ($=2\pi\nu$) is the angular frequency of the vibrational *normal mode* characterized by the eigenvector ξ and D is the so-called "dynamical matrix". It is quite easy to prove that:

$$D = M^{-1/2} D_o M^{-1/2} \tag{4}$$

where M is the diagonal mass matrix and D_o is a force constant matrix, whose elements are the *second derivatives of the potential energy* with respect to the atomic shifts. Therefore, if such derivatives of the potential energy are known, by diagonalizing D it is quite easy to obtain the vibrational frequencies and the eigenvectors describing the modes. It is quite important to consider that the eigenvectors of D contain *phase information.*

For a crystal, the problem is very similar, although the procedure followed in solving it is instead rather different, for evident reasons. Such a situation occurs because, whereas for a molecule the number of atoms N is a reasonable figure, for a crystal N is instead of the order of magnitude of Avogadro's number. Therefore *an extremely large number of different vibrational modes are present in the crystal,* and the procedure we have seen for molecules would lead to an utterly unreasonable order of D. For this reason, according to the Born-von Karman lattice-dynamical procedure, particular solutions of the equations of motion can be considered in the form of waves of wavelength λ:

$$v(\kappa,t) = -V_o(\kappa) \exp2\pi(-ivt + iqr_\kappa) \tag{5}$$

where q is the so-called "wave vector" ($|q| = 1/\lambda$) and r_κ defines the position of the atom κ. Usually r_κ is given in unit-cell (fractional) coordinates, whereas the components of q are given instead in reciprocal-lattice units. By developing the pertinent mathematical expressions, we arrive at the following *lattice dynamical* equation, exactly corresponding to the one we have already seen for molecules:

$$\omega^2\xi = D\xi \qquad \text{where } D = M^{-1/2}D_o M^{-1/2} \tag{6}$$

Here the elements of D_o are those of the force constant matrix as above, each one multiplied by $\exp2\pi(iqr_\kappa)$ and summed over all the translation-equivalent atoms in the lattice. The order of D is now equal to 3N, where N is however no longer of the order of magnitude of Avogadro's number, but is only equal to the *number of atoms in the primitive unit cell,* a quite reasonable figure. The drawback of the situation is that the *calculations must be repeated* for different values of the wave vector q within a reciprocal unit cell, equivalent to one "Brillouin zone". The density of sampled points is a delicate argument and a convenient strategy may shorten the computing time very considerably [11, 30].

It should always be kept in mind that for a crystal the number of vibrational modes is *enormous*. For each definite value of **q**, there are however only 3N modes, where N is the number of atoms in the primitive unit cell. For **q** = 0, there are three frequencies equal to zero, corresponding to translation of the whole crystal; the rest of the frequencies (not necessarily all) may correspond to Raman or infra-red active frequencies, depending upon the *symmetry representation* carried out by each mode. For values of **q** different from zero, there are instead no active modes either in the Raman or in the infrared spectra. An important feature for all the **q** = **0** modes (also mentioned as the modes at the origin of the Brillouin zone) is that there is no phase difference between the motion of atoms related to each other by unit-cell translations. Since Raman and infrared spectra *are only an insignificant part of the whole vibrational spectrum,* they can be quite important for a number of different purposes, but contrary to a quite widespread belief *no deduction concerning properties such as ADP's or thermodynamic functions can be reasonably inferred from such a limited information.* Therefore, for instance, the alleged correlations of the amplitude of libration of a certain molecular group (such as NO_2 *etc.*) or of isolated ions (such as CO_3^{2-} in carbonates) in the crystal with some observed Raman- or IR-active frequencies in the vibrational spectra are *not significant,* although in many cases they may look plausible. Contrarily to the usual "dogmas" of crystallography , for the vibrational modes relative to non-zero **q**, at any instant the positions of atoms related by unit-cell translations are no longer related by such translations, leading to non- Bragg diffuse scattering. In some aspects, the situation has points in common with incommensurate crystal structures.

On similar grounds, and for any value of **q**, also the instantaneous positions of atoms related to each other by rotational symmetry are no longer exactly so, although - as in the previous case- this difference can be described as involving phase only. Here, a wider diffusion among crystallographers of group theory, with a special emphasis on representations, leading to a detailed comparison of the concepts of symmetry as seen by crystallographers and spectroscopists, could be fundamental.

From well-known theory, since ADP's can be expressed as $U(\kappa) = \langle u(\kappa)u(\kappa)^T \rangle$, and by writing the displacement $u(\kappa)$ as a function of each vibrational mode in the crystal (see, for instance, [49]), the following expression can be deduced:

$$U(\kappa)=1/(Nm_\kappa)\Sigma_{jq}E_j(q)/(2\pi\nu)^2 e(\kappa \mid jq)e^*(\kappa \mid jq)^T \qquad (7)$$

where $E_j(q) = h\nu_j(q)\{^1/_2 +[\exp(h\nu_j(q)/kT)-1]^{-1}\}$ is the average energy of the mode (jq), $e(\kappa \mid jq)$ is the mass-adjusted polarization vector of the atom κ (which is the part of the eigenvector corresponding to the coordinates of κ); the summation is carried out with respect to all possible vibrational modes in the crystal, *i.e.* to each vibrational mode j for a certain value of **q** and then all over the Brillouin zone.

Such a procedure is *the correct way to evaluate ADP's theoretically,* provided of course that the motion is harmonic; a very similar expression, where $e(\kappa \mid jq)$ and $e^*(\kappa' \mid jq)^T$ are relative to different atoms, leads to the correlation tensors $U(\kappa\kappa') = \langle u(\kappa)u(\kappa')^T \rangle$, which are essential for bond-length correction in the most general case [23,24,44].

Lattice-dynamical calculations of ADP's can also provide arguments of notable weight in interpreting or solving several problems. For instance, in some garnets such as pyrope $Mg_3Al_2Si_3O_{12}$ or almandine $Fe_3Al_2Si_3O_{12}$ the disordered atom (Mg or Fe, respectively) is the only one whose observed ADP's are much greater than the corresponding calculated ones, indicating disorder [35] (see also below). For fayalite Fe_2SiO_4 and tephroite Mn_2SiO_4 the experimental values of the ADP's of Fe and Mn, respectively, have implied a re-interpretation of the vibrational spectra [34]. Similarly, in coesite, a modification of silica stable at high pressure, where one 180° O-Si-O angle is present in the structure, the experimental values of ADP's for all the atoms involved are very satisfactorily reproduced by such harmonic models [39], thereby showing such an angle to be real and not a sort of average in space of several positions rotated with respect to one another, as some authors seemed instead to believe.

In general, for lattice-dynamical procedures, the main problems are the following:

1) Difficulty of calculations, which could be quite long.
2) Availability of potential-energy functions (transferable or not?) in order to formulate reasonable force-constant matrices.

On the other hand, very useful results can be achieved, as for instance:

1) Essential information for solving the problem of thermal-motion correction of bond lengths, in the most general case.
2) Interpretation of vibrational spectra (Raman, IR, phonon dispersion curves)
3) Improved crystal-structure models.
4) Temperature-dependent estimates of thermodynamic functions starting from crystal-structure information (or even *a priori*, if a reasonable model can be formulated)
5) Empirical potentials derived from fit not only to crystal structure parameters, but also to vibrational frequencies and to values of thermodynamic functions.

The difficulties (and the length) of calculations, although rather notable, are facing a marked reduction in the near future. Such a perspective takes place either because of the very notable and rapid evolution of available computers, and also because of the possibility of developing adequate routines.

Since the order of the dynamical matrix **D** can be quite large [6N in the most general case, since the dynamical matrices are complex, unless $q=0$, and N as we have seen includes all the atoms in the (primitive) unit cell and not only those in the "asymmetric unit"], considerable computing time can be spared for molecular crystals if the calculations are carried out on normal-coordinate basis of the isolated molecule [1,2,14,18,25,46]; there are also interesting applications of such a technique to evaluate the perturbations occurring on passing from the isolated molecule to the one packed in the crystal, with important consequences in deriving the force field and thermodynamic properties [12,15]. The particular case of the molecule as a rigid body can be considered as the simplest possibility of this kind [4,27,28], and if only the lowest-frequency internal modes can be included in the complete lattice-dynamical treatment the extension to non-rigid bodies comes out quite easily. For a rigid body, in fact, the order of the dynamical matrices is equal to 6 times the number of molecules in the unit cell in the most general case, a quite reasonable figure, and for non-rigid bodies, the number of internal modes mixing with lattice vibrations rarely exceeds one third of the number of atoms in the molecule [7,14,16,17,18].

The treatment on molecular basis also leads to an interpretation of the motion which is easier to describe, and can be linked on one hand to the rigid body Schomaker-Trueblood (here onwards "ST") treatment [45], and on the other hand to the "exclusively crystallographic" non-rigid-body interpretations. Besides rigid molecules, where such a procedure directly leads to lattice-dynamical estimates of T, L, and S tensors, useful examples of the situation can be also provided by some molecules containing several phenyl rings, such as for instance o-terphenyl and tetraphenylmethane [3,43].

In these cases, using lattice dynamics [7,17] a notable agreement between theoretical and experimental results has been achieved (see Table 1):

Table 1. Examples of ADP's as B^{ij} ($Å^2 \times 10^4$) for carbon atoms in tetraphenylmethane (observed data from [43]; calculated data from [7].The temperature factor is in the form: $T = exp(-B^{ij} h_i h_j)$)

Atom	B^{11}	B^{22}	B^{33}	B^{12}	B^{13}	B^{23}
C0 obs	51(1)	51(1)	115(3)	0	0	0
cal	53	53	112	0	0	0
C1 obs	56(1)	60(1)	111(3)	-5(1)	10(2)	0(2)
cal	55	58	122	-2	0	-5
C2 obs	71(1)	83(2)	129(4)	-5(1)	-4(2)	-6(2)
cal	62	73	134	-6	-8	-7
C3 obs	83(2)	111(2)	134(4)	-25(2)	1(2)	-20(2)
cal	81	86	155	-19	-8	-23

In Table 2, calculated and observed ADP data for an ionic inorganic compound (mineral) are reported. The agreement is satisfactory, especially keeping in mind that the "true" uncertainty of experimental results is greater than the estimated standard deviation. Here, the only notable disagreement concerns the U^{22}'s of all the atoms at any temperature, and differences are almost always the same. This disagreement is very probably due to an error in the absorption correction for "obs2" [37]. Also the important contribution of zero-point motion is evident, since it generally amounts to more than one third of the corresponding room-temperature value.

Table 2. ADP's as U^{ij} ($\times 10^4$) at different temperatures for andalusite Al_2OSiO_4. For all the observed values (obs1 by Pilati *et al.* [37]; obs2 by Winter and Ghose [51] z.p. = zero-point contribution), the reported standard deviations are of the order of or lower than the last digit. The same denominations of atoms as in these works is maintained.

Atom and T(K)		U^{11}	U^{12}	U^{13}	U^{22}	U^{23}	U^{33}
Al1 (298)	calc	68	18	0	43	0	28
	obs1	69	17	0	51	0	37
	obs2	65	16	0	92	0	36
	z.p.	28	6	0	21	0	16
Al2 (298)	calc	37	-2	0	33	0	40
	obs1	36	0	0	45	0	46
	obs2	28	0	0	82	0	44
	z.p.	19	-1	0	18	0	20
Si (298)	calc	31	-1	0	31	0	30
	obs1	34	0	0	39	0	42
	obs2	22	0	0	76	0	39
	z.p.	15	0	0	15	0	15
OA (298)	calc	63	-10	0	35	0	38
	obs1	63	-5	0	47	0	51
	obs2	55	-9	0	88	0	47
	z.p.	33	-4	0	21	0	24
Al1(1073)	calc	231	64	0	141	0	86
	obs2	265	84	0	212	0	92
Al2(1073)	calc	119	-7	0	104	0	130
	obs2	102	-3	0	158	0	124
Si(1073)	calc	97	-4	0	98	0	95
	obs2	102	3	0	136	0	106
OA(1073)	calc	200	-34	0	101	0	113
	obs2	188	-30	0	158	0	127

The availability of empirical potential energy functions surely is a serious argument: a very important property of such functions should be the transferability from one substance to another, at least within a group of similar compounds. Among organic crystals, there seem to be no problems, since a number of transferable potentials either as intramolecular valence-force fields (VFF's) or also intermolecular have long been available. Especially the former ones have been developed and used by spectroscopists, whereas the latter ones (as "van der Waals potentials") have been developed especially by crystallographers, mostly in connection with crystal structure modelling [47,48]. It has been noticed that such intermolecular potentials, although often derived on different grounds, provide satisfactory results when applied to lattice vibrations (see, for instance, [9,10,15]). It should also be noticed that since the lowest frequency modes provide the dominant contributions in determining the values of ADP's and thermodynamic functions (see below), and since such modes can be described mainly as "lattice" modes, in these calculations intermolecular force fields are the most important, often by far.

For inorganic crystals, most of which are ionic and where molecules do not exist, the situation is more difficult, for many reasons. A first one is the much greater number of possibilities, arising from all possible combinations of a quite relevant number of different atoms and of different schemes of bonding, far greater than for organic compounds; a second one is linked to the atomic charge, which usually is much higher than for molecular compounds. Furthermore, Coulombic interactions do not fall to zero even at relevant distances and therefore the so-called "Ewald sums" carried out in the reciprocal lattice should be used instead, a lengthy process. Moreover, the presence of charge generates other charges by electrostatic induction, and- which is still more serious- for this reason charges vary during atomic motion, thereby possibly requiring more advanced models, where the external electron cloud does not necessarily follows the core (such as, for instance, "shell" models) and therefore accounting in some way for polarizability. For a long time, all these difficulties have restricted application of lattice dynamics only to very simple compounds, and transferability of empirical potentials has often been denied, although not always clearly.

However, recent experience with a considerable number of ionic compounds, mainly silicates and oxides, but also carbonates, has shown empirical force fields to be essentially transferable, much more than it had been assumed so far [26, 30-40]. Quite often, rather than to the models, the inadequacies are instead due to misinterpretation of the spectra to be fitted: such a misinterpretation may be serious, especially for the lowest frequency modes, which are the most important. On these grounds not only a satisfactory agreement between experimental frequencies and those obtained using empirical potentials could be attained even for minerals (see, for instance,[41]); quite often it was even possible to predict the Raman- or IR-spectra (as well as the phonon dispersion curves and ADP's) of a substance, starting from crystal structure data only and using potentials derived from fitting the vibrational spectra of a number of other compounds, not including the substance in question [39]: examples concerning diopside, a chain silicate, and one garnet, grossular [35,36] are reported in Table 3.

Table 3. Examples of Raman-active vibrational frequencies (cm^{-1}) for diopside CaMgSi$_2$O$_6$ and grossular Ca$_3$Al$_2$Si$_3$O$_{12}$. Observed data from [21] and [52]; calculated data from [35] and [36]

diopside CaMgSi$_2$O$_6$ A$_g$	obs	cal	grossular Ca$_3$Al$_2$Si$_3$O$_{12}$ T$_{2g}$	obs	cal
	140	129		178	177
	182	193		238	225
	235	211		246	244
	256	246		278	261
	n.o.	306		330	334
	326	327		349	356
	360	356		383	391
	390	394		478	492
	508(?)	526		577(?)	552
	530	557		629	631
	n.o.	770		n.o.	667
	858	906		850	877
	1016	969		n.o.	920
	1050	998		1007	992

Besides interpreting vibrational spectra, such calculations are closely connected with crystal-structure modelling. In a certain sense, considering vibrational frequencies does add completeness to a certain crystal-structure model. Moreover, because a certain crystal structure does not necessarily correspond to an energy minimum, but rather to a free energy minimum, the possibility of also deriving entropy and free energy for the model (see below) provides a definite possibility of improving results. However, there is something else still more fundamental, which has been neglected too often.

In deriving a plausible model for a crystal structure, sometimes the energy minimization procedure is carried out by considering first derivatives only, until such derivatives are zero at the presumed minimum. However, in order to check whether the final result is indeed a minimum, the *second derivatives matrix* of potential energy *should also be considered*. It is easy to see that such a matrix is just $\mathbf{D}_o$ (see above): there is, accordingly, a close connection with the dynamical matrix for $\mathbf{q} = \mathbf{0}$, at least up to the point that the eigenvalues of $\mathbf{D}$ should be all positive, thereby excluding imaginary frequencies.

On expanding such an argument, it follows that only in order to be sure of the physical significance of the "static" model, a procedure essentially identical with the calculation of vibrational frequencies all over the Brillouin zone is necessary [19]. Therefore, such simple theoretical reasoning shows that even *having to do with a "static" model should also include, more or less implicitly, the evaluation of the whole vibrational spectrum of the substance.*

Another important point is the *evaluation of thermodynamic functions*. The vibrational partition function Z_v can be considered for such purposes, applied to each normal mode, or, better, to the phonon density of states: see, for instance,[13,19,42]. From these considerations, *temperature-dependent* expressions can be deduced for the vibrational energy E_v, the specific heat c_v, the entropy S, and the free energy A= E-TS:

Table 4. Values of thermodynamic functions (J/mol.K) at different temperatures for aragonite $CaCO_3$ [40]. Observed values of thermodynamic functions from [20].

T(K)	c_p obs	c_p cal	c_v cal	Sobs	Scal.
260	74.7	75.7	75.4	77.4	78.2
280	78.4	78.3	78.0	83.0	84.0
298	81.2	80.6	80.2	88.0	88.9
400	92.6	91.2	90.7	113.7	114.1
500	100.0	99.0	98.3	135.2	135.4
600	106.0	104.8	104.0	153.9	154.0
700	111.3	109.1	108.2	170.7	170.5

Such evaluations are definitely superior to those obtained using Einstein or Debye models.

A detailed lattice-dynamical calculation including simultaneous comparison of ADP's and thermodynamic functions with the corresponding experimental data has provided evidence for order-disorder transitions taking place at low temperature in two garnets, *i.e.* pyrope $Mg_3Al_2Si_3O_{12}$ and almandine $Fe_3Al_2Si_3O_{12}$ [35], contrasting a diffuse opinion, which explained the situation in terms of large-amplitude anharmonic motion. The disorder involves the Mg and the Fe atoms, respectively, and such a situation is evident also in the ADP's, since for these atoms only the observed values are much greater than their corresponding lattice-dynamical estimates. Whereas on considering ADP's only there is no ground for distinguishing between these two possibilities, however, the value of entropy above 240K is almost equal to the theoretical estimates, provided an additional contribution of 34.5 J/mol.K is considered for each substance. Such a contribution is exactly the one corresponding to random distribution of the Mg (or Fe) atom originally in the *24c* special position (in the *Ia3d* space group) over four different near-by sites of lower symmetry. If such a coincidence is due to a mere chance, then (apart from other reasons) it would be difficult to explain why the same difference is observed for two different cases, and, moreover, why the calculated values of the specific heat C_p at room temperature (298 K) are in good agreement with the corresponding experimental values (318.6 against 325.3, and 332.3 against 342.8 J/mol.K for pyrope and almandine, respectively).

68

References

1. Bonadeo, H. and Burgos, E. (1982) Lattice dynamical calculations of the mean square amplitudes of crystalline biphenyl, *Acta Crystallogr.* **A38**, 29-33.

2. Bonadeo, H. and Taddei, G. (1973) Calculation of dispersion relations and frequency distribution of crystalline benzene, *J.Chem.Phys.* **58**, 979-984.

3. Brown, G.M. and Levy, H.A. (1979) *o*-Terphenyl by neutron diffraction, *Acta Crystallogr.* **B35**, 785-788.

4. Cochran, W. and Pawley, G.S. (1964) The theory of diffuse scattering of X-rays by a molecular crystal, *Proc.Roy.Soc.* **A280**, 1-22.

5. Filippini, G. and Gramaccioli, C.M. (1981) Deriving the equilibrium conformation in molecular crystals by the quasi-harmonic procedure: some critical remarks, *Acta Crystallogr.* **A37**, 335-342.

6. Filippini, G. and Gramaccioli, C.M. (1984) Lattice-dynamical calculations for tetracene and pentacene, *Chem.Phys.Lett.* **104**, 50-53.

7. Filippini, G. and Gramaccioli, C.M. (1986) Thermal motion analysis in tetraphenylmethane: a lattice-dynamical approach, *Acta Crystallogr.* **B42**, 605-609.

8. Filippini, G. and Gramaccioli, C.M. (1989) Benzene crystals at low temperature: A harmonic lattice-dynamical calculation, *Acta Crystallogr.* **A45**, 261-263.

9. Filippini, G., Gramaccioli, C.M., Simonetta, M. and Suffritti, G.B. (1973) Lattice-dynamical calculations on some rigid organic molecules, *J.Chem.Phys.* **59**, 5088-5101.

10. Filippini, G., Gramaccioli, C.M., Simonetta, M. and Suffritti, G.B. (1974) On some problems connected with thermal motion in molecular crystals and a lattice-dynamical interpretation, *Acta Crystallogr.* **A30**, 189-196.

11. Filippini G, Gramaccioli, C.M., Simonetta M and Suffritti, G.B. (1976) Lattice-dynamical application to crystallographic problems: consideration of the Brillouin-zone sampling, *Acta Crystallogr.* **A32**, 259-264.

12. Filippini, G., Simonetta, M. and Gramaccioli, C.M. (1984) A simplified force field for non-planar vibrations in aromatic polycyclic hydrocarbons, *Molec.Phys.* **51**, 445-449.

13. Gramaccioli, C.M. (1987) Spectroscopy of molecular crystals and crystallographic implications, *Intern.Rev.Phys.Chem.* **6**, 4, 337-349.

14. Gramaccioli, C.M. and Filippini, G. (1983) Lattice-dynamical evaluation of temperature factors in non-rigid molecular crystals: a first application to aromatic hydrocarbons, *Acta Crystallogr.* **A39**, 784-791.

15. Gramaccioli, C.M. and Filippini, G. (1984) Lattice-dynamical calculations for orthorhombic sulfur: a non-rigid molecular model, *Chem.Phys.Letters* **108**, 585-588.

16. Gramaccioli, C.M. and Filippini, G. (1985) Thermal motion for non-rigid molecules in crystals: symmetry of the generalized mean-square displacement tensor **W**, *Acta Crystallogr.* **A41**, 356-361.

17. Gramaccioli, C.M. and Filippini, G. (1985) Thermal motion analysis in *o*-terphenyl: a lattice-dynamical approach, *Acta Crystallogr.* **A41**, 361-365.

18. Gramaccioli, C.M., Filippini, G. and Simonetta, M. (1982) Lattice-dynamical evaluation of temperature factors for aromatic hydrocarbons, including internal

molecular motion: a straightforward systematic procedure, *Acta Crystallogr.* **A38**, 350-356

19. Gramaccioli, C.M. and Pilati, T. (1992) in: Thermodynamic data: systematic and estimation (ed. S.K. Saxena) *Advances in Phys.Geochem.* **10**, 239-263. New York: Springer.

20. Holland, T.J.B. and Powell, R. (1990) An enlarged and updated internally consistent thermodynamic dataset, *J.Metamorph.Geol.* **8**, 89-124.

21. Hofmeister, A.M. and Chopelas, A. (1991) Thermodynamic properties of pyrope and grossular from vibrational spectroscopy, *Amer.Mineral.* **76**, 880-891.

22. Johnson, C.K. (1969) Addition of higher cumulants to the crystallographic structure-factor equation: A generalized treatment for thermal-motion effect, *Acta Crystallogr.* **A25**, 187-194.

23. Johnson, C.K. (1970) Generalized treatments for thermal motion. In *Thermal neutron diffraction* (ed. B.T.M. Willis) pp.132-160. Oxford University Press.

24. Johnson, C.K. (1970) An introduction to thermal-motion analysis. *In Crystallographic computing* (ed. F.R. Ahmed) pp 207-219 and 220-226. Munksgaard, Copenhagen.

25. Neto, N., Righini, R., Califano, S. and Walmsley, S.H. (1978) Lattice dynamics of molecular crystals using atom-atom and multipole-multipole potentials, *Chem.Phys.* **29**, 167-179.

26. Patel, A., Price, G.D. and Mendelssohn M.J. (1991) A computer simulation approach to modelling the structure, thermodynamics and oxygen isotope equilibria of silicates,*Phys Chem Minerals* **17**, 690-699.

27. Pawley, G.S. (1967) A model for the lattice dynamics of naphthalene and anthracene, *Phys.Status Solidi* **20**, 347-360.

28. Pawley, G.S. (1968) Anisotropic temperature factors and screw rotation coefficients from a lattice dynamical viewpoint, *Acta Crystallogr.* **B24**, 485-486.

29. Pilati, T., Bianchi, R. and Gramaccioli, C.M. (1990) Evaluation of atomic displacement parameters by lattice-dynamical calculations. Efficiency in Brillouin-zone sampling, *Acta Crystallogr.* **A46**, 485-489.

30. Pilati, T., Bianchi, R. and Gramaccioli, C.M. (1990) Lattice-dynamical estimation of atomic thermal parameters for silicates: Forsterite α-Mg_2SiO_4, *Acta Crystallogr.* **B46**, 301-311.

31. Pilati, T., Demartin, F., Cariati, F., Bruni, S and Gramaccioli C.M. (1993) Atomic thermal parameters and thermodynamic functions for chrysoberyl ($BeAl_2O_4$) from vibrational spectra and transfer of empirical force fields, *Acta Crystallogr.* **B49**, 216-222.

32. Pilati, T., Demartin, F. and Gramaccioli, C.M. (1993) Atomic thermal parameters and thermodynamic functions for corundum (α-Al_2O_3) and bromellite (BeO): a lattice-dynamical estimate, *Acta Crystallogr.* **A49**, 473-480.

33. Pilati, T., Demartin, F. and Gramaccioli, C.M. (1994) Thermal parameters for α-quartz: a lattice-dynamical calculation, *Acta Crystallogr.* **B50**, 544-549.

34. Pilati, T., Demartin, F. and Gramaccioli, C.M. (1995) Thermal parameters for minerals of the olivine group: their implication on vibrational spectra, thermodynamic functions and transferable force fields, *Acta Crystallogr.* **B51**, 721-733.

35. Pilati, T., Demartin, F. and Gramaccioli, C.M. (1996) Atomic displacement parameters for garnets: A lattice-dynamical evaluation, *Acta Crystallogr.* **B52**, 239-250.

36. Pilati, T., Demartin, F. and Gramaccioli, C.M. (1996) Lattice-dynamical evaluation of atomic displacement parameters of minerals and its implications: The example of diopside, *Amer. Mineral.* **81**, 811-821.

37. Pilati, T., Demartin, F. and Gramaccioli, C.M. (1997) Transferability of empirical force fields in silicates: Lattice-dynamical evaluation of atomic displacement parameters and thermodynamic properties for the Al_2OSiO_4 polymorphs, *Acta Crystallogr.* **B53**, 82-94.

38. Pilati, T., Demartin, F. and Gramaccioli, C.M. (1997) Lattice-dynamical evaluation of thermodynamic properties and atomic displacement parameters for beryl using a transferable force field, *Amer. Mineral.* **82**, 1054-1062.

39. Pilati, T., Demartin, F. and Gramaccioli, C.M. (1998). Lattice-dynamical evaluation of atomic displacement parameters for coesite from an empirical force field with implications on thermodynamic properties, *Phys.Chem.Minerals*, **25**, 152-159.

40. Pilati, T., Demartin, F. and Gramaccioli, C.M. (1998). Lattice-dynamical estimation of atomic displacement parameters in carbonates: calcite and aragonite $CaCO_3$, dolomite $CaMg(CO_3)_2$ and magnesite $MgCO_3$, *Acta Crystallogr.* **B54**, 515-523.

41. Price, G.D., Parker, S.C. and Leslie, M. (1987) The lattice dynamics of forsterite, *Mineral Mag*, **51**, 157-170.

42. Price, G.D., Parker, S.C. and Leslie, M. (1987) The lattice dynamics and thermodynamics of the Mg_2SiO_4 polymorphs, *Phys Chem Minerals* **15**, 181-190.

43. Robbins, A., Jeffrey, G.A., Chesick, J.P., Donohue, J.C., Cotton, F.A., Frenz, B.A. and Murillo, C.A. (1975) A refinement of the crystal structure of tetraphenylmethane: Three independent redeterminations, *Acta Crystallogr.* **B31**, 2395-2399.

44. Scheringer, C. (1972) A lattice-dynamical treatment of the thermal motion bond length corrections, *Acta Crystallogr.* **A28**, 616-619.

45. Schomaker, V. and Trueblood, K.N. (1968) On the rigid-body motion of molecules in crystals, *Acta Crystallogr.* **B24**, 63-76.

46. Taddei, G., Bonadeo, H., Marzocchi, M.P. and Califano, S. (1973) Calculation of crystal vibrations of benzene, *J.Chem.Phys.* **58**, 966-978.

47. Williams, D.E. (1965) Nonbonded potential parameters derived from crystalline aromatic hydrocarbons,*J.Chem.Phys.* **45**, 3770-3778.

48. Williams, D.E. (1967) Nonbonded potential parameters derived from crystalline aromatic hydrocarbons, *J.Chem.Phys.* **47**, 4680-4684.

49. Willis, B.T.M. and Pryor, A.W. (1975) *Thermal Vibrations in Crystallography.*, Cambridge University Press.

50. Winkler B., Dove, M.T. and Leslie, M. (1991) Static lattice energy minimization and lattice dynamics calculations on aluminosilicate minerals. *Amer Mineral.* **76**, 313-331.

51. Winter, J.K. and Ghose, S. (1979) Thermal expansion and high-temperature crystal chemistry of the Al_2SiO_4 polymorphs, *Amer.Mineral.* **64**, 573-586.

52. Zulumyan, N.O., Mirgorodskii, A.P., Pavinich, V.F. and Lazarev, A.N. (1976) Study of calculation of the vibrational spectrum of a crystal with complex polyatomic anions: Diopside $CaMgSi_2O_6$, *Optika i Spektroskopiya* **41**, 1056-1064.

MOLECULAR MECHANICS MODELING OF TRANSITION METAL COMPOUNDS

PETER COMBA

Anorganisch-Chemisches Institut, Universität Heidelberg
Im Neuenheimer Feld 270, 69120 Heidelberg, Germany

1. Introduction

The art of chemistry is to understand the properties of substances, to know how new compounds can be made and how they interact with the environment in the solid, in solution and in the gas phase, to understand what their behavior toward substrates is and how they interact with magnetic fields, light and any other form of energy. The understanding of properties of chemical compounds may be based on their structures, the three-dimensional arrangement of the atoms of a molecule or a molecular assembly in space. The determination, prediction, understanding and interpretation of structures and their interdependence with macroscopic physical properties are based on models and theories. These may be more or less generally applicable, accurate, efficient, rigorous and scientifically appealing but they are never right or wrong in a general sense. Models are based on observed patterns, regularities in experimentally detected properties, and they are usually restricted in their applicability to the set of data from which they were derived. Hence, the quality of a model is given by the accuracy of its predictions and interpretations in the area where it is applied.

2. Structure and Strain

Structural distortions, sterically hindered reaction centers, steric crowding and strained molecules are very intuitive, useful and simple empirical properties and concepts that relate structures with stabilities and reactivities. The underlying models are based on the visualization of structures and the important question is whether empirical rules can be found, allowing a quantification of distortion, strain and steric crowding. A number of models have been developed that are based on simple geometrical rules. One of the most successful concepts in this area is that of cone angles, where the steric demand of a ligand, usually a phosphine, is related to thermodynamic properties and reactivities [1]. With the increasing bulk of the phosphorus substituents, steric crowding and strain build up. Are phosphine compounds with the smallest possible substituent, H-atoms, strain-free? The answer to this question may not be of importance in the concept of

71

J.A.K. Howard et al. (eds.),
Implications of Molecular and Materials Structure for New Technologies, 71–86.
© 1999 *Kluwer Academic Publishers. Printed in the Netherlands.*

cone angles, but the question of what undistorted, relaxed, strain-free bond lengths, valence angles, van-der-Waals distances etc. are is as interesting as it is difficult or impossible to answer. Cobalt(III) hexaamine compounds, for example, with minimum and maximum values of the Co-N bond distances of 1.94 Å and 2.05 Å, respectively, have been reported [2]. Are these extreme values compressed and elongated, respectively, or are they simply short and long? How much energy cost is involved in changing a particular bond from a putative ideal value to that observed in a particular structure? Molecular mechanics models try to answer these questions.

3. Molecular Mechanics

The fundamental assumption of molecular mechanics is that the positions of the atoms in a molecule are determined by forces between each atom and all the others. Generally, and following intuitive chemical concepts, there is a distinction between forces involving pairs of atoms (covalent, H-bonding, electrostatic or van-der-Waals interactions), three atoms (valence angle forces), four and more atoms (forces involving the torsion around bonds, out-of-plane bending and others; see Figure 1).

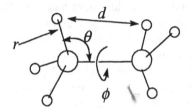

Figure 1. The classical molecular mechanics concept.

The nuclear positions in a (strained) structure are the result of a compromise that is based on the minimized sum of all energies resulting from these forces (E_{total}, Eqn. 1). Minimization of E_{total} (the strain energy of a molecule in a particular geometry) by rearrangement of the nuclear positions, starting from an assumed configuration, leads to an optimized structure and the corresponding minimized strain energy. Thus, E_{total} depends on the structure of the molecule. Typical potential energy functions for various terms are given in Eqns. 2 - 11.

Additional potentials that involve, for example, a combination of bond stretching and angle bending terms (so called cross terms) are sometimes added to account for the fact that these motions are not strictly decoupled [3]. Other functions, developed specifically for transition metal compounds, are discussed below.

$$E_{total} = \sum_{molecule} (E_b + E_\theta + E_\varnothing + E_{nb} + E_\sigma + \cdots) \qquad (1)$$

$$E_b = \tfrac{1}{2} k_b (r_{ij} - r_o)^2 \qquad (2)$$

$$E_b^{morse} = D(1 - \exp(-\alpha(r - r_o)))^2 - D \qquad (3)$$

$$E_\theta = \tfrac{1}{2} k_\theta (\theta_{ijk} - \theta_o)^2 \qquad (4)$$

$$E_\varnothing = \tfrac{1}{2} k_\varnothing (1 + \cos(m(\varnothing_{ijkl} + \varnothing_{off}))) \qquad (5)$$

$$E_{nb} = A \cdot e^{-Bd_{ij}} - C \cdot d_{ij}^{-6} \qquad (6)$$

$$E_\delta = \tfrac{1}{2} k_\delta \delta^2 \qquad (7)$$

$$E_\varepsilon = \frac{q_i \, q_j}{\varepsilon \, d_{ij}} \qquad (8)$$

$$E_{hb} = F \cdot d_{ij}^{-12} - G \cdot d_{ij}^{-10} \qquad (9)$$

$$E_\theta^M = \tfrac{1}{8} k_{LML} \cdot \sin^2(2\theta) \qquad (10)$$

$$E_\delta = \tfrac{1}{2} k_\gamma (\gamma - \gamma_o) \qquad (11)$$

The basic ideas of force field calculations may be traced back to 1930 [4] and the first applications to inorganic coordination compounds were reported around 50 years ago [5]. For many years now molecular mechanics has been a well-established tool in organic chemistry [6]. Also, computational studies of biological systems and methods for drug design often involve force field calculations [7]. For a long time, the number of transition metal elements, their variety in terms of oxidation and electronic states, the range of possible coordination numbers and the plasticity of coordination geometries, electron transfer between donors and metal centers and specific electronic effects, such as Jahn-Teller distortions, *trans*-influences, spin-crossover and allogonism were believed to be intractable by the simple approach of molecular mechanics (see separate section below). The development of a number of new techniques in the last decade has lead to force field models that now are powerful tools for inorganic chemists, and this is documented in a series of review articles [8].

A number of algorithms are used to minimize strain energies. Most fall into two classes: first-derivative techniques (steepest descent, conjugate gradient, Powell) and second-derivative methods (block-diagonal Newton-Raphson, full-matrix Newton-Raphson) [6,9]. The former methods converge rather slowly, particularly if the structures are close to the energy minimum, the latter are computationally more demanding (determination and storing of the second derivative matrix) but convergence is much more efficient. Also, the second derivatives are needed to compute vibrational frequencies and to apply mathematically precise constraints [8a,10]. Often it is advisable to optimize structures with a combination of methods, starting with the first-derivative, followed by second-derivative algorithms.

The optimization of a structure generally leads to a geometry that is similar to that of the starting structure, that is, there usually is no conformational interconversion, and the result depends on the conformation chosen when the molecule was drawn. There is no general and conclusive method to find the global energy minimum structure and the conformational equilibrium involving all possible low energy structures [8a]. A number of methods are used to deal with this problem, ranging from manual and automated grid searches to Monte Carlo and molecular dynamics methods [8a,11].

4. Force Fields

The optimized structure and the corresponding strain energy of a given conformation of a molecule depend on the type of potential energy terms, the functional form of the potentials and their parameterization. Examples for potential energy functions are given in Eqns. 2 – 11 and, together with a corresponding set of parameters, this is an example of a force field. As a whole, such a force field is empirical, that is, it is based on an (as large as possible) set of experimental data. It is, therefore, not unexpected that a number of different combinations of energy terms, of functional forms for the potentials and corresponding parameter sets have been developed and successfully used. From this, there are three obvious and important emerging properties of force fields:

- *Range of applicability*: Force fields are based on sets of experimental data, refined to reproduce experimental structural, thermodynamic and vibrational spectroscopic data with high accuracy (for restrictions in terms of vibrational frequencies, see paragraph on minimizers above). It follows that the range of applicability and the quality of the results depends strongly on the type of data used to set up the force field (structural, thermodynamic and/or vibrational spectroscopic data; range of structural distortions). An important fact is that molecular mechanics is an interpolative method, that is, a force field for cobalt(III) hexaamines, for example, that has been developed on the basis of Co^{III}-N distances from 1.95 Å to 1.98 Å may not be used without caution for compounds that are supposed to have Co^{III}-N distances larger than 2.0 Å [2c,8a]. Another critical point which follows from the fact that force field calculations are interpolations is that force fields which have been set up for structural modeling may not produce acceptable results when thermodynamic properties (e.g. conformational equilibria) or vibrational spectra are computed: structures are related to the minima of potential energy functions, energies to the steepness of the potentials (first derivatives) and vibrational frequencies to the curvature (second derivatives) [8a]. The quality of the force fields used in organic chemistry has often been related to the accuracy of predicted vibrational frequencies. For a number of reasons, in inorganic molecular modeling, these have not often been computed and compared to experimental data. Also, accurate experimental thermodynamic data for inorganic compounds are not always readily available. Therefore, in inorganic molecular mechanics, force fields are primarily based on structural data. Since the parameterization of the organic part of the molecules (ligand backbone) has often been adapted from force fields used in organic chemistry, and since all parameters within a well tuned force field are interdependent (see below), these force fields are expected to and have been shown to accurately reproduce thermodynamic data [8a-c].

- *Transferability:* Each parameter value in a force field (force constants, ideal distances, ideal angles etc.) depends on the set of functions and the functional form and shape of each potential in the force field. It follows that force field parameters are interdependent and not transferable. The two important consequences are that (i) when any function or parameter in an existing force field is changed all other parameters have to be checked and some might need to be adjusted; (ii) parameters may not be exchanged between different force fields, that is, published parameters from other force fields may only be included to complete a force field if they are considered as initial guesses for careful tuning based on experimental data [8a].

- *Force fields and reality:* From the discussion above, it emerges that force field parameters are not related to reality: A molecular mechanics program, that is, a force field together with the strain energy minimization algorithm, may be used to reproduce, predict and/or interpret structural, thermodynamic and vibrational spectroscopic molecular properties, but the parameter values of the force field are not necessarily scientifically meaningful [8a-c]. Depending on the set of potential energy functions (Eqns. 1 – 11) and the experimental structures used to tune a cobalt(III) hexaamine force field, for example, the range of force constants (k_b, see Eqn. 2) and ideal Co^{III}-N distances (r_0, see Eqn. 2) presented in Table 1 have been found to accurately reproduce certain sets of structures [8a,12,13,14,15]. It follows that, from molecular mechanics force field parameters, it is not warranted to get accurate information on strain-free bond distances and bond strengths. Consider, for example, the C-C distance in ethane (see also Fig. 1): The parameter value for the strain-free C-C distance r_0 depends on both the parameter value of the force

TABLE 1. Variation of published ideal Co^{III}-N distances (r_0) and force constants (k_b)

r_0	k_b	ref
1.905	1.750	12
1.927	2.25	13
1.915	2.00	13
1.915	1.75	13
1.925	1.70	14
1.950	2.25	15

constant k_b of the C-C bond and on the functional form and parameterization of the van-der-Waals interactions (H···H repulsion, see Eqn. 6). Due to the fact that observed structures are generally the result of a compromise between a number of preferences, similar to these discussed for ethane, it is not generally possible to determine experimentally strain-free distances and angles (see also the discussion on bond elongation and compression in the section "structure and strain", above).

An interesting and generally successful approach to model the angular geometry of transition metal coordination compounds is to replace the angle bending potential (Eqn. 4) around the metal center by 1,3-nonbonded interactions (points-on-a-sphere or Urey-Bradley model, see below) [8a,12,16]. Here, the repulsion between the donor atoms is used to model their angular distribution. However, this also leads to a significant metal-donor elongation, and this may be adjusted by a decrease of the parameter for the ideal metal-donor distances r_0. Therefore, force fields that are based on a points-on-a-sphere approach generally have smaller strain-free metal-donor distances [8a-c].

In conclusion: conventional molecular mechanics is an interpolative method, force field parameters are interdependent and not transferable and they are not necessarily related to reality.

Although molecular mechanics force field parameters are not strictly related to observables, starting values for fitting procedures to experimental data are usually obtained from experimentally observed data. Ideal or strain-free distances or angles are assumed from theoretical considerations (180°, 120° or 109.5° angles for sp, sp^2, or sp^3 hybrids) or averaged structural data (due to secondary effects - e.g. repulsion of substituents, see above - starting values for ideal bond distances are usually reduced by approximately 10% of those observed in average in relatively unstrained experimental structures). Upper limits for force constants may be obtained from Eqn. 12 and similar equations for bending and other interactions, where v is the fundamental vibrational

$$v = \frac{\left(\dfrac{k_b}{\mu}\right)^{1/2}}{2\pi c} \qquad (12)$$

frequency of the corresponding bond, taken from spectra, c is the speed of light and μ is the reduced mass of the two atoms involved in the bond. Theoretically more accurate values may be obtained by normal coordinate analysis. However, independently of the approach used for the calculation of the force constant, due to the fundamental difference between spectroscopic (molecule based) and molecular mechanics (molecule independent) force fields, these are generally overestimated by as much as 45 % [8a]. This discrepancy is, to some extent, due to the fact that in most normal coordinate analyses, nonbonded interactions other than 1,3- and 1,4-interactions are excluded.

An additional fact that has not been discussed in detail, but is obvious from the discussion above, is that only strain energy *differences* may be used to discuss thermodynamic properties. A high value of the strain energy of a computed structure does not necessarily indicate a highly strained, unstable structure but the energy difference between to conformations of a molecule may under certain conditions be used to analyze the relative stability.

Apart from conventional force fields, there have been reports on rule-based, more general, universal force fields [17]. In conventional force fields, specific parameter sets need to be developed for each class of compounds, that is, parameters are based on atom types. For example, there are specific and different parameterization schemes for carbon atoms in aliphatic and aromatic carbohydrates, in carboxylates, ethers, amines etc. In rule-based force fields the parameterization may be more general and atom-based. Simple, empirical rules are used to build-up a conventional specific parameter set for each molecule from the atom-based parameters. Universal force fields have the advantage that, in principle, any new compound may be refined, but usually the accuracy is inferior to that obtained when using conventional force fields [17]. The VALBOND model is an interesting new approach, using a rule-based force field, that, in many areas of chemistry, has produced results of high accuracy [8e,8h,18] and this will be discussed below in more detail.

5. Environment

There are three terms that are often neglected in conventional molecular mechanics calculations:

- Entropic contributions to the free energy, other than statistical terms, are excluded. Strain energies may be assumed to be related to enthalpies (ΔH; see, however, other neglections, such as solvation, ion-pairing and specific electronic factors, below). There are partition functions that allow the computation of the vibrational (S_{vib}), transitional (S_{trans}), rotational (S_{rot}) and statistical (S_{stat}) contributions to the entropy (Eqns. 13 – 16; ν_i are the vibrational energy levels, T is the temperature, m is the molecular weight, p is the pressure, ABC is the product of the moments of inertia, σ is the rotational symmetry number, ρ is the number of ways in which a given isomer may be formed, i.e. all these parameters are available for and from conventional force field calculations [8a,19]).

$$S_{vib} = R \sum_{i=1}^{3N-6} \frac{h\nu_i / kT}{\exp\left(\dfrac{h\nu_i}{kT} - 1\right)} - \ln\left(1 - \exp\left(-\frac{h\nu_i}{kT}\right)\right) \tag{13}$$

$$S_{trans} = R\left\{\frac{3}{2}\ln m + \frac{5}{2}\ln T - \ln p - 3.664 + \frac{5}{2}\right\} \tag{14}$$

$$S_{rot} = R \ln\left(\frac{8\pi^2(8\pi^3 ABC)^{\frac{1}{2}}(kT)^{\frac{3}{2}}}{\sigma\, h^3}\right) + \frac{3}{2}R \tag{15}$$

$$S_{stat} = R \ln \rho \tag{16}$$

There are two problems related to the neglect of entropic terms: (i) "entropic strain" may lead to some distortion of a computed structure, that is the entropic terms (Eqns. 13 – 16) need to be involved in the structure optimization (Eqn. 1); (ii) there might be a significant contribution of the entropic terms to the minimized energy E_{total}, leading to errors in computed isomer distributions in a set of conformers when entropies are neglected [8a,19,20].

- Specific electronic terms, such as *cis*- and *trans*-influences and Jahn-Teller effects are difficult to include in conventional molecular mechanics programs. These are discussed in a separate section below.

- Interactions between "naked" molecules and their environment. This last problem is now discussed in some detail.

Ion-pairing and solvation in the liquid phase and crystal lattice effects in the solid state are usually not included explicitly when structural, thermodynamic and vibrational spectroscopic properties of molecular compounds are computed. There is no fundamental reason why these intermolecular forces may not be included in a structural optimization procedure. The required potentials (mainly van-der-Waals, electrostatic and H-bonding terms) and the corresponding parameterization schemes are usually available (Eqns. 1 – 11; a possible restriction is the determination of the charge distribution but there are approaches to deal with this potentially difficult task [7c,8a,21]). The main problem is the computational expense when a crystal lattice or a solvated and ion-paired molecule need to be optimized, but relevant techniques have been developed and are readily available [7c,20c,22].

There are good reasons for the fact that acceptable agreement between experimentally observed and computed properties is usually observed when the refinement is based on a model that does not include the environment. The important fact is that molecular mechanics is an interpolative method and the force field is generally based on data obtained from solid state or solution studies. That is, the environment is inherently present in the force field. Even if the environment is not explicitly included in the structural optimization, it is not appropriate to discuss such results as those of gas-phase calculations; it might be more appropriate to denote isolated molecules that are optimized with a force field that is based on "real" structures, thermodynamics and spectra as "naked" or "interactionless" [23].

There are structural, thermodynamic and vibrational spectroscopic differences between gas-phase molecules and those in solution and solid phases (distortions by crystal lattice effects, selective outer-sphere effects used in chromatography and extractions, stereoselective crystallization, *e.g.* for racemate separations). The interactions that are neglected when isolated, interactionless molecules are optimized are nonbonded interactions. Van-der-Waals interactions are attractive at long distances (decreasing with approximately d^{-6}), the hydrogen bonding energy decreases with approximately d^{-10} and that of electrostatic attraction decreases with d^{-1}. Therefore, an isolated molecule in a lattice (or solvent sheath) must experience some pressure that may lead to a contraction of the structure. This is built-in in force fields optimized for interactionless but not for gas-phase molecules [20c]. It follows that the molecular mechanics optimization of structures in the gas phase must be based on force fields that are tuned to gas-phase structures. Also, since intermolecular interactions are inherently present in conventional force fields, different parameterization schemes have to be used in computations where entire lattices or solvated and ion-paired molecules are optimized. It emerges that in these latter cases, force fields which are based on interactionless gas phase molecules might be appropriate.

The forces that are included in force fields for interactionless molecules are isotropic. That is, in general, optimized structures are more symmetrical than those of experimentally observed molecules. Therefore, for evaluating the quality of an optimized structure averaged structural data for parameters that are symmetrically related in the more symmetrical computed structure must be used. For modeling anisotropic distortions by the lattice (or solvent or counter ion) the whole lattice (or solvent sheath and ion pairs) must be optimized.

6. Transition Metal Compounds

The large number of transition metal ions and their broad range of available oxidation states, coordination numbers and electronic ground states require an extensive set of force field parameters for the computation of transition metal compounds. However, the major problem is due to the partially filled d-orbitals which are responsible for electronic effects that generally are difficult to model with conventional force field methods [8a-c, 24]. The approaches available for dealing with this problem may be grouped into three types of methodology:

- Hybrid methods, where the metal center and the coordination sphere are computed with quantum mechanical methods, whereas the rest is computed by conventional molecular mechanics [24,25].

- Enhanced molecular mechanics methods, that is, approaches that include conventional as well as general electronic terms in the set of potential energy functions [8e,8g,8h,18,26].

- Electronically doped conventional force field methods [8a-c,12b-c, 20c].

A number of combined quantum mechanical/molecular mechanics methods have been reported and used successfully [24,25]. These include methods where the quantum mechanical part is based on ab-initio or semi-empirical quantum mechanics, and various techniques have been described to deal with the interface between the quantum mechanically and the molecular mechanically computed parts of the molecule.

The CLFSE/MM approach is an enhanced molecular mechanics model that includes a ligand field electronic term in the basic expression for the strain energy (Eqn. 17, see also Eqn. 1) [8g,26].

$$E_{total} = \sum_{molecule} \left(E_b + E_\theta + E_\phi + E_{nb} + E_\sigma + CLFSE \right) \qquad (17)$$

CLFSE is the cellular ligand field stabilization energy, and the cellular ligand field model is a variant of the angular overlap model (AOM), which is a simplified molecular orbital model for the d-orbital part of transition metal ions, that is, a ligand field approach with a different, MO-based, parameterization scheme [27]. Thus, the CLFSE/MM method has a general d-orbital electronic term (one-electron d-levels, obtained by crystal field matrix diagonalization) included in the force field. One of the main advantages is that the CLFSE/MM approach allows the computation of electronically distorted compounds (for example, high-spin and low-spin d^8 nickel(II) and Jahn-Teller distorted d^9 copper(II) systems) with a single set of force field parameters, each [8g,26].

The VALBOND model is a molecular mechanics based approach that uses qualitative valence bond theory, that is, hybridized localized bonds, to define molecular shapes [8e,h,18]. Approximate hybridizations are derived from Lewis structures and adjusted with a generic algorithm based on Bent's rule [28]. From these refined hybridizations and a set of generalized hybrid orbital strength functions the molecular mechanics angle bending energy terms are set up. Thus, the VALBOND model uses a generic, atom based force field that includes an electronic angle bending function, and it has been used successfully for p- and d-block molecules, including hypervalent compounds, metal hydrides and metal alkyls [8e,h,18]. So far, it has not been applied to classical coordination compounds.

The MOMEC force field is based on a conventional molecular mechanics model with potential energy functions and parameter sets that have been adapted specifically for transition metal coordination compounds (see Eqns. 1 – 11) [12,29]. The main features of MOMEC are:

- The optimization of the coordination geometry is based on a points-on-a-sphere model, that is, in the basic form of the force field [12a], donor-metal-donor potentials are replaced with 1,3-nonbonded interactions (van-der-Waals term). Another possibility for modeling 1,3-non-bonded interactions is by electrostatic interactions; this is less straightforward since it involves the assignment of partial charges, see above. The points-on-a-sphere model has the advantage that the only metal-dependent parameters are those involved in the metal-donor stretching potential (metal-donor-ligand backbone potentials are assumed to be metal-ion independent [8a,b,12a], and force constants for the torsional potentials around the metal center are usually set to zero [8a,b,12a]). The refined structures are generally in good agreement with experimental data [20a,30].

82

- The points-on-a-sphere model does not directly account for metal-ion-based electronic effects [8a,30]. One of the reasons why it still leads to reasonably accurate predictions is that the metal-ion preferences are not as strong as those of the organic ligand backbones [8a,b]. In a few specific examples, where the directionality of the d-orbitals is of importance, this leads to significant inaccuracies. These have been corrected by a perturbation to the points-on-a-sphere model with a harmonic sine function (minima at 90° and 180°), using ligand-field based generic force constants $k_{LML'}$ (see Eqn. 10) [12b]. Since the original force field has been fitted on the basis of a donor-atom donor-atom repulsion model the resulting angle bending force constants are, as expected, much smaller than those in other force fields that do not involve 1,3-nonbonded interactions. This is an illustration that force field parameters are not related to real physical observables (see above) [12b].

- For four-coordinate metal ions the points-on-a-sphere model leads to tetrahedral coordination geometry. There are a number of possibilities to enforce planarity [8a]. A plane twist function has been introduced that allows the modelling of planarity with a harmonic potential (Eqn. 11) and with a generic force constant k_γ that depends on the ligand field strength (Fig. 2) [31]. An advantage of this approach is that, with constraints applied to the twist angle γ, it is possible to scan the energy along a twist coordinate. The plane twist function may also be applied to modeling of the Bailar twist of hexacoordinate compounds and of the torsional barrier of metallocenes [8f,31]. The recently developed technique of applying mathematically precise constraints to the sum of the internal coordinates, based on Lagrange multipliers, may also be used to enforce planarity of four coordinate compounds [32].

Figure. 2. Application of the plane twist function.

- An effect that, in general, has not been appreciated until recently is that upon coordination of a ligand molecule to a metal center, there is some electron redistribution which leads to changes of bond strengths within the ligand backbone (Table 2) [8a,12c]. Although, for typical coordination compounds (for example for hexaamine complexes), there are only minor effects, an improved MOMEC force field has been developed that uses different parameterization schemes for metal-free and coordinated ligands [12c].

TABLE 2. Average differences in bond distances for metal-free and coordinated ligands

	metal-free	coordinated
c- C- N – M	1.47	1.49
M - N – C - C -	1.51	1.50

- The structure optimization of Jahn-Teller distorted compounds is a difficult problem to solve with conventional molecular mechanics, especially if no assumption on the direction and extent of the distortion is made. MOMEC uses a method that is based on a harmonic first-order model to compute the Jahn-Teller stabilization energy, and this is computed conventionally (Fig.3) [33].

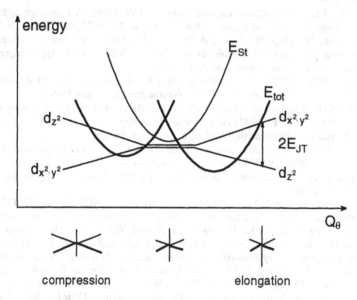

Figure 3. Molecular mechanics and the Jahn-Teller effect.

Acknowledgments

Our studies are supported by the German Science Foundation (DFG), the Funds of the Chemical Industry (FCI) and the VW-Stiftung. I am grateful for that, for the excellent work done by my coworkers, whose names appear in the references, and to Karin Stelzer for her help in preparing the manuscript.

84

References

1. 1a) Tolman, C. A. (1970) Phosphorus ligand exchange equilibria on zerovalent nickel. A dominant role for steric effects, *J. Am. Chem. Soc.* **92**, 2956; b) Tolman, C. A. (1977) Steric effects of phosphorus ligands in organometallic chemistry and homogeneous catalysis, *Chem. Rev.* **77**, 313; c) Stahl, L. and Ernst, R.D. (1987) Equilibria studies involving ligand coordination to 'open' titanocenes: Phosphine and pentadienyl cone angle influences and the existence of these electron-deficient molecules, *J. Am. Chem. Soc.* **109**, 5673.

2. a) Bernhardt, P.V., Hambley, T.W. and Lawrance, G.A. (1989) 6,13-diamino-6,13-dimethyl-1,4,8,11-tetra-azacyclotetradecane, L7, a new, potentially sexidentate polyamine ligand. Variable co-ordination to cobalt(III) and crystal structure of the complex [Co(L7)]Cl$_2$[ClO$_4$], *J. Chem. Soc., Dalton Trans.*, 1059; b) Geue, R.J., Hanna, J., Höhn, A., Qin, C.J., Ralph, S.F., Sargeson, A.M. and Willis, A.C. (1997) Steric effects in redox reactions and electron transfer rates, in: S. Isied (Ed.) *Electron Transfer Reactions*, ACS Symp. Ser.; c) Comba, P. and Sickmüller, A. F. (1997) The solution structures of a pair of stable cobalt(III) hexaamine conformers, *Angew. Chem.* **109**, 2089; *Angew. Chem. Int. Ed. Engl.* **36**, 2006.

3. a) Allinger, N.L. (1977) Conformational analysis. 130. MM2. A hydrocarbon force field utilizing V1 and V2 torsional terms, *J. Am. Chem. Soc.* **99**, 8127; b) Allinger, N.L., Geise, H.J., Pyckhout, W., Paquette, L.A. and Galluci, J.C. (1989) Structure of norbornane and dodecahedrane by molecular mechanics calculations (MM3), x-ray crystallography, and electron diffraction, *J. Am. Chem. Soc.* **111**, 1106.

4. a) Andrews, D.H. (1930), The relation between the Raman spectra and the structure of organic molecules, *Phys. Rev.* **36**, 544; b) Westheimer, F.H. and Mayer, J.E. (1946) The theory of the racemization of optically active derivatives of dephenyl, *J. Chem. Phys.* **14**, 733; c) Westheimer, F.H. (1947) A calculation of energy of activation for the racemization of 2,2´-dibromo-4,4´-dicarboxydiphenyl, *J. Chem. Phys.* **15**, 252.

5. a) Mathieu, J.-P. (1944) Dichroisme circulaire, liaisons de valence et synthèse asymétique secondaire, *Ann. Phys.* **19**, 335; b) Corey, E.J. and Bailar, J.C. (1959) The stereochemistry of complex inorganic compounds. XXII. Stereospecific effects in complex ions, *J. Am. Chem. Soc.* **81**, 2620.

6. Burkert, U. and Allinger, N.L. (1982) *Molecular Mechanics*, ACS Monograph 177.

7. a) Warshel, A. (1991) *Computer Modeling of Chemical Reactions in Enzymes and Solutions*, Wiley, New York.; b) Doncet, J.-P. and Weber, J. (1996) *Computer-aided Molecular Design*, Academic Press, London; c) Banci, L. und Comba, P. (1997) *Molecular Modeling and Dynamics of Bioinorganic Systems*, Kluwer Academic Publishers, Dordrecht.

8. a) Comba, P. and Hambley, T. W. (1995) *Molecular Modeling of Inorganic Compounds*, VCH, Weinheim; b) Comba, P. (1993) The relation between ligand structure, coordination stereochemistry and electronic and thermodynamic properties, *Coord. Chem. Rev.* **123**, 1; c) Comba, P. (1996) Modeling of structural and spectroscopic properties of transition metal compounds, in Gans W., Amann, A. and Boeyens, J. C. A. (eds.) *Fundamental Priciples of Molecular Modeling*, Plenum Press, New York, 167; d) Hay, B.P. (1993) Methods for molecular mechanics modeling of coordination compounds, *Coord. Chem. Rev.* **126**, 177; e) Landis, C.R., Root, D.M. and Cleveland, T. (1995) Molecular mechanics force field for modeling inorganic and organometallic compounds in: Lipkowitz, K.B. and Boyd, D.B. (eds.) *Reviews in Computational Chemistry*, Vol. 6, VCH, New York; f) Comba, P. (1997) Modeling of structures and molecular properties of transition metal compounds - toward metalloprotein modeling in: Banci, L. and Comba, P. (eds.) *Molecular Modeling and Dynamics of Bioinorganic Systems*, Kluwer Academic Publishers, Dordrecht; g) Deeth, R.J.; Munslow, I.J.; Paget, V.J. (1997) A novel molecular mechanics strategy for transition metals bound to biological molecules in: Banci, L and Comba, P. (eds.) *Molecular Modeling and Dynamics of Bioinorganic Systems*, Kluwer Academic Publishers, Dordrecht; h) Landis, C.R., Firman, T.K., Cleveland, T. and Root, D.M. (1997) Extending molecular

mechanics methods to the description of transition metal complexes and bond-making and - breaking processes in: Banci, L and Comba, P. (eds.) *Molecular Modeling and Dynamics of Bioinorganic Systems*, Kluwer Academic Publishers, Dordrecht.

9. a) Press, W.H., Flannery, B.P., Tenkolsky, S.A. and Vetterling, W.T. (1988) *Numerical Recipes in C*, Cambridge University Press, Cambridge; b) Schlick, T. (1992) Optimization methods in computational chemistry in: K.B. Lipkowitz und D.B. Boyd (eds.) *Reviews in Computational Chemistry*, Vol. 3, VCH, New York; c) Boyd, R.H. (1968) Methods for calculation of the conformation of minimum potential-energy and thermodynamic functions of molecules from empirical valence-force potentials - application to the cyclophanes. *J. Chem. Phys.* **49**, 2574.

10. a) Van de Graaf, F. and Baas, J.M.A. (1980) The implementation of constraints in molecular mechanics to explore potential energy surfaces, *Recl. Trav. Chim. Pays-Bas* **99**, 327; b) Hambley, T.W. (1987) Strain energy minimization study of the mechanism of, and the barrier to, conformational interconversion in five-membered diamine chelate rings *J. Comput. Chem.* **8**, 651.

11. a) Saunders, M., Houk, K.N., Wu, Y.-D., Still, W.C., Liptow, M., Chang, G. und Guida, W.C. (1990) Conformations of cycloheptadecane. A comparison of methods for conformational searching *J. Am. Chem. Soc.* **112**, 1419; b) Chang, G., Guida, W.C. and Still, W.C. (1989) An internal coordinate Monte Carlo method for searching conformational space. *J. Am. Chem. Soc.* **111**, 4379; c) Saunders, M. (1987) Stochastic exploration of molecular mechanics energy surfaces. Hunting for the global minimum, *J. Am. Chem. Soc.* **109**, 3150; d) Haile, J.M. (1992) *Molecular Dynamics Simulations*, Wiley, New York.

12. a) Bernhardt, P. V. and Comba, P. (1992) Molecular mechanics calculations of transition metal complexes, *Inorg. Chem.* **31**, 2638; b) Comba, P.; Hambley, T. W. and Ströhle, M. (1995) The directionality of d-orbitals and molecular mechanics calculations of octahederal transition metal compounds, *Helv. Chim. Acta* **78**, 2042; c) Bol, J. E., Buning, C., Comba, P., Reedijk, J. and Ströhle, M. (1998) Molecular mechanics modeling of the organic backbone of metal-free and coordinated ligands, *J. Comput. Chem.*, **19**, 512.

13. Comba, P. and Hambley, T. W. (1999) Molecular mechanics of inorganic compounds: a Tutorial in Computational Chemistry.

14. Snow, M.R. (1970) Structure and conformational analysis of coordination complexes. The alpha alpha isomer of chlorotetraethylenepentaaminecobalt(III), *J. Am. Chem. Soc.* **92**, 3610.

15. Brubaker, G.R. and Johnson, D.W. (1983) Cobalt(III) complexes with the disubstituted tetraamine ligand 1,10-diamino-2,9-dimethyl-4,7-diazadecane, *Inorg. Chem.* **22**, 1422.

16. Hambley, T. W, Hawkins, C. J., Palmer, J. A. and Snow, M. R. (1981) Conformational analysis of coordination compounds. XI. Molecular structure of tetraammine-{(+-)-pentane-2,4-diamine}cobalt(III) dithionate, *Aust. J. Chem.* **34**, 45.

17. a) Rappé, A. K., Casewit, C. J., Colwell, K. S., Goddard III, W. A. and Skiff, W. M. (1992) UFF, a full periodic table force field for molecular mechanics and molecular dynamics simulations, *J. Am. Chem. Soc.* **114**, 10024; b) Casewit, C. J., Colwell, K. S. and Rappé, A. K. (1992) Application of a universal force field to main group compounds, *J. Am. Chem. Soc.* **114**, 10046; c) Rappé, A.K., Casewit, C. J. and Colwell, K. S (1993) Application of a universal force field to metal complexes *Inorg. Chem.* **32**, 3438.

18. Root, D.M., Landis, C.R. and Cleveland, T (1993) Valence bond concepts applied to the molecular mechanics description of molecular shapes. 1. Application to nonhypervalent molecules of the p-block, *J. Am. Chem. Soc.* **115**, 4201.

19. a) De Hayes, L.J. and Busch, D.H. (1973) Conformational studies of metal chelates. I. Intra-ring strain in five- and six-membered chelate rings, *Inorg. Chem.* **12**, 1505; b) Gollogly, J.R., Hawkins, C.J. and Beattie, J.K. (1971) Conformational analysis of coordination compounds. IV. Conformational energies and activation energies for ring inversion of ethylenediamine complexes, *Inorg. Chem.* **10**, 317; c) Hilleary, C.J., Them, T.F. and Tapscott, R.E. (1980) Stereochemical studies on diastereomers of tris(2,3-

butanediamine)cobalt(III), *Inorg. Chem.* **19**, 102; d) Comba, P.; Hambley, T. W. and Zipper, L. (1988) Energy minimized structures and calculated and experimental isomer distributions in the cobalt(III) hexaamine system Co(trap)$_2^{3+}$ (trap = 1,2,3-propanetriamine), *Helv. Chim. Acta* **71**, 1875.

20. Comba, P. and Sickmüller, A. F. (1997) Modeling of the redox properties of hexaaminecobalt(III/II) couples, *Inorg. Chem.* **36**, 4500; b) Comba, P. and Jakob, H. (1997) Redox potentials of tetraaminecopper(II/I) couples, *Helv. Chim. Acta* **80**, 1983; c) Comba, P. (1998) The importance of intra- and intermolecular weak bonds in transition metal coordination compounds, in: W. Gans and J. C. A. Boeyens (eds) *Intermolecular Interactions*, Plenum Press, New York, p 97.

21. a) Cox, S. and Williams D.E. (1981) Representation of the molecular electrostatic potential by a net atomic charge model, *J. Comput. Chem.* **2**, 304; b) Chirlian, L. E. and Miller-Francl, M. (1987) Atomic charges derived from electrostatic potentials: A detailed study, *J. Comput. Chem.* **8**, 894-905; c) Abraham, R. J., Griffiths, L. and Loftus, P. (1982) Approaches to charge calculations molecular mechanics, *J. Comput. Chem.* **3**, 407-416; d) Rappé, A. K. and Goddard, W. A. III (1991) Charge equilibration for molecular dynamics simulations, *J. Phys. Chem.* **95**, 3358-3363.

22. (a) J. Sabolovic and K. Rasmussen (1995) In vacuo and in crystal molecular-mechanical modeling of copper(II) complexes with amino acids, *Inorg. Chem.* **34**, 1221; (b) L. Glasser, Packing molecules and ions into crystals, in *Fundamental Principles of Molecular Modeling* Gans, W., Amann, A. and Boeyens, J.C.A. (eds.), Plenum Press, New York (1995).

23. Comba, P. (1994) Solution structures of coordination compounds, *Comm. Inorg. Chem.* **16**, 133.

24. Bersuker, I.B. (1996) *Electronic Structures and Properties of Transition Metal Compounds: Introduction to the Theory,* Wiley, New York.

25. a) Maseras, F. and Morokuma, K. (1995) IMOMM: A new integrated *ab initio* plus molecular mechanics geometry optimization scheme of equilibrium structures on transition states, *J. Comput. Chem.* **16**, 1170; b) Tchougréeff, A.L. (1997) The effective crystal field methodolgy as used to incorporate transition metals into molecular mechanics in: Banci, L and Comba, P. (eds.) *Molecular Modeling and Dynamics of Bioinorganic Systems*, Kluwer Academic Publishers, Dordrecht.

26. a) Burton,V.J.; Deeth, R.J. (1995) Molecular modeling for copper(II) centers, *J. Chem. Soc., Chem. Comm.*, 8407-8415.

27. a) Schaeffer, C. E. and C. K. Jorgensen (1965) The angular overlap model, an attempt to revive the ligand field approaches, *Mol. Phys.* **9**, 401; b) Schäffer, C.E. (1974) The non-additive ligand field, *Theor. Chim. Acta* **34**, 237. (c) Larsen, E. and La Mar, G. N. (1974) The angular overlap model, *J. Chem. Educ.* **51**, 633; e) Gerloch, M., Harding, J.H. and Woolley, R.G. (1981) The context and application of ligand field theory, *Struct. Bonding (Berlin)* **46**, 1.

28. Bent, H. (1961) An appraisal of valence-bond structures and hybridization in compounds of the first-row element, *Chem. Rev.* **61**, 275-311.

29. Comba, P., Hambley, T. W., Lauer, G. and Okon, N. (1997) *MOMEC97, a molecular modeling package for inorganic compounds*, Lauer & Okon Chemische Verfahrens- & Softwareentwicklung, Heidelberg, Germany, e-mail: CVS-HD@T-Online.de.

30. Comba, P. (1989) Coordination geometries for hexaamine cage complexes, *Inorg. Chem.* **28**, 426.

31. Comba, P., Lauer, G., Lienke, A. and Okon, N., publication in preparation.

32. Comba, P., Lauer, G., Okon, N. and Remenyi, R. (1999) The computation of cavity shapes, sizes and plasticicies, *J. Comput. Chem., in press.*

33. Comba, P. and Zimmer, M. (1994) Molecular mechanics and the Jahn-Teller effect, *Inorg. Chem.* **33**, 5368.

MODELING STRUCTURAL, SPECTROSCOPIC AND REDOX PROPERTIES OF TRANSITION METAL COMPOUNDS

PETER COMBA
Anorganisch-Chemisches Institut, Universität Heidelberg
Im Neuenheimer Feld 270, 69120 Heidelberg, Germany

1. Introduction

The visualization of structures of chemical compounds is probably the most frequent activity of chemists. With each synthesis, spectrum, mechanistic problem, reaction rate, stability and catalytic, biological and pharmaceutical activity we have a picture of the corresponding molecular or supramolecular structure in our mind. Simple rules, based on experience and models, help us to predict probable connectivities, distances and angles; observed reactivities, stabilities and spectroscopic data are used to confirm a structural model or to suggest another. Steric crowding, preorganization, self-assembly, molecular shapes, selectivity, electronic effects and steric strain are properties and principles that interrelate structural factors with molecular properties, stabilities, reactivities and electronics. Molecular modeling is a computational tool, based on models, to visualize and quantify structures and to derive molecular and bulk properties.

Crystal structure analysis, based on X-ray or neutron diffraction studies, has become a routine tool for the accurate and detailed structural characterization of chemical compounds. Of some concern is the fact that solution structures are not generally identical to those in the solid state and that, therefore, the corresponding thermodynamic properties, reactivities and spectra might change significantly. This is specifically so for coordination compounds involving relatively labile transition metal cations (for example zinc(II), copper(II), cobalt(II)), where solvent molecules and/or counter-anions might lead to ligand exchange, and where the plasticity of the coordination center may lead to angular distortions of the chromophore with corresponding changes in the electronic ground state. Since, very often, solution rather than solid state properties are of importance, solid state structural information may not generally be relevant.

Among the solution structural characterization techniques, EXAFS and NMR experiments (primarily NOE and paramagnetic NMR techniques) have been found to yield data that are, compared with other solution spectroscopic methods, most directly

87

J.A.K. Howard et al. (eds.),
Implications of Molecular and Materials Structure for New Technologies, 87–100.
© 1999 *Kluwer Academic Publishers. Printed in the Netherlands.*

related to nuclear coordinates. However, the resulting structural information is by no means as complete and as unambiguously interpreted as that obtained from crystallographic studies. With these and other spectroscopic methods (see below) that yield data which are strongly coupled to structural parameters, the computation of structures in combination with the corresponding spectra has become a useful approach for the determination of the solution structural behavior.

The computation of molecular structures with quantum mechanical methods has the advantage that the optimization involves electronic wave functions. Thus, in comparison with empirical approaches, the underlying model is scientifically more appealing, the structure optimization is less heavily parameterized and therefore more general and, as a result, charge distributions and electronic energy levels are also available. The main advantage of empirical force field calculations is that they are computationally much less expensive. This is of importance when large molecules or molecular assemblies are studied (metalloproteins, crystal lattices, solvated compounds) or when large series of structures are optimized (conformational search, molecular dynamics). Most problems related to the molecular mechanics modeling of transition metal compounds have been solved to the extent that now molecular mechanics can be successfully used in coordination chemistry [1]. In this chapter, I will concentrate on structure optimizations by classical force field calculations.

For simplicity, structural optimization usually involves interactionless molecules, that is, the environment (crystal lattice, solvent, ion-pairs) is neglected. For molecular mechanics computations this does not lead to serious problems since the parameterization of force fields is generally based on experimental data of solids or solutions, that is, averaged environmental effects are built into the force field and errors from not explicitly including intermolecular interactions are minimal [2]. This also emerges from studies where solution structures and properties are determined and interpreted with a combination of molecular mechanics and the simulation of solution spectra (see below).

Properties of chemical compounds in general, and specifically of transition metal compounds, that may be related to data from molecular mechanics (structures, strain energies, vibrational frequencies), spectra simulations and electronic calculations (single point quantum mechanical computations or ligand field calculations for transition metal compounds), based on the computed structures, include metal ion and stereoselectivities, conformational analyses, redox potentials and electron transfer rates, IR, EPR, NMR and electronic spectra and magnetic properties (see Fig. 1). Data from NMR spectra (primarily using Karplus relations) and NOE experiments have been widely used for structural refinement of biomolecules [3] and also for low molecular weight transition metal compounds [1a]. More recently, similar methods for the structure determination of paramagnetic metalloproteins have been developed and successfully used [4]. From NMR experiments, direct structural parameters are available, and these may be used as constraints when structural models are built and refined and when their dynamic behavior is studied by molecular dynamics.

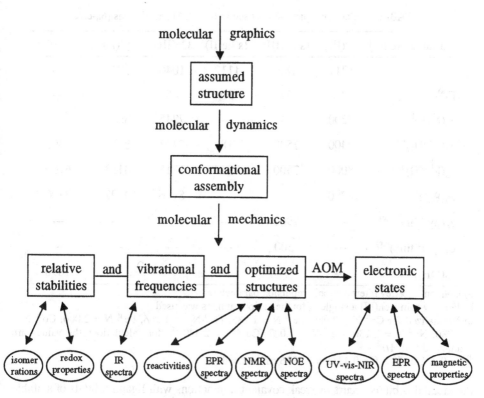

Figure 1. Molecular modeling and the relation between structures and molecular properties.

From all other properties shown in Fig. 1 there is less extensive and less direct structural information available, that is, the computational structure optimization and, in the case of force field calculations, the quality of the force field is of increasing importance. In this chapter I will concentrate on three of these latter techniques that have been used to compute ligand field spectroscopic properties (MM-AOM), redox potentials (MM-Redox) and EPR spectra, based on structures optimized by force field calculations (MM-EPR).

2. The Computation of Molecular Properties

MM-AOM Term energies of transition metal ions are dependent on the metal center, donor atoms and their distances to and angular distributions around the metal center. The angular overlap model (AOM) is based on a simplified MO treatment of the metal-ligand interaction and has been developed for the analysis of electronic d-d transitions, EPR spectra and magnetic properties [5]. The basis of the AOM is that a perturbation of

TABLE 1. Electronic parameters used in the AOM calculations [6a-c,7]

Parameter [cm^{-1}]	Cr(III)	1s Fe(III)	1s Fe(II)	Co(III)	Ni(II)	Cu(II)^d
F_2[a]	1211	786	1112	1046	1243	---
F_4[a]	101	57	90	85	79	---
$e_\sigma(NH_3)$[b,c]	7200	---	---	7245	3582	--
$e_\sigma(RNH_2)$[b,c]	7400	7500	7188	7433	3857	6400
$e_\sigma(R_2NH)$[b,c]	8000	7500	7344	7715	4133	6700
$e_\sigma(R_3N)$[b,c]	8700	---	---	8186	4592	7100
$e_\sigma(pyridine)$[b,c]	---	7900	---	---	---	---
$e_\pi(pyridine)$[b,c]	---	-500	---	---	---	---
$e_\sigma(OH_2)$[c]	---	---	---	---	---	1090

a) Condon-Shortley interelectronic repulsion parameters
b) For low spin iron(III) a single e_σ for the various amines was used
c) Normalized for Cr^{III}-N = 2.080, Fe^{III}-N = 1.985, Fe^{III}-N$_{py}$ = 1.968, Fe^{II}-N = 2.000, Co^{III}-N = 1.980, Ni^{II}-N = 2.130, Cu^{II}-N = 2.027, Cu^{II}-O = 2.440 Å; for calculations the values are adjusted with $1/r^6$
d) $k = 0.7$, $\xi = -580$, $K = 0.43$, $P = 0.0036$ cm^{-1}, $\alpha^2 = 0.74$

the metal d-orbitals occurs by weak covalent interactions with ligand orbitals of suitable symmetry. The destabilization of the d-orbitals is proportional to a two-atom overlap integral, and this is parameterized with e_σ, e_π and e_δ values. The application of AOM calculations to predict electronic properties, based on computed structures, involves a constant and transferable set of e-parameters for a given set of compounds, modified exclusively by structural differences. This is not a priori a given property of the AOM model, and transferability of ligand field parameters has been discussed controversially [6]. We have successfully used transferable parameter sets, adjusted by $1/r^n$ (r is the metal-donor distance, n=5,6) and an empirical correction for misdirected valences (Table 1) [6a-c,7].

MM-Redox. Reduction potentials are related to the free energy of the corresponding electron transfer process ($\Delta G° = -nFE°$), where $\Delta G°$ for transition metal coordination compound couples involves terms that are related to the ionization potential of the free metal ion, the complexation and the solvation of the reduced and oxidized forms of the couples. The basis for using strain energy differences ΔU_{strain} between the oxidized and the reduced forms of the complexes is that some of these terms may be neglected. It emerges that

$$E° = \frac{f}{-nF} \cdot \Delta U_{strain} \qquad (1)$$

where f is a function that depends on steric factors, that is, f is assumed to be linearly dependent on the strain energy difference ΔU_{strain}. With $f = 1$ the slope of the correlation between $E°$ and ΔU_{strain} is 96.5 kJ Mol^{-1}V^{-1} and the intercept is 0.0 [8]. The applicability of this approach depends on the relative importance of the structural difference between the oxidized and the reduced forms of the couples with respect to the neglected terms. For hexaaminecobalt(III/II) couples, good correlations with slopes of 45-65 kJ Mol^{-1}V^{-1} (depending on the type of amine ligand) [8a,b], and for tetraaminecopper(II/I) couples, slopes of 60-70 kJ Mol^{-1}V^{-1} have been observed [8c].

MM-EPR. EPR spectra of weakly dipole-dipole coupled dicopper(II) compounds can be simulated using the spin Hamiltonian parameters (anisotropic g- and A-values of each copper(II) center), line shape parameters and four geometric parameters that are available from the computed or observed structure (r, τ, η and ξ, i.e. the copper-copper distance, the angle between the two z-axes of the g-Tensors, the angle between the y-axis, transformed to the xy plane of one of the copper centers, and the vector between the two metal centers, and the angle between the z-axis of one of the copper centers and the vector between the two metal centers, respectively) [9]. The combination of molecular mechanics and the simulation of EPR spectra of weakly coupled dinuclear copper(II) compounds has been used to determine solution structures of dicopper(II) compounds [10].

MM-AOM, MM-Redox, MM-EPR and MM-Isomer (not discussed in this chapter, see for example [11]) have been used successfully to determine solution structures. The same algorithms may also be applied to predict molecular properties of unknown compounds, that is, for the design of new compounds with specific properties or for the interpretation of unexpected experimental data. Usually, there is a series of stable conformations in solution, each one with significantly different spectroscopic and redox properties. The observed spectra and redox potentials are then a weighted average of all species present. It is quite common that the different spectra and potentials may not be resolved, and in many cases, due to poor resolution and due to the inaccuracy of the modeling procedures, computed spectra and redox potentials of the most abundant isomer alone have been used for comparison with the corresponding experimental data.

Two specific examples will now be discussed in detail. Others have been presented in some recent review articles [1a-c,f,12 and references therein].

3. Example 1: Spectroscopic and redox properties of copper(II) compounds of tetraazamacrocyclic ligands with variable ring sizes

Tetraazamacrocyclic ligand copper(II) compounds usually have 4+1 or 4+2 chromophores with four amine donors building a (distorted) square planar arrangement with the copper(II) center and one or two more distant axial donors from solvent molecules, counter-anions or pendant groups of the macrocyclic ligands. In these systems, the distortion of the CuN$_4$ chromophore is primarily a function of the macrocyclic ring-size, with fourteen-membered macrocycles of the type of cylam

92

(1,4,8,11-tetraazacyclotetradecane) being most suitable for copper(II). With smaller rings, copper(II) is usually five-coordinate with a square pyramidal coordination geometry, and with larger rings, a tetrahedral distortion leads to the accommodation of the usual Cu-N distances of approximately 2.0 Å. The observed structures of a series of relevant complexes are shown in Fig. 2 and the corresponding structural and spectroscopic parameters are presented in Table 2 [6c, 8c,13]. The correlation of the reduction potentials for copper(II/I) couples with the strain energy differences is shown in Fig. 3.

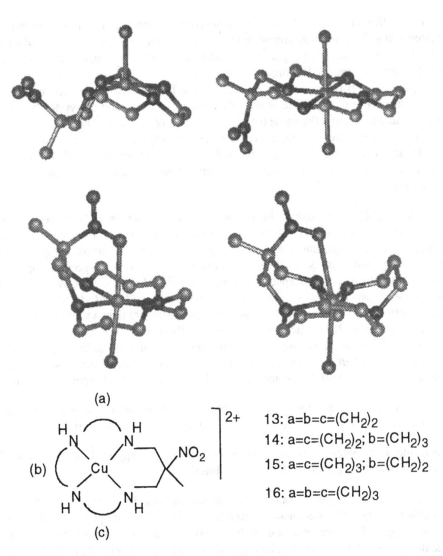

(a)

(b)

(c)

$$\left[\text{...} \right]^{2+}$$

13: $a=b=c=(CH_2)_2$
14: $a=c=(CH_2)_2$; $b=(CH_2)_3$
15: $a=c=(CH_2)_3$; $b=(CH_2)_2$
16: $a=b=c=(CH_2)_3$

Figure 2. Calculated structures of $[Cu([13]N_4)]^{2+}$, $[Cu([14]N_4)]^{2+}$, $[Cu([15]N_4)]^{2+}$ and $[Cu([16]N_4)]^{2+}$ [14a,b].

TABLE 2: Experimentally determined and calculated (italics) structural and spectroscopic parameters (MM-AOM in brackets) of a series of copper(II) tetraamines [6c]

Compound[a]	Cu-N^{av}(Å)	Cu–X^{av}(Å)		θ^{av}(deg)[b]	E$_{xy}$ (cm^{-1})
[Cu([13]N$_4$)]$^{2+}$	2.014	1	(2.507)[c]	28.0	18020 (*19530*)
	1.999		*2.195*	*21.6*	*18640*
[Cu([14]N$_4$)]$^{2+}$	2.006	2	2.558	0.3	20830 (*21200*)
	2.019		*2.549*	*1.1*	*20400*
[Cu([15]N$_4$)]$^{2+}$	2.037	2	2.512	11.5	18180 (*18840*)
	2.038		*2.493*	*17.2*	*18340*
[Cu([16]N$_4$)]$^{2+}$	2.029	2	2.643	35.3	17240 (*18080*)
	2.008		*2.682*	*42.5*	*17850*

a) 13-to-16-membered tetraazamacrocyclic ligand copper(II) compounds shown in Fig. 2.
b) tetrahedral twist angle (square planar: $\theta=0°$, tetrahedral: $\theta=90°$)
c) experiment: X = Cl; modeling: X = OH$_2$

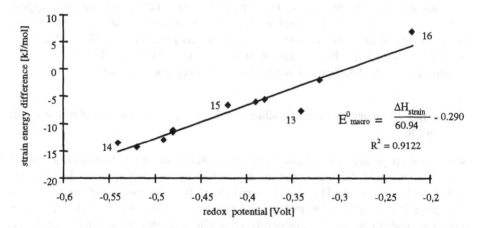

Figure. 3. Correlation of the strain energy difference ΔU_{strain} of copper(II/I) tetraamine couples with macrocyclic tetraamine ligands (see Table 2 for the four complexes shown in Fig. 2, labelled 13, 14, 15, 16) with the experimentally determined reduction potentials (*vs.* NHE) in aqueous solution [8c].

94

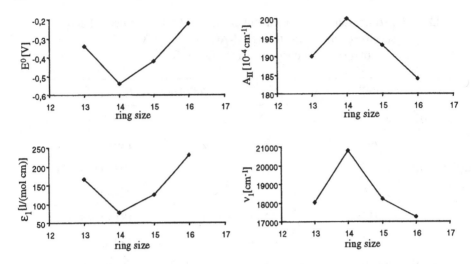

Figure 4. Correlation of the reduction potential $E°$, the EPR hyperfine parameter $A_{\parallel}$, the d-d transition v_1 and first extinction coefficient ε_1 with the macrocycle ring size.

Fig. 4 indicates that the d-d transition, the hyperfine coupling (EPR) and the redox potential have, as expected, a common dependency on the macrocyclic ring-size. The fact that neither the Cu-N distances nor the tetrahedral twist angles or axial bonding alone are directly correlated to the observed properties is not unexpected and suggests that only a careful and combined analysis of all parameters allows a meaningful interpretation of the observed trends. The computed parameters in Table 2 and the correlations shown in Fig. 3 and 4 indicate that MM-AOM (UV-vis and EPR parameters) and MM-Redox lead to predictions of acceptable accuracy.

4. Example 2: Structure determination of a dicopper(II) compound of a new cyclic octapeptide

Novel cyclic peptides, isolated from bacteria, fungi and marine organisms that contain unusual amino acids have attracted much attention due to their specific chemical and bacteriological properties [14]. Metal complexes of these ligands are of interest due to their possible role in their biosynthesis and metal ion accumulation [14a,15]. However, only few transition metal compounds of cyclic peptides have been structurally characterized so far [14a,16], and the reported solution structure of [Cu$_2$(PatN)] (PatN = cyclo(Ile-ser-(gly)-Thz-Ile-Thr-(Gly)-Thz)) is the first example of a fully characterized metal complex of a synthetic cyclic peptide [17]. PatN has structural features that are similar to those of the naturally occurring Patellamide A and the hydrolyzed form of Asciadiacyclamide (for structures see Fig. 5).

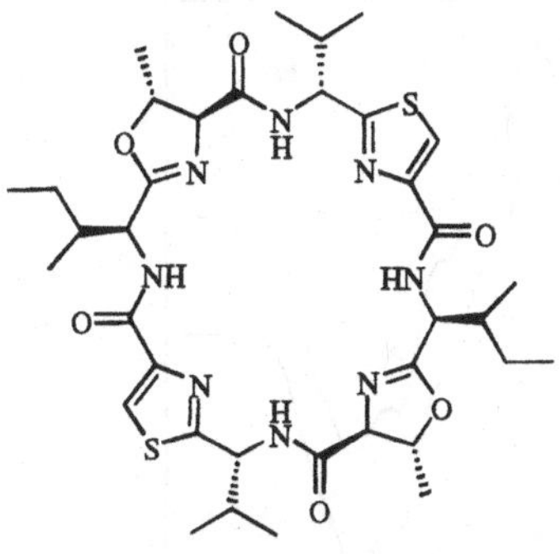

Ascidiacyclamide

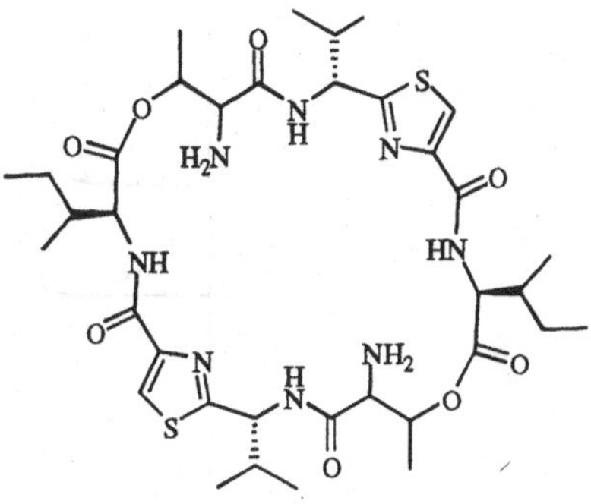

hydrolyzed Ascidiacyclamide

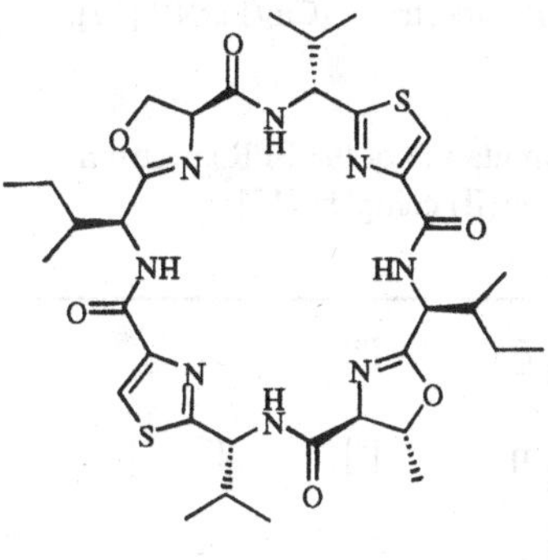

Patellamide A

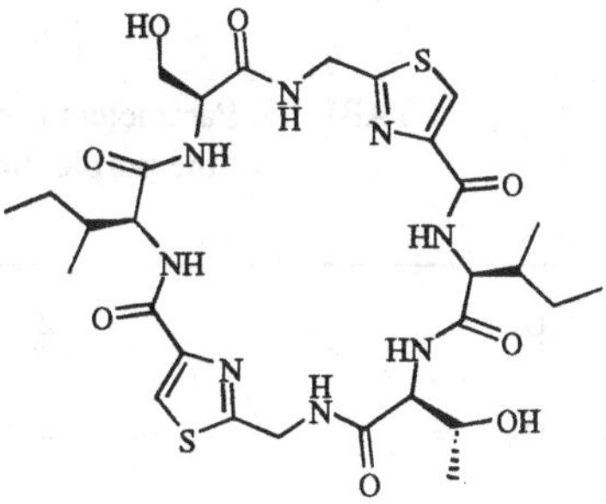

PatN

Figure 5. Structure of the synthetic cyclic octapeptide PatN and of three natural products

The most stable species isolated from the reaction in methanol of PatN with copper(II) and base is a purple dicopper(II) compound. A thorough spectroscopic analysis of this compound and four other species (three mononuclear and a blue dinuclear compound), involving ion-spray mass spectroscopy, UV-vis, IR, CD and multifrequency EPR spectroscopy, indicate that the purple dicopper(II) compound has each of the two copper(II) ions bound to one thiazolyl N-donor, two amide N-donors and three oxygen atoms from water. The experimentally observed EPR spectrum is shown together with its simulation in Fig. 6, and the parameters used for the simulation are presented in Table 3.

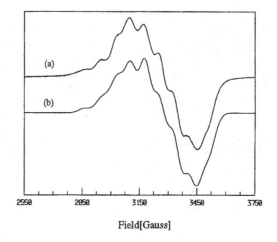

Figure 6. Observed (a) and simulated (b) EPR spectra of [Cu₂(PatN)] [17].

TABLE 3: Parameters used for the simulation of the EPR spectrum
of the purple dinuclear copper(II) complex [17].

$g_\parallel$	2.260	ξ	[°]	88
$g_\perp$	2.085	η	[°]	25
$A_\parallel$ $[10^4\,cm^{-1}]$	82	τ	[°]	35
$A_\perp$ $[10^4\,cm^{-1}]$	77	$r_{Cu\text{-}Cu}$	[Å]	5.89
$W_\parallel$ [Gauss]	28			
$W_\perp$ [Gauss]	28			

Based on the structural parameters, a model of the dicopper(II) complex was constructed, and molecular dynamics calculations lead to three low energy conformations of the dicopper(II) compound ([Cu₂(PatN1)], [Cu₂(PatN2)], [Cu₂(PatN3)]). These were optimized, and EPR spectra were simulated, based on the corresponding structural parameters (for structural parameters, see Table 4). For only one of the structures were the simulated and the observed spectrum in reasonably good agreement. The corresponding structure was that with the lowest strain energy (see Table 4; the experimentally observed and simulated spectra ([Cu₂(PatN)] and [Cu₂(PatN1)], respectively) are shown in Fig. 7, and the computed solution structure ([Cu₂(PatN1)]) is given in Fig. 8).

TABLE 4: Angles, Cu-Cu-distances and strain energies of the calculated structures compared with the parameters from the EPR simulation of the purple dinuclear copper(II) complex [Cu$_2$(PatN)] [17].

	ξ [°]	τ [°]	η [°]	$r_{Cu\text{-}Cu}$ [Å]	Strain energy [kJ/mol]
[Cu$_2$(PatN1)]	75	31	26	5.28	86.0
[Cu$_2$(PatN2)]	71	34	20	5.42	94.8
[Cu$_2$(PatN3)]	43	2	17	4.96	125.8
EPR-Simulation	88	35	25	5.98	-

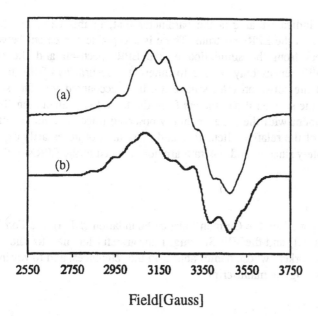

Field[Gauss]

Figure 7. Experimentally observed EPR spectrum of [Cu$_2$(PatN)] (a) and simulated spectrum of Cu$_2$(PatN1)] (b) [17].

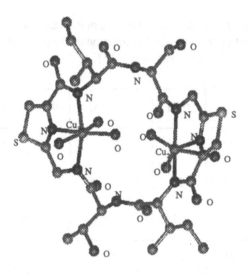

Figure 8. Solution structure of [Cu$_2$(PatN)] (computed structure of [Cu$_2$(PatN1)]) [17].

Model calculations indicate that ξ is the parameter that, in this case, has the most significant influence on the EPR spectrum. There is acceptable agreement between the structural parameters from the simulation of the EPR spectrum and the structure optimization. The differences may be due to inherent inaccuracies of the force field and/or the fact that the solvation effects have not been accounted for. The simulated spectrum, based on the structural parameters from the force field calculation (Fig. 7) is in acceptable agreement with the experimentally observed trace. The main differences are due to changes of the relative intensities, and these are not necessarily expected to be computed accurately since g- and A-strain are not included in the EPR simulation.

Acknowledgments

Our studies are supported by the German Science Foundation (DFG), the Fonds of the Chemical Industry (FCI) and the VW-Stiftung. I am grateful for that, for the excellent work done by my coworkers whose names appear in the references, and to Karin Stelzer for her help in preparing the manuscript.

References

1. a) Comba, P. and Hambley, T. W. (1995) *Molecular Modeling of Inorganic Compounds,* VCH, Weinheim; b) Comba, P. (1993) The relation between ligand structure, coordination stereochemistry and electronic and thermodynamic properties, *Coord. Chem. Rev.* **123**, 1; c) Comba, P. (1996) Modeling of structural and spectroscopic properties of transition metal compounds, in Gans W., Amann, A. and Boeyens, J. C. A. (eds.) *Fundamental Principles of*

Molecular Modeling, Plenum Press, New York, 167-187; d) Hay, B.P. (1993) Methods for molecular mechanics modeling of coordination compounds *Coord. Chem. Rev.* **126**, 177; e) Landis, C.R., Root, D.M. and Cleveland, T. (1995) Molecular mechanics force field for modeling inorganic and organometallic compounds in: Lipkowitz, K. B. and Boyd, D. B. (eds.) *Reviews in Computational Chemistry*, Vol. 6, VCH, New York; f) Comba, P. (1997) Modeling of structures and molecular properties of transition metal compounds - toward metalloprotein modeling in: Banci, L. and Comba, P. (eds.) *Molecular Modeling and Dynamics of Bioinorganic Systems*, Kluwer Academic Publishers, Dordrecht; g) Deeth, R.J., Munslow, I. J. and Paget, V. J. (1997) A novel molecular mechanics strategy for transition metals bound to biological molecules in: Banci, L and Comba, P. (eds.) *Molecular Modeling and Dynamics of Bioinorganic Systems*, Kluwer Academic Publishers, Dordrecht; h) Landis, C.R., Firman, T.K., Cleveland, T. and Root, D.M. (1997) Extending molecular mechanics methods to the description of transition metal complexes and bond-making and -breaking processes in: Banci, L. and Comba, P. (eds.) *Molecular Modeling and Dynamics of Bioinorganic Systems*, Kluwer Academic Publishers, Dordrecht.

2. Comba, P. (1998) Molecular mechanics modeling of transition metal compounds in: Howard, J.A.K. and Allen, F.H. (eds.) *Implications of Molecular and Materials Structure for New Technologies*, Kluwer Academic Publishers, Dordrecht.

3. a) Warshel, A. (1991) *Computer Modeling of Chemical Reactions in Enzymes and Solutions*, Wiley, New York; b) Doncet, J.-P. and Weber, J. (1996) *Computer-aided Molecular Design*, Academic Press, London; c) Banci, L. und Comba, P. (1997) *Molecular Modeling and Dynamics of Bioinorganic Systems*, Kluwer Academic Publishers, Dordrecht; d) Wüthrich, K. (1986) NMR of proteins and nucleic acids, Wiley, New York; e) Wüthrich, K. (1989) The development of nuclear magnetic resonance spectroscopy as a technique for protein structure determination, *Acc. Chem. Res.* **22**, 36.

4. a) Bertini, I. and Rosato, A. (1997) Solution structures of proteins containing paramagnetic metal ions in: Banci, L. und Comba, P. (eds.) *Molecular Modeling and Dynamics of Bioinorganic Systems*, Kluwer Academic Publishers, Dordrecht; b) Banci, L. and Gori-Savellini, G. (1997) Molecular dynamics calculations on metalloproteins in: Banci, L. und Comba, P. (1997) *Molecular Modeling and Dynamics of Bioinorganic Systems*, Kluwer Academic Publishers, Dordrecht.

5. a) Schaeffer, C. E. and C. K. Jorgensen (1965) The angular overlap model, an attempt to revive the ligand field approaches, *Mol. Phys.* **9**, 401; b) Schäffer, C.E. (1974) The non-additive ligand field, *Theor. Chim. Acta* **34**, 237; (c) Larsen, E. and La Mar, G. N. (1974) The angular overlap model *J. Chem. Educ.* **51**, 633; e) Gerloch, M., Harding, J. H. and Woolley, R.G. (1981) The context and application of ligand field theory, *Struct. Bonding (Berlin)* **46**, 1.

6. a) Bernhardt, P. V. and Comba, P. (1993) Prediction and interpretation of electronic spectra of transition metal complexes *via* the combination of molecular mechanics and angular overlap model calculations, *Inorg. Chem.* **32**, 2798; b) Comba, P. (1994) Prediction and interpretation of EPR spectra of low spin iron(III) complexes with the MM-AOM method, *Inorg. Chem.* **33**, 4577; c) Comba, P., Hambley, T. W., Hitchman, M. A. and Stratemeier, H. (1995) Interpretation of electronic and EPR spectra of copper(II)amine complexes - A test of the MM-AOM Method, *Inorg. Chem.* **34**, 3903; d) Glerup, J., Monsted, O. and Schäffer, C.E. (1980). Transferability of ligand field parameters and nonlinear ligation in chromium(III) complexes. *Inorg. Chem.* **19**, 2855; e) Gerloch, M. and Woolley, R. G. (1981) Is the angular overlap model chemically significant?, *J. Chem. Soc., Dalton Trans.*, 1714; f) Vanquickenborne, L. G., Coussens, B., Postelmans, D., Ceulemans, A. and Pierloot, K. (1991). Electronic structure of $Cr(NH_3)_6^{3+}$ and substituted chromium(III) ammine compounds, *Inorg. Chem.* **30**, 2978.

7. a) Börzel, H., Comba, P., Pritzkow, H. and Sickmüller, A. (1998) Preparation, structure and electronic properties of a low spin iron(II) hexaamine compound, *Inorg. Chem.*, **37**, 3853 ; b) Comba, P., Hilfenhaus, P. and Nuber, B. (1997) The structure of an isomeric pair of copper(II) tetraamines in the solid state and in solution, *Helv. Chim. Acta* **80**, 1831.

8. a) Comba, P. and Sickmüller, A. F. (1997) The solution structures of a pair of stable cobalt(III) hexaamine conformers, *Angew. Chem.* **109**, 2089; *Angew. Chem., Int. Ed. Engl.* **36**, 2006; b) Comba, P. and Sickmüller, A. F. (1997) Modeling of the redox properties of hexaaminecobalt(III/II) couples, *Inorg. Chem.* **36**, 4500; c) Comba, P. and Jakob, H. (1997) Redox potentials of tetraaminecopper(II/I) couples, *Helv. Chim. Acta* **80**, 1983.

9. Smith, T. D, Boas, J. F. and Pilbrow, J. R. (1974) An electron spin resonance study of certain vanadyl polyaminocaboxylate chelates formed in aqueous and frozen aqueous solutions, *Aust. J. Chem.* **27**, 2535.

10. a) Bernhardt, P. V., Comba, P. Hambley, T. W., Massoud, S. S. and Stebler, S. (1992) Determination of solution structures of binuclear copper(II) complexes, *Inorg. Chem.* **31**, 2644; b) Comba, P. and Hilfenhaus, P. (1995) One step template synthesis and solution structures of bis(macrocyclic) octaamine dicopper(II) complexes *J. Chem. Soc., Dalton Trans.*, 3269. c) Comba, P., Hambley, T. W., Hilfenhaus, P. and Richens, D. T. (1996) Solid state and solution structures of two structurally related dicopper complexes with markedly different redox properties, *J. Chem. Soc., Dalton Trans.*, 533; d) Brudenell, S. J., Bond, A. M., Spiccia, L., Comba, P. and Hockless, D. C. R. (1998) Structural, EPR and electrochemical studies of binuclear copper(II) complexes of bis(pentadentate) ligands derived from bis(1,4,7-triazacyclononane) macrocycles, *Inorg. Chem.*, **37**, 4389; e) Comba, P., Gavrish, S. P., Hay, R. W., Hilfenhaus, P., Lampeka, Y. D., Lightfoot, P. and Peters, A. (1999) Analysis and interpretation of significant structural differences of dinuclear complexes (M = Ni(II), Cu(II)) of a bismacrocyclic ligand, *Inorg. Chem., accepted.*

11. a) Comba, P., Maeder, M. and Zipper, L. (1989) Energy minimized structures and calculated and experimental isomer distributions in the cobalt(III) hexaamine system CoL_2^{3+} with the chiral facially coordinating triamine L = 1,2,4-butanetriamine, *Helv. Chim. Acta* **72**, 1029; b) Comba, P., Jakob, H., Keppler, B. K. and Nuber, B. (1994) Solution structures and isomer distributions of bis(ß-diketonato) complexes of titanium(IV) and cobalt(III), *Inorg. Chem.* **33**, 3396.

12. a) Comba, P. (1998) The importance of intra- and intermolecular weak bonds in transition metal coordination compounds, in: W. Gans and J. C. Boeyens (eds.) *Intermolecular Interactions*, Plenum Press, New York, p.97; b) Comba, P. (1994) Solution structures of coordination compounds, *Comm. Inorg. Chem.* **16**, 133.

13. Comba, P., Curtis, N. F., Lawrance, G. A., O'Leary, M. A., Skelton, B. W. and White, A H. (1988) Comparison of thirteen- to sixteen-membered tetraazacyclo-alkane copper(II) complexes derived from template synthesis involving nitroethane and formaldehyde. Crystal structures of (10-methyl-10-nitro-1,4,8,12-tetraazacyclopentadecane)copper(II) and (3-methyl-3-nitro-1,5,9,13-tetraazacyclohexadecane)copper(II) perchlorate, *J. Chem. Soc., Dalton Trans.*, 2145.

14. (a) Michael, J.P. and Pattenden, G. (1993) Marine metabolites and metal ion chelation: The facts and the fantasies, *Angew. Chem.* **105**, 1-23; (b) Lewis, J.R. (1989) Muscarine, oxazole and peptide alkaloids and other miscellaneous alkaloids, *Natural Products Reports* **6**, 503; (c) Faulkner, D.J. (1988) Marine natural products, *Natural Products Reports* **5**, 613; (d) Krebs, H.C. (1986) Recent developments in the field of marine natural products with emphasis on biologically active compounds, *Fortschr. Chem. Org. Naturst.* **49**, 151; e) Rosen, M.K. and Schreiber, S. L. (1992) *Angew. Chem., Int. Ed. Engl.* **31**, 384-400; (f) Evans, D. A. and Ellman, J. A. (1991) *J. Am. Chem. Soc.* **11**, 1063-1672; (g) Fusetani, N., Sugawara, T. and Matsunaga, S. (1991) *J. Am. Chem. Soc.* **113**, 7811-7812.

15. a) Carlisle, D. B. (1968) Vanadium and other metals in ascidians, *Proc. Royal Soc. B.* **171**, 31-42; (b) Kustin, K. and Weaver, R. (1990) Vanadium, a biologically relevant element, *Adv. Inorg. Chem.* **35**, 81-115.

16. (a) van den Brenk, A. L., Fairlie, D. P., Hanson, G. R., Gahan, L. R., Hawkins, C. J. and Jones, A. (1994) Binding of copper(II) to the cyclic octapeptide patellamide D., *Inorg. Chem.* **33**, 2280-2289; b) van den Brenk, A. L., Byriel, K. A., Fairlie, D. P., Gahan, L. R., Hanson, G. R., Hawkins, C. J., Jones, A., Kennard, C. H. L., Moubaraki, B. and Murray, K. S. (1994) Crystal structure, electrospray ionization mass spectrometry, electron paramagnetic resonance and magnetic susceptibility study of $[Cu_2(ascidH_2)(1,2-\mu-CO_3)(H_2O)_2]$ $\cdots$ $2H_2O$, the bis(copper(II)) complex of ascidiacyclamide (ascidH$_4$), a cyclic peptide isolated from the ascidian lissoclinum-patella, *Inorg. Chem.* **33**, 3549-3557; c) van den Brenk, A. L., Fairlie, D. P., Gahan, L. R., Hanson, G. R. and Hambley, T. W. (1996) A novel potassium-binding hydrolysis product of ascidiacyclamide: A cyclic octapeptide isolated from the ascidian, *Lissoclinum patella, Inorg. Chem.* **35**, 1095-1100.

17. Comba, P., Cusack, R. M., Fairlie, D. P., Gahan, L. R., Hanson, G. R., Kazmaier, U. and Ramlow, A. (1998), The solution structure of a copper(II) compound of a new cyclic octapeptide by EPR spectroscopy and force field calculations, *Inorg. Chem.* **37**, 6721.

MOLECULES IN CRYSTALS - WHAT MAKES THEM DIFFERENT?

KERSTI HERMANSSON

Inorganic Chemistry, The Ångström Laboratory, Uppsala University, Box 538, S-75121 Uppsala, Sweden

1. Introduction: From molecule to crystal

The goal of molecular design is to create molecular aggregates and materials with special, tailored properties. Such an endeavour will be more successful the more we know about molecular interactions and how the properties of a molecule change in different surroundings. As long as a molecule keeps its essential identity in the bound state, it is appealing to retain the concept of 'a molecule' even within a crystal or a liquid, because it relates so closely to well known chemical concepts. However, the properties of many molecules, including the water molecule, are known to be sensitive to their surroundings and their states of aggregation. For instance, the average $O\cdots O$ distance in the gaseous water dimer is 2.98 Å, while in liquid water it is only ~2.85 Å, and in normal ice ~2.75 Å. This gradual shortening of intermolecular bond distances is a manifestation of the cooperative nature of the intermolecular bonding, where certain geometrical arrangements give rise to large changes in molecular properties. In this paper, I will use the water molecule as an example, because water is a good and important representative of molecules which exhibit long-ranged interactions. Such molecules affect their neighbours, but the interaction also leads to changes in the molecule itself - a feature highly relevant in the context of molecular design.

In this lecture we will discuss the development of the 'in-crystal' molecular properties as the system is assembled from its constituent components ("*Molecules in crystals - what makes them different?*"). Most of the results presented are based on quantum-mechanical *ab initio* calculations (for finite or infinite systems). Experimental data for molecules in crystals generally display the combined effect of all interactions: short- and long range interactions of different types (ionic, dipolar, van der Waals, etc.), etc. One particularly compelling aspect of theoretical calculations is that they make it possible to investigate tailored model complexes. This facility allows us to focus on different parts/types of the molecular surroundings individually and investigate their separate contributions to the total "sum interaction". A similar kind of enlightening division is usually difficult to achieve by experimental methods.

In the condensed phases the water molecule is found in a variety of diverse environments and has been amply studied with a multitude of experimental techniques. Interesting examples with different characteristics and technologically important properties are therefore easy to find. Here we will focus on a few representative examples of water molecules in surroundings with very different bond strengths: the ionic $LiOH \cdot H_2O$ and

J.A.K. Howard et al. (eds.),
Implications of Molecular and Materials Structure for New Technologies, 101–117.
© 1999 *Kluwer Academic Publishers. Printed in the Netherlands.*

102

LiClO$_4$·3H$_2$O crystals, hydrogen-bonded ice structures and, finally, the β-quinol·H$_2$O clathrate crystal structure, where the water molecule resides as a guest molecule in a large cage (**Fig. 1**). The host-guest interaction energy in this clathrate structure is about 20 kJ/mol per guest molecule. This energy is rather small, but large enough to stabilize the clathrate structure and prevent the compound from transforming into the α-quinol structure, which turns out to be the stable polymorph in the absence of guest molecules. The properties we will study are electron density, molecular charges and dipole moments, electric field gradients at the O and H nuclei and, finally, the intramolecular OH stretching vibrations.

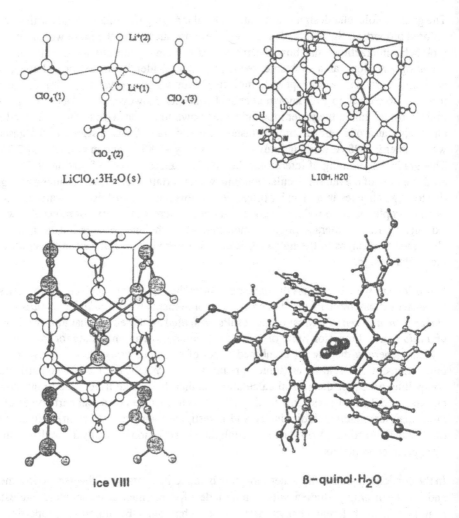

Figure 1. Experimentally determined structures for some of the crystals discussed in this paper.

2. Quantum-chemical calculations of extended systems

Theoretical quantum-mechanical calculations are always affected by systematic errors (limited basis set, limited treatment of electron correlation effects, neglect of temperature-dependent effects, etc.). One usually hopes, and often also tries to demonstrate, that the *change* or *shift* of a property is much less affected by error than the absolute value itself, provided one and the same method has been used for the situations compared. This will be the underlying understanding in the results I present here, but for the uncredulous reader I refer to the references provided together with each investigation; these references give more detail concerning the methods used.

Provided the chemical system is small enough, quantum-mechanical ab initio calculations yield molecular properties and intermolecular interactions to practically any desired degree of accuracy. But, as implied above, for larger systems the effect of the surroundings has to be incorporated by means of approximate methods; various models have been designed, such as "continuum" and "reaction field" models, the isolated "supermolecule" approach, and combinations thereof. For crystals with long-ranged interactions the crystal field has to be taken into account. The crystal results I present in this paper were obtained by two different approaches: (*i*) the water molecule and all its nearest neighbours were treated as a supermolecule, and the rest of the crystal was included implicitly, via a set of atomic point charges constructed so as to mimic the Madelung potential at every point within a region enclosing the supermolecule, or (*ii*) the crystal was treated as an infinite periodic 3-dimensional lattice. There are a number of 3-D periodic Density Functional Theory (DFT)-based programs in use in the scientific community today, but only one based on the Hartree-Fock approach, namely the program package CRYSTAL [1], which was used for all the crystal-orbital calculations presented in this paper. The finite systems were computed with the HONDO [2], Gaussian92 and Gaussian94 programs [3].

The majority of the energy calculations presented here have been performed at the Restricted Hartree-Fock (*RHF*) level with or without electron correlation taken into account (via the second-order Møller-Plesset perturbation formalism (*MP2*) [4]). The basis sets were generally of *DZP* quality [5,6].

3. Electron density redistribution

We will start by discussing the molecular electron distribution, $\rho(\mathbf{r})$, which is a sensitive probe of the strength and nature of intermolecular interaction. Many experimental electron density investigations based on X-ray diffraction measurements have been performed for molecular crystalline systems (see, forexample, [7] and references therein).

104

3.1. WATER CHAINS

Fig. 7 (presented further down) shows a series of puckered $(H_2O)_n$ chains of different lengths. We have studied the convergence of various molecular properties in these chains as a function of chain length [8]. The water molecules all lie in the same plane, and the O-H···O angles are 180°. Our intention was to create a "quasi-crystalline" infinite chain with a geometry as close as possible to the "infinite chains" found in different ice polymorphs, but still keep the system as simple as possible. The effect of the intermolecular interaction is clearly demonstrated by the resulting optimized O···O hydrogen-bond distances, which range from 3.021 Å (in the dimer) to 2.846 Å (for the middlemost molecules in the 12-chain) to 2.840 Å (the infinite chain).

Fig. 2 shows $\Delta\rho(\mathbf{r}) = \rho_{total}(\mathbf{r}) - \Sigma\rho_{molecules}(\mathbf{r})$, the so-called interaction density, for some of the shorter chains and for the infinite chain. The isolated molecules subtracted in these

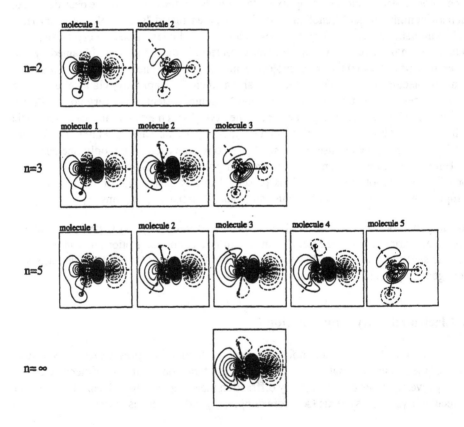

Figure 2. Calculated difference density maps, $\Delta\rho(\mathbf{r}) = \rho_{total}(\mathbf{r}) - \Sigma\rho_{molecules}(\mathbf{r})$, around each water molecule in some H-bonded water chains (the geometries are shown in Fig. 7). The sections are taken in the planes of the water molecules, which all lie in the same plane (see text). Solid lines denote electron excess, dashed lines electron loss, and the zero contour is omitted. Contour levels at ±0.01 e/Å³.

plots have the same geometries as in the optimized complex, and the maps thus display the direct electron redistribution caused by the intermolecular interaction. For the dimer, the two $\Delta\rho(\mathbf{r})$ maps show that both molecules are polarised in the same direction by the H-bond, with an electron flow along the H-bond donating OH bond in molecule *1* (the left-most molecule, the H-bond donor) and along both OH bonds in molecule *2* (the H-bond acceptor).

High-level experimental single-crystal electron density maps are seldom accurate to a resolution larger than ± 0.05 e/Å^3, which, by the way, is often the contour level interval chosen to present experimental $\Delta\rho(\mathbf{r})$ maps. Comparison with the infinite chain shows that such a convergence has been reached for the middlemost molecule for a chain length of $n= 5$, and we suggest that for comparative experimental/computational electron density studies of chemical systems where H-bonds are the strongest bonds present, such as in the ices, it might be sufficient to take only nearest and next-nearest water neighbours into account in the *ab initio* calculations. Fig. 2 also shows that $\Delta\rho(\mathbf{r})$ converges much faster for the end molecules than for the molecules within the chains.

3.2. CLUSTERS AND CRYSTALS

The interaction density for an isolated tetrahedron-like $(H_2O)_5$ cluster, taken out of the ice VIII structure, is plotted in **Fig. 3a**. This structure has been determined experimentally by neutron diffraction [9], and it is those positions that were used in the calculations for Fig. 3. The central water molecule is here surrounded by four H-bonded neighbours at an $R(\text{O}\cdots\text{O})$ distance of 2.879 Å, all oriented so as to increase the polarisation of the central water molecule. Consequently we are perhaps not surprised to find that the $\Delta\rho(\mathbf{r})$ features are more pronounced for the tetrahedron than for both the 3-chain and the 5-chain in Fig. 2 (Note the different contour intervals used).

The effect of the next-nearest neighbours and the rest of the crystal field (summarized here as 'the crystal-field effect') is seen by comparing **Figs. 3a** and *b*, and is found to be modest.

The situation is rather different in the LiOH·H_2O crystal. Here the water molecule is strongly bound by four ionic nearest neighbours (see Fig. 1): two water - Li$^+$ interactions of 1.981 Å (determined by neutron diffraction [10]) and two quite strong hydrogen bonds donated by the water molecule to the hydroxide ions (HOH$\cdots$OH$^-$ bond length of 1.685 Å). The resulting ab initio-computed [11] $\Delta\rho(\mathbf{r})$ map for the water molecule in its cluster of nearest neighbours (**Fig. 3c**) shows the same overall features as for the ice tetrahedron, but much more pronounced. If we include the crystal field in the calculations (**Fig. 3d**), the electron density redistribution is reduced to about half. This occurs because ionic crystals are built up of alternating positive and negative ions, so the next-nearest neighbours of the water molecule will partly "neutralize" the effect of the nearest neighbours, and so on. The final experimental electron density maps obtained from X-ray diffraction for LiOH·H_2O [10] show a fairly modest electron redistribution around water and we can now explain why this is so: the intermolecular interaction from the nearest and the next-nearest neighbours counteract each other.

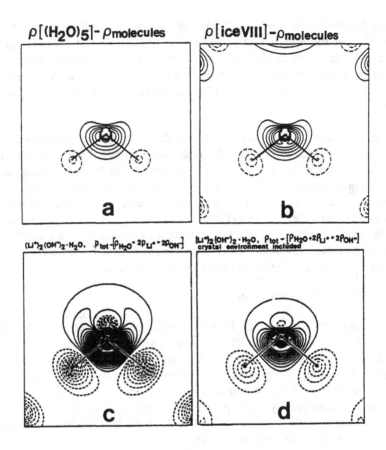

$\rho[(H_2O)_5] - \rho_{molecules}$ $\rho[\text{ice VIII}] - \rho_{molecules}$

a   b

$(Li^+)_2(OH^-)_2 \cdot H_2O,$ $\rho_{tot} - [\rho_{H_2O} + 2\rho_{Li^+} + 2\rho_{OH^-}]$

$(Li^+)_2(OH^-)_2 \cdot H_2O,$ $\rho_{tot} - [\rho_{H_2O} + 2\rho_{Li^+} + 2\rho_{OH^-}]$ crystal environment included

c   d

Figure 3. Calculated difference density maps, $\Delta\rho(\mathbf{r}) = \rho_{total}(\mathbf{r}) - \Sigma\rho_{molecules}(\mathbf{r})$, through the water molecules in the ice VIII and LiOH·H$_2$O crystal structures. In (a) and (c) only the nearest neighbours have been included in the calculation, in (b) and (d) also long-range crystal effects are taken into account. Contour levels at ±0.05 e/Å^3.

4. Molecular dipole moment and charges

For polar crystals (*i.e.* crystals consisting of ions or dipolar molecules) the size of the molecular dipole moment ($\mu_{molecule}$) and the atomic charges to a large part determine the lattice energy and are thus vital to the crystal's stability and properties. At a more practical level, the molecular and atomic charges and the higher electrostatic moments of a molecule all have important implications in 'computational chemistry modelling' (Monte Carlo, Molecular dynamics, molecular mechanics calculations). The electrostatic model used in molecular dynamics simulations, for example, often consists of atomic partial charges which are kept fixed regardless of the geometrical arrangement of the molecules in the simulation. The success of such a model and the choice of the best average dipole moment depend on the strength and range of the intermolecular interactions, and it is thus important to have a realistic estimate of the size of these effects.

Several different strategies have been used in the literature to try to incorporate the effect of charge redistributions from many-body (many-molecular) interactions during an MC or MD simulation. So, for example, there are models which incorporate one or several molecular sites possessing an isotropic or anisotropic polarizability which react to the electric field created by the surroundings. A second class of models which go beyond pair-wise interaction functions use analytical potential energy functions involving clusters of three or even more molecules; such analytical three- and four-body functions are usually derived by a fitting procedure based on ab initio interaction energies generated for a large number of triplet (or larger cluster) geometrical configurations. A third approach to remedy the lack of many-body interactions is the 'ab initio MD' approach, e.g. the Car-Parrinello type simulations (see, for example, [12] and references therein), where the nuclear equations of motion are solved using forces obtained not from analytical potential functions, but from electronic quantum-mechanical calculations performed during the course of the MD simulation. Whichever approach is considered, however, it is important to know and understand the range and effect of intermolecular interaction on the molecular properties, since that knowledge will help determine the most feasible strategy to pursue.

4.1. MOLECULAR DIPOLE MOMENT

We have computed the net dipole moment for each molecule in the water chains using the Mulliken population analysis inherent in the CRYSTAL package. The total molecular dipole moment here originates from two contributions: the local atomic monopoles (charges q_i) provide a contribution equal to $\Sigma \, q_i \cdot (r_i - r_0)$ where i is O, H1 and H2, and r_0 the center-of-mass for each molecule, and the second contribution originates from the local atomic dipoles which is obtained by summing up the three μ_i vectors for O, H1 and H2.

The magnitude of the dipole moment for the isolated water molecule is 2.214 D [8] at the computational level employed here. The total molecular dipole moment for each molecule along the 12-chain is shown in **Fig. 4a**. The water molecules at the ends are much less polarized than those at the middle of the chain. The dipole moment of the middlemost water molecule in the 5-chain is 12% larger than in the free molecule and in the infinite chain ~15% larger than in the free molecule. $\Delta\mu$ converges more quickly with chain length for the end molecules than for the middlemost molecules in the chains.

In the dimer the dipole moment of molecule 1 is enhanced by 5.1% compared to the isolated water molecule and the dipole moment of molecule 2 is enhanced by 4.7%, *i.e.* they are quite similar. On the other hand, the electron density maps in Fig. 2 demonstrate rather large differences between the two molecules. It is clear that both properties are needed to obtain a full picture of the effect of the electron redistribution.

Fig. 5 shows how the molecular dipole moment in the ice VIII crystal is built up as we go from monomer $\Rightarrow$ trimer chain $\Rightarrow$ tetrahedral cluster $\Rightarrow$ ice VIII crystal. All nearest neighbours contribute towards the increased polarisation of the central water molecule. Neighbours further away also help polarise the water molecule but, as we also saw in the electron density maps, the crystal-field effect is fairly modest.

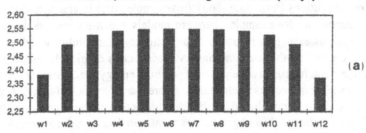

(a)

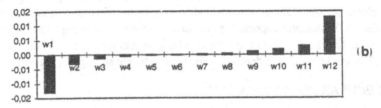

(b)

Figure 4. Calculated molecular dipole moments (*a*) and molecular charges (*b*) for each individual water molecule along an optimized $(H_2O)_{12}$ chain (the chain geometry is the same as those shown in Fig. 7)

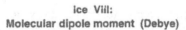

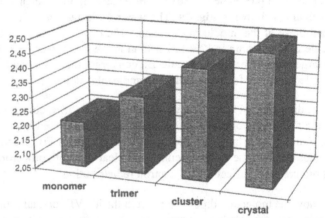

Figure 5. Calculated molecular dipole moment for the (central) water molecule in the H_2O, $(H_2O)_3$, $(H_2O)_5$ fragments taken out from the ice VIII crystal structure (cf. Fig. 1). The rightmost column shows the result from the full-crystal calculation.

We conclude that the water molecules in the chains and in the clusters are substantially polarised by their environment, and, indeed, computer simulations (Monte Carlo, molecular dynamics) of liquid water in the literature based on pairwise additive potentials with fixed atomic charges have often required exaggerated monomer dipoles of up to 2.4 D to reproduce experimental data with any success [13-14].

4.2. MOLECULAR CHARGES

The net molecular charge for each molecule in the 12-chain was calculated by summing the atomic Mulliken charges and is shown in **Fig. 4b**. All the chains we have studied follow the same pattern. There is a net charge transfer 'from right to left' in the chains, *i.e.* from the terminal H-bond acceptor to the terminal H-bond donor. The end molecules build up a significant net charge while all the other molecules remain virtually neutral. In an earlier study of H-bonded water chains with fixed hydrogen-bond distances (2.84 Å) we found a similar charge transfer effect from one end of the chain to the other [15].

These results have interesting implications for quantum-mechanical calculations of liquids or solids based on the cluster approximation: the results may be heavily affected by the electric field created by charge transfer from one side to the other across the small cluster. This is an unphysical effect since in reality there is no abrupt limit between the nearest neighbours and the rest of the environment. In a study of OH frequency shifts in liquid water [16] using point charge-embedded water pentamer clusters, the effect (read: error) from the "cluster dipole" on the OH vibrational frequency for the central water molecule was estimated to about 50 cm^{-1}. A larger supramolecular cluster would move the border between properly treated atoms and point-charge atoms further away from the central water molecule and reduce the cluster dipole and its accompanying, unwanted electric field.

5. Electric field gradient shifts

The electric field gradient (EFG) tensor at the atomic nuclei can be determined from NMR and NQR experiments which measure the quadrupole coupling constant (QCC) of nuclei which possess electric-quadrupole moments. The largest component of the EFG tensor at a point r' essentially depends on the quantity $\rho(r)/|(r-r')|^3$ integrated over all volume elements in space, and is therefore sensitive to the local environment around the point where it is probed.

The EFG tensor has five independent components. Diagonalization of the EFG tensor yields the principal components. Here we will only discuss the largest principal component, V_{33}, at the O and H nuclei because at least the latter has been found to show interesting correlations with H-bond strength. V_{33} and QCC are proportional to each other, with the proportionlity constant depending on the electric quadrupole moment of the nucleus studied. The high sensitivity of the deuteron quadrupole coupling constant, QCC(^{2}H), to small variations in hydrogen-bond strength and the high precision with which it can be determined experimentally makes it a very valuable quantity in hydrogen-bond analyses. This was first pointed out by Chiba in 1964 [17]. Berglund *et al.* [18] have correlated the electric field gradient parameters with other well known hydrogen-bond parameters, such as OH intramolecular stretching vibrational frequencies and the O···O hydrogen-bond distance.

It is known that also $QCC(^{17}O)$ is affected by the H-bonding environment. NMR and NQR measurements show that $QCC(^{17}O)$ in different ice polymorphs is reduced by ~35% relative to the vapour phase [19]. We have studied also the correlation of $QCC(^{17}O)$ with hydrogen-bond strength and bonding environment using ab initio methods [20].

Fig. 6 shows how $V_{33}(H)$ (called 'EFG(H)' in the figure) and $V_{33}(O)$ ('EFG(O)' in the figure) in ice VIII are built up from different neighbour interactions We find that [20]:

Both $V_{33}(H)$ and $V_{33}(O)$ are sensitive probes of the H-bonded environment around water molecules. Both are downshifted due to the interaction with the environment.

$V_{33}(H)$ is affected by (i) the electric field from the neighbours, (i) the polarisation of the water molecules due to this field, and (iii) the intramolecular geometry changes (brought about by the intermolecular interaction). For $V_{33}(O)$ the third factor is almost negligible.

The next-nearest neighbours and the rest of the crystal field in ice has a marked effect on the both the oxygen and hydrogen field gradients. A cluster of the five water molecules in a tetrahedral configuration is too small to be used for predictions of the V_{33} value. However, it gives a reasonable estimate of the V_{33} downshift in liquid water.

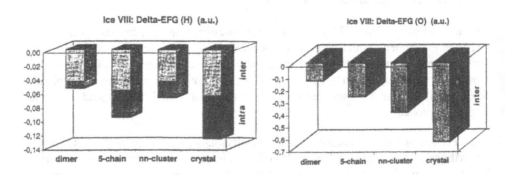

Figure 6. Calculated V_{33} shifts induced by the environment at the H and O nuclei of the (central) water molecule in $(H_2O)_2$, $(H_2O)_5$ chain, $(H_2O)_5$ tetrahedral fragments taken out from the ice VIII crystal structure. In both figures the rightmost column shows the result from the full-crystal calculation. For H the V_{33} shifts have been separated into intra- and intermolecular contributions.

6. Intramolecular frequency shift

The sensitivity of intramolecular vibrational stretching frequencies to intermolecular interactions is utilized in a range of commercial and scientific applications. Infrared and Raman measurements have been performed on a large number of crystalline hydrates and aqueous solutions, and with today's instrumental resolution it is possible to detect frequency shifts as small as a few wavenumbers. As we will see, however, the frequency shifts of even weakly bound water molecules are usually much larger.

6.1. CALCULATION OF VIBRATIONAL SPECTRA

The OH vibrational frequencies reported here refer to uncoupled vibrations as obtained experimentally from isotope-isolated measurements, *i.e.* they correspond to the OH vibration of an HDO molecule in an environment where all other water molecules present are D_2O molecules. Because of the large mass difference between H and D, the resulting OH vibrational mode is almost entirely a "one-legged" OH stretching vibration, decoupled from other vibrations. The result is a cleaner spectrum, which is also easier to interpret in terms of intermolecular interactions. Moreover, as an extra bonus, the anharmonic vibrational problem becomes easier to treat theoretically this way, since we can now use a one-dimensional oscillator to model the vibration. The vibrational frequencies reported here are either harmonic frequencies obtained directly from the normal coordinate analysis inherent in the Gaussian program, or they were obtained from separate anharmonic vibrational calculations [21] subsequent to the generation of the ab initio potential energy curves. In general, whenever intramolecular X-H bonds are discussed, it is important to include the anharmonicity contribution, since both the absolute frequencies and the frequency shifts are greatly affected by the anharmonicity. Nevertheless, many *trends*, remain similar for harmonic and anharmonic X-H shifts.

6.2. WATER CHAINS AND CLUSTERS

Fig. 7 shows calculated spectra from all the water molecules in all the chains from $n=2$ to $n=10$. All the hydrogen-bond distances were optimized, as mentioned earlier. We find that the (harmonic, uncoupled) frequencies are grouped into four groups: (*i*) The non-H bonded OH bonds at the right end of the chains (Group 1 in Fig. 7), (*ii*) a combined group consisting of the water molecules at the left end of all the chains (Group 2 in Fig. 7), and those which are next-closest to the right end (Group 3), (*iii*) the broad group of water molecules "in the interior of" the chains (Group 4), and (*iv*) the H-bonded OH bond of water *1* in the dimer. The water molecule in the middle of the 12-chain (not included in the plot) is downshifted ~200 cm^{-1} relative to the free-water frequency.

Fig. 8 shows how the OH frequency for a water molecule in an ice-like cluster is build up of its constituents [15]. The downshift in the dimer in Fig. 8*a* is large. Also H-bonded neighbours on the oxygen side of the central water induce frequency downshifts. We have seen from electron density maps (not shown here) that the H-bond-accepting neighbour on the "opposite" OH bond causes an electron density transfer over to the vibrating OH

112

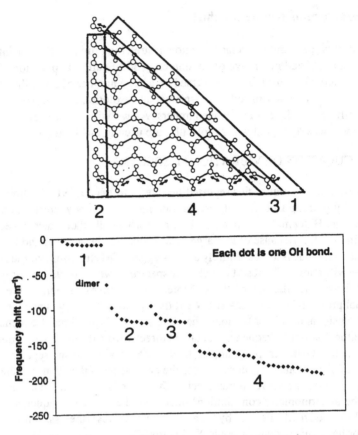

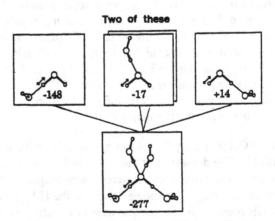

Figure 7. Calculated harmonic, uncoupled OH frequency shifts induced by the environment for each water molecule in the geometry-optimized $(H_2O)_n$ chains shown in the upper part of the figure. The frequencies were calculated one at a time, and the vibrational modes studied are indicated by arrows in the upper figure.

Figure 8. Calculated anharmonic, uncoupled OH frequency shifts induced in the central water molecule by its surroundings in a water tetrahedron and its trimer fragments. The R(O···O) distance was kept at 2.84 Å.

bond and we believe this to be the reason for the frequency upshift found for that bond in Fig. 8c. The total downshift for the cluster is substantial, much larger than the sum of the contributing terms. We will return to this question in the next section.

6.3. CATION-WATER CLUSTERS

A number of mono-, di- and trivalent cations in aqueous solution have been studied by vibrational spectroscopy (see *e.g.* Refs. 22-27). The uncoupled OH vibration of water molecules in the first hydration shell of Li^+, for example, undergoes an average frequency downshift of -270 cm^{-1} [27] compared to the isolated-water frequency, while for the first hydration shell around an Mg^{2+} ion, the redshift is -450 cm^{-1} [22].

Fig. 9 shows the calculated uncoupled, anharmonic stretching vibrational frequency for HDO molecules in a number of $Mg^{2+}(H_2O)_n$ complexes at the *RHF(DZP)* level [28]. It is interesting to note that the frequency shifts in the cation-water complexes are highly non-additive. Let us study complex 6 and its two 'constituent complexes' 2 and 5. The

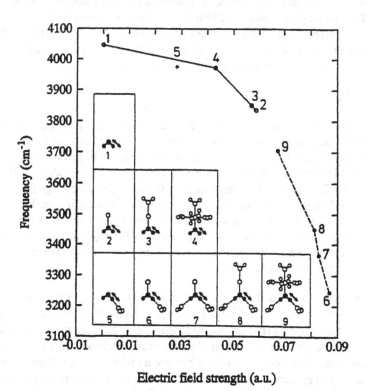

Figure 9. Calculated anharmonic, uncoupled OH frequency for water vibrations in nine different water and $Mg^{2+}(H_2O)_n$ systems. For each system the vibration studied is indicated by an arrow in the molecular drawing. For each system the electric field strength at the equilibrium position of the vibrating H atom was calculated using nominal charges for the surrounding atoms (-0.80 *e* for O, +0.40 for H and +2.0 for Mg).

water dimer in complex *5* is downshifted by -70 cm^{-1} and the cation-water complex in *2* by -200 cm^{-1}, yet the total downshift for 'the sum complex' in *6* is as large as -800 cm^{-1}. The reason for the large non-additivity of the frequency shifts is hinted at in Fig. 9, where the frequencies are plotted against the electric field strength at the vibrating H nucleus. The plot shows that the various complexes lie on a curved line (approximately harmonic), which immediately implies that the field increment caused by a certain neighbour depends on how many neighbours are already present around the vibrating water, or, differently said, "where on the bent curve in Fig. 9 we start". The non-linearity of the frequency vs. field curve thus "explains" the non-additivity of the frequency shifts.

Let's look at some other interesting complexes in Fig. 9. The frequency downshift for the Mg^{2+}(H$_2$O) complex (complex *2*) is about -200 cm^{-1}. The addition of a second neighbour on the other side of the cation ion (complex *3*), results in a frequency upshift compared to *1*, mostly due to charge transfer from the water molecules to the ion. In Mg^{2+}(H$_2$O)$_6$ (complex *4*) the resulting downshift compared to the free water molecule is only -70 cm^{-1}. Adding two second-shell hydrogen-bonded water molecules to complex *4* will give us complex *9* with a rather large downshift, in agreement with our discussion in the previous paragraph (the second-shell water which binds to the non-vibrating H atom in *9* actually has only a small effect on the frequency shift, as seen from Fig. 8). Thus the difference between complexes *4* and *9* is much larger than the difference between *1* and *5*, giving us yet another example of the frequency non-additivity effect.

6.4. CRYSTALS

Five crystalline hydrates are discussed here. The experimental (isotope-isolated) O-H frequency shifts here range from one of the very smallest observed in ionic crystalline hydrates (-150 cm^{-1} in LiClO$_4$·3H$_2$O [29]) to the very largest (-930 cm^{-1} in LiOH·H$_2$O [29]). The different O-H vibrations in ice II and ice IX are downshifted by about -310 to -370 cm^{-1} [30,31]. The O···O distance in these ice modifications lie in the range 2.75 - 2.84 Å. No experimental frequency value exists for the β-quinol clathrate structure. **Fig. 10** shows our calculated frequency shifts (uncoupled, anharmonic) with respect to the monomer frequency as we go from dimer ⇒ tetrahedral nearest-neighbour cluster ⇒ crystal for all but the clathrate (the data in the figure are taken from Refs. 32-34). The last column in each sequence is the experimental shift.

A comparison of the second and third columns confirms that for the ionic compounds the nearest neighbours and the crystal field have opposite effects, while for the ices they cooperate. The bar for LiOH·H$_2$O is only sketched. This is to illustrate that in a fictitious (Li$^+$)$_2$(OH$^-$)$_2$·H$_2$O complex the interaction of the water molecule with its environment is so strong that the water would deprotonate and become a hydroxide ion instead. This is seen by the potential energy curve denoted H$_2$O(Li$^+$)$_2$(OH$^-$)$_2$ in the bottom figure in Fig. 10. With the crystal field present, however, the potential energy curve has a minimum to the left, corresponding to an intact water molecule, but the curve is shallow at the minimum, indicating a low-lying vibrational ground state, i.e. a large frequency downshift compared to the isolated water molecule.

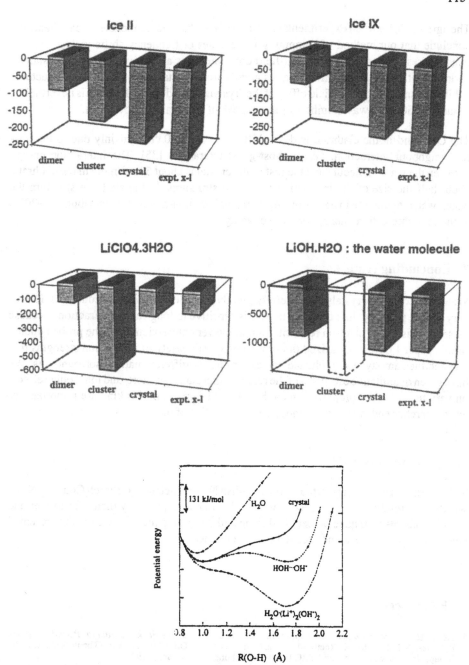

Figure 10. Calculated and experimental anharmonic, uncoupled OH frequency shifts induced in a water molecule by its surroundings in four crystals (cf. Fig. 1). For each crystal the four columns are as follows. 1st: calculated dimer fragment, 2nd: calculated nearest-neighbour cluster, 3rd: calculated full-crystal environment, 4th: experimental full-crystal environment. The potential energy curves for LiOH·H₂O in the bottom figure are all calculated.

116

The agreement between experimental and calculated absolute frequencies is excellent. We conclude that our method of computing frequencies or frequency shifts yields very satisfactory results and that the crystal-field contribution is always important in crystalline hydrates, although the magnitude varies greatly from crystal to crystal. The ice structures exhibit the smallest crystal-field effect of the crystals studied in Fig. 10, but it is nevertheless as large as 50 to 70 wavenumbers or about 25% of the total OH downshift.

The OH bond in the clathrate is downshifted by about 50 cm^{-1}, mainly due to the nearest-neighbour interaction, *i.e.* the host-guest interaction [35]. The interaction energy between the water molecule and the rest of the crystal is about 5 kcal/mole in the clathrate, about half the size of the interaction in the ice structures. In the clathrate structure the water-water interaction in neighbouring cages makes a non-negligible contribution (~20%) to the total interaction energy for large coverages.

7. Concluding remarks

Water molecules in crystals are generally greatly modified by their environments. Even a very 'local' property, like the EFG tensor, is sensitive to long-range interactions because these give rise to an electron redistribution in the very close vicinity of the probed nuclei. In this paper I have emphasized the broad range and variety of interaction strengths that one and the same type of molecule can experience in different natural sorroundings. We have given quantitative estimates of the relative importance of short- and long-range effects on various molecular properties in the belief that estimates of this kind are important for any researcher endeavouring to modify the properties of molecular materials.

8. Acknowledgements

This research has been supported by the Swedish Natural Science Research Council (NFR) which is gratefully acknowledged. I would also like to express my thanks to former and present students who have helped producing and discussing many of the results presented here. Their names are acknowledged in the reference list.

9. References

1. *CRYSTAL 95. User's Manual. An ab initio All-Electron LCAO-Hartree-Fock Program for Periodic Systems,* R. Dovesi, V.R. Saunders, C. Roetti, M. Causà, N.M. Harrison, R. Orlando and E. Aprà (Theoretical Chemistry Broup, University of Turin, and SERC Daresbury Laboratory), December 1996.

2. M. Dupuis, J. Rys and H.F. King (1976), *J. Chem. Phys.* **65**, 111; M. Dupuis, A. Farazdel, S.P. Karna and S.A. Maluendes, In: *MOTECC, Modern Techniques in Computational Chemistry,* Ed. E. Clementi, ESCOM, Leiden (1990).

3. Gaussian 92/DFT, Revision G.4, M. J. Frisch, G. W. Trucks, H. B. Schlegel, P. M. W. Gill, B. G. Johnson, M. W. Wong, J. B. Foresman, M. A. Robb, M. Head-Gordon, E. S. Replogle, R. Gomperts, J. L. Andres, K. Raghavachari, J. S. Binkley, C. Gonzalez, R. L. Martin, D. J. Fox, D. J. Defrees, J. Baker, J. J. P. Stewart, and J. A. Pople, Gaussian, Inc., Pittsburgh PA, 1993.

4. C. Møller and M.S. Plesset (1934), *Phys. Rev.* **46**, 618.

5. T.H. Dunning, Jr. and P.J. Hay, in *Modern Theoretical Chemistry* Vol. 3: *Methods of Electronic Structure Theory*; Ed.: Henry F. Schaefer III, p. 1, Plenum Press, New York (1977).

6. A.J. Thakkar, T.Koga, M. Saito and R.E. Hoffmeyer (1993), *Int. J. Quantum Chem. Quantum Chem.Symp.* **27**, 343.

7. P. Coppens, *X-ray Charge Densities and Chemical Bonding*, Oxford University Press (1997).

8. K. Hermansson and M. Alfredsson. J. Chem. Phys. In press.

9. W.F. Kuhs, J.L. Finney, C.Vettier and D.V. Bliss (1984), *J. Chem. Phys.* **81**, 3612.

10. K. Hermansson and J.O. Thomas (1982), *Acta Crystallogr.* **B38**, 2555.

11. K. Hermansson and S. Lunell (1982), *Acta Crystallogr.* **B38**, 2563.

12. K. Laasonen, M. Sprik, M. Parrinello and R. Car (1993), *J. Chem. Phys.* **99**, 9080.

13. J. Caldwell, X.D. Liem and P.A. Kollman (1990), *J. Am. Chem. Soc.* **112**, 9144.

14. M. Sprik (1991), *J. Chem. Phys.* **95**, 6762.

15. L. Ojamäe and K. Hermansson (1994), *J. Phys. Chem.* **98**, 4271.

16. K. Hermansson, S. Knuts and J. Lindgren (1991), *J. Chem. Phys.* **95**, 7486.

17. T. Chiba (1964), *J. Chem. Phys.* **41**, 1352.

18. B. Berglund, J. Lindgren and J. Tegenfeldt (1978), *J. Mol. Struct.* **43**, 179.

19. D. T. Edmonds, S. D. Goren, A. A. L. White and W. F. Sherman (1977), *J. Magn. Reson.* **27**, 35.

20. M. Alfredsson and K. Hermansson. Chem. Phys. In press.

21. M.J. Wojcik, J.Lindgren and J. Tegenfeldt (1983), *Chem. Phys. Lett.* **99**, 112.

22. O. Kristiansson, A. Eriksson, J. Lindgren (1984), *Acta Chemica Scand.* **A38**, 613.

23. H. Kleeberg, G. Heinje and W. Luck (1986), *J. Phys. Chem.* **90**, 4427.

24. G. Zundel and J. Fritsch, In: *The Chemical Physics of Solvation*, Vol. 2, Eds.: R.R. Dogonadze, E. Kalman, A. Kornyshev and J. Ulstrup, Chapter 2: Interactions in and Structures of Ionic Solutions and Polyelectrolytes. Infrared Results. Elsevier, Amsterdam (1986).

25. P.-Å. Bergström, J. Lindgren, M. Read and M. Sandström (1991), *J. Phys. Chem.* **95**, 7650.

26. W. Mikenda (1986), *Monatshefte fur Chemie* **117**, 977.

27. J. Lindgren, K. Hermansson and M.J. Wojcik (1993), *J. Phys. Chem.* **97**, 3712.

28. K. Hermansson, J. Lindgren and M.M. Probst (1995), *Chem. Phys. Lett.* **233**, 371.

29. B. Berglund, J. Lindgren and J. Tegenfeldt (1978), *J. Mol. Struct.* **43**, 169.

30. J.E. Bertie and E. Whalley (1964), *J. Chem. Phys.* **40**, 1646.

31. J.E. Bertie and F.E. Bates (1977), *J. Chem. Phys.* **67**, 1511.

32. L. Ojamäe and K. Hermansson (1992), *J. Chem. Phys.* **96**, 9035.

33. S. Knuts, L. Ojamäe and K. Hermansson (1993), *J. Chem. Phys.* **99**, 2917.

34. K. Hermansson and L. Ojamäe (1995), *Solid State Ionics* **77**, 34.

35. K. Hermansson. To be published.

A NEW *AB INITIO* POWDER METHOD AND PROFILE REFINEMENT IN MATERIALS DESIGN: APPLICATION TO POLYMER ELECTROLYTES

PETER G. BRUCE, GRAHAM MacGLASHAN and YURI G. ANDREEV
School of Chemistry, University of St. Andrews, St. Andrews, Fife KY16 9ST, UK

1. Introduction

Crystallography is an enabling science and as such its influence on chemistry has been incalculable. This article describes the critical role that crystallography is playing in a key area of solid state coordination chemistry (polymer electrolytes) [1-4]. Crystallographic studies have led to new strategies for the design of solid coordination compounds with polymeric ligands which as a result display substantially higher ionic conductivity than was previously possible opening the way to applications of these polymeric coordination compounds as solid polymer electrolytes in devices. The demands placed on crystallography often act as a catalyst for the development of crystallographic methods. In order to tackle the crystallographic problems in the area of polymer electrolytes, powder diffraction methods have been vital. Our interest in such methods has led us to develop a particularly powerful approach to *ab initio* structure determination from powder diffraction data, which combines simulated annealing with chemical constraints and full profile fitting.

2. Polymer electrolytes

Salts may be dissolved in liquid ethers and extending the chain length of the latter so that the solvent becomes a polymer does not change fundamentally the energetics of solvation. Dissolution is driven primarily by coordination of the cations by the ether

119

J.A.K. Howard et al. (eds.),
Implications of Molecular and Materials Structure for New Technologies, 119–133.

oxygens associated with the polymer chains. In the vast majority of cases the polymer host contains the ethylene oxide unit $(CH_2\text{-}CH_2\text{-}O)$ which is a particularly good solvating group [1,2]. Although a variety of polymer architectures have been studied we will concentrate in this article on the archetypal polymer solvent, poly(ethylene oxide) $[(CH_2CH_2O)_n]$ usually abbreviated to PEO. Many salts may be dissolved in poly(ethylene oxide) as shown in Fig. 1.

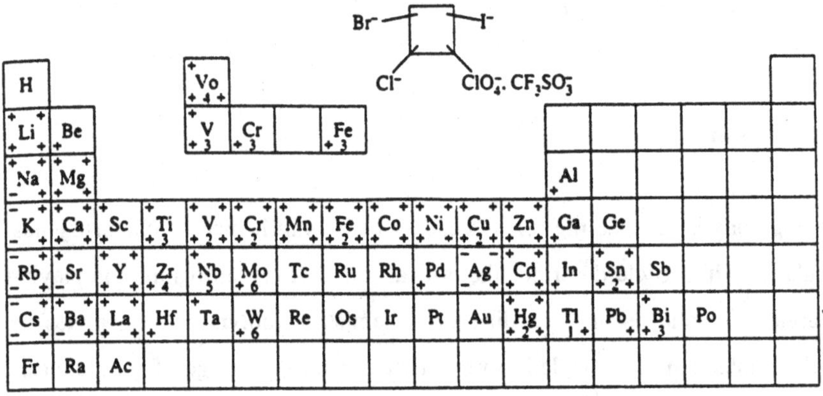

Figure 1. Part of the periodic table in which the elements marked with + form salts that are complexed by poly(ethylene oxide) and those with – do not. (Reproduced by permission of World Scientific from Armand & Gauthier (1989).)

The resulting polymer-salt complexes have obvious analogies to the crown ethers. Almost nothing is known of the structural chemistry of the many thousands of polymer-salt complexes that have been prepared; this interesting area of coordination chemistry deserves further structural characterisation. However there is another important reason for our interest in the structural chemistry of these compounds. These materials combine uniquely solid, yet flexible, mechanical properties with ionic conductivity and as such are of considerable interest as electrolytes for the development of all-solid-state devices such as rechargeable lithium batteries, electrochromic displays and smart windows [4]. For many of these applications a high level of conductivity is essential and much effort has been expended over the last twenty years in order to improve the conductivity. Polymer electrolytes may be prepared as crystalline or amorphous materials and it turns out that ionic conductivity is largely confined to the amorphous phase above its glass transition temperature in which state the polymer chains are in motion and can facilitate ion transport by constantly creating new sites into which the ions may migrate. Based on this understanding many polyether solvents have been designed with low T_g in order to enhance ionic conductivity. However it became clear that the level of ionic conductivity which could be achieved by lowering T_g had reached a plateau and that to make a further substantial advance a different approach to materials design would be necessary. In many areas of chemistry a knowledge of structure is essential to an understanding, and hence ultimately optimisation, of physical properties. The paucity of knowledge concerning the structural chemistry of polymer electrolytes was a major barrier in the field and its removal was essential to achieve a substantial advance in our understanding of ion transport and ultimately exploitation of polymer electrolytes. Of course it is the structure of the amorphous materials which is of direct interest since ionic conductivity occurs in this state, however such studies are notoriously difficult and the level of information that has been obtained, despite much excellent effort over the years, has remained somewhat limited. We decided to take a different approach and to develop, through crystallography, a knowledge of the structural chemistry by first establishing the structures of the crystalline materials and then using this as the basis for interpreting spectroscopic studies of analogous amorphous phases. In this article we will show that powder crystallography has provided a substantial advance in our knowledge of the structural chemistry of polymer electrolytes overturning long held views in the field and that by combining these results with infra red studies on both crystalline and amorphous polymer electrolytes it has been possible to obtain unprecedented knowledge of the structural chemistry in the amorphous state [5]. This has in turn led to new design strategies for polymer electrolytes involving the organisation of polymer chains using liquid crystal approaches which, recent results indicate, are offering substantial improvement in the level of ionic conductivity.

The challenge of developing the structural chemistry of polymer electrolytes reduces to two problems. The first is establishing the crystal structures of several model systems and the second extrapolating this information into the amorphous state. Both aspects present formidable difficulties. As is exemplified in Fig. 2a, the quality of single crystal data obtained from the vast majority of polymer electrolytes is inadequate for structure solution. In contrast to the single crystal data, powder diffraction data of high quality may be obtained from many polymer electrolytes (Fig. 2b).

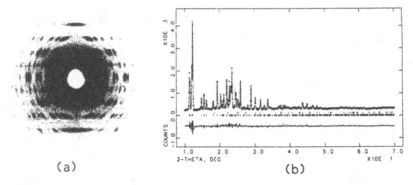

(a)   (b)

Figure 2. Fibre pattern of PEO₄:KSCN (a). Powder X-ray diffraction pattern of PEO₃:NaClO₄(b).

In cases where models of structures are already available, the established technique of full profile refinement using the Rietveld method may be used, however, given that the structural chemistry of polymer electrolytes is largely unexploited we must employ *ab initio* methods of structure solution from powder diffraction data if significant progress is to be made. In a succeeding section the structural chemistry of some important polymer electrolytes is described and it is shown that there is a very close connection between the structures of the crystalline and amorphous phases. It is further indicated that these structures have led to new design strategies for polymer electrolytes with higher ionic conductivity.

3. Simulated Annealing Method of Structure Solution

3.1 BASICS OF THE METHOD

In attempting to solve the crystal structures of polymer electrolytes it became evident that the established methods of ab initio structure solution from powder diffraction data were insufficient for the task. As a result we have developed an approach to structure solution which builds on earlier work by Newsam *et al* [6], combining simulated annealing (SA) with full profile fitting and chemical constraints. Although an origin of motivation could be found in our study of polymer electrolytes, the development of the methodology is of considerable interest in its own right and has many ramifications beyond the field of polymer electrolytes where the only access to crystallography is through powder diffraction.

Similar to the Rietveld method [7] of structure refinement, the objective of the SA method of *ab initio* structure solution is to find the atomic arrangement which gives the best fit, in the least-squares sense, of calculated and experimental powder diffraction patterns. It is achieved by minimising the figure-of-merit function (χ^2) represented by the sum of weighed residuals. The difference between structure refinement and structure solution in this case lies in the minimisation method.

Minimisation is carried out in a multidimensional non-linear space of variable parameters which contains multiple minima of various depths. The refinement procedure is based on iterative linearisation of χ^2 in the neighbourhood of a minimum with parameter increments being determined by the gradient value. Thus the refinement succeeds only if the starting structural model is close to the true structure. Attempts to find the true structure with an arbitrarily chosen initial model using this approach are destined to fail because the gradient method would always lead to the nearest minimum of the χ^2 which is unlikely to coincide with the global (deepest) minimum. Such minimisation may be compared to fast cooling of a molten metal which leads to a high energy (amorphous) structure corresponding to a local minimum.

In order to reach the lowest energy (crystalline) state, cooling of the molten metal must be performed slowly. Such a regime allows both uphill and downhill variations of the energy, which enables the ensemble to escape local minima in order to find the atomic arrangement corresponding to the lowest possible energy. Similar idea forms the basis of the SA method of global minimisation [8] of the χ^2 function in the process of structure solution. During the solution, assigning random stepwise increments to parameter values generates a series of trial structures. For each new trial structure a corresponding powder pattern is generated and compared to the experimental pattern by calculating χ^2_{new}. The new structural model is accepted if either $\chi^2_{new} < \chi^2_{old}$ or if $exp(-(\chi^2_{new} - \chi^2_{old})/S < R$, where R is a randomly chosen value between 0 and 1. The value of S determines the marginal uphill step of the χ^2 function. It serves as an analogue of temperature in the molten metal example. The larger S the more uphill steps are accepted thus reducing the possibility of becoming trapped in a local minimum and ensuring a comprehensive investigation of parameter space. A gradual reduction of S after generating a sufficient number of random trial atomic arrangements permits progression towards the deepest minimum of the figure-of-merit function corresponding to the true structure.

Another distinctive feature of the SA method in comparison with the more well-known refinement procedures, is a different approach to the introduction of structural constraints when dealing with molecular structures. The known inter-atomic connectivity in molecules is used successfully to narrow the parameter space during structure refinements by adding punishing terms to the profile figure-of-merit function [9]. These terms increase the χ^2 value when inter-atomic distances, calculated *via* variable crystallographic coordinates of the constituent atoms, in the model go beyond the well-established limits for bond lengths and bond angles. The same approach when applied to structure solution is troublesome. The main problem arises from the efficiency of a SA algorithm, which is computationally expensive because of the Monte Carlo fashion of assigning parameter increments while generating trial models. As a result constraining by punishment inevitably involves generating a vast number of chemically unreasonable structural models, which must be rejected because of implausible bond lengths and bond angles. In order to obviate this problem, instead of varying crystallographic coordinates of individual atoms belonging to a molecular fragment we vary stereochemical parameters, bond lengths, bond angles and torsion angles, together with positional (crystallographic coordinates of the reference atom) and orientation (Eulerian angles) parameters of the fragment as a whole in the unit cell. This approach preserves all the flexibility of the model, in which coordinates of each atom from a molecular moiety are independent variables, while allowing us to consider only stereochemically reasonable structures because the values of variable parameters are randomly chosen only within pre-determined ranges. In order to exploit the usual form of calculating structure factors we have developed a versatile method of transforming from stereochemical variables to crystallographic coordinates of individual atoms belonging to a molecule of any shape. This algorithm is based on splitting a molecule into a sequence of three-dimensional chains. It also incorporates rotation of bonds around a chosen axis. In addition to the above constraints, the number of evaluated trial structures is reduced by rejecting models in which the distance between atoms belonging to separate molecules becomes less than the sum of their Van der Waals radii and the models, in which the continuity of a molecular moiety (*e.g.* PEO chain) is violated at the junctions of neighbouring asymmetric units. A detailed account of the SA algorithm for structure solution from powder data together with a geometrical description of molecules using bond lengths, bond angles and torsion angles is given in [10].

3.2 EXAMPLE

The previously unknown crystal structure of the polymer:salt complex poly(ethylene oxide):$NaSO_3CF_3$ was solved *ab initio* from a powder diffraction pattern collected on a laboratory X-ray diffractometer STOE STADI/P using Cu $K\alpha_1$ radiation and a position-sensitive detector. Indexing and subsequent refinement of lattice parameters yielded the monoclinic cell with $a=9.8491(4)$ Å, $b=12.7862(5)$ Å, $c=5.7750(3)$ Å, $\beta=90.638(5)°$. Judging from systematic absences of reflections the space group was identified as $P2_1/c$. Density considerations indicated that there are four formula units in the unit cell. On this basis the structure solution was carried out assuming that all atoms in the cell occupy 4e general positions and the asymmetric unit comprises a single formula unit. This being so, the content of the asymmetric unit was divided in three parts, Na^+ cation, triflate $SO_3CF_3^-$ anion and the basic repeat unit CH_2-CH_2-O of the polymer chain. The crystallographic coordinates of the atoms belonging to the triflate were described using S-C, S-O, C-F bond lengths, C-S-O, S-C-F, O-S-O, F-C-F bond angles, and the O-S-C-F torsion angle. The description for the ethylene oxide unit included C-C, C-O bond lengths and the C-C-O bond angle. In addition, each of the molecular fragments required a dedicated set of three Eulerian angles and the crystallographic coordinates of the reference atom, which was sulphur for the triflate and the carbon adjacent to the ether oxygen for the chain.

The initial trial structure (Fig. 3a) was chosen randomly and as can be seen from Fig. 3b, provided no match of the calculated to the observed diffraction pattern. During

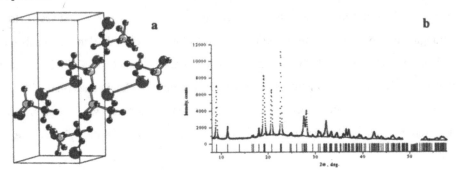

Figure 3. Starting trial structure (a) and profile fit (b).

the subsequent SA run 27 parameters were varied to fit the pattern. It should be noted that in this instance additional constraints were applied to reduce the number of variables. All like bond lengths (*e.g.* S-O1, S-O2, S-O3) and bond angles (*e.g.* O1-S-O2, O1-S-O3) in the triflate were set to be equal to each other and varied simultaneously. After evaluating about 140,000 trial configurations the SA procedure converged producing a structural model (Fig. 4a) which after refinement by the Rietveld method gave a reasonable profile fit (Fig. 4b). However all attempts to improve the fit further failed, with the best χ^2 equal to 6 and a noticeable misfit in the 2θ range from 35 to 45°. The model placed fluorines, rather than the more negatively charged oxygens of the

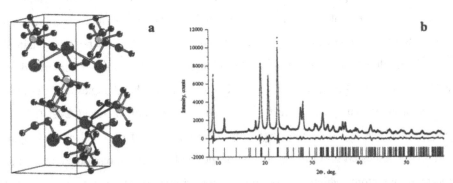

Figure 4. Final structure (a) and profile fit (b) with over-constrained triflate model.

triflate anion, adjacent to the sodium cation and did not ensure coordination of the sodiums by the ether oxygens thus making the model unreasonable. In addition, insufficient separation of adjacent Na^+ ions and complex values of amplitudes of mean-squared displacements for some of the atoms provided further evidence indicating the inappropriateness of the structural model. Successful structure determination was achieved after removing the constraint that all like bond lengths and bond angles in the triflate were equal. A new SA minimisation was performed allowing all such lengths and the angles to vary independently thus making the model fully flexible and increasing the number of variable parameters to 37. The structural model produced by the second SA run (Fig. 5a) revealed 6-fold coordination of the Na^+ cation by equidistant oxygens from the triflate and the chain and provided an excellent match between the observed and calculated patterns (Fig. 5b) with $\chi^2=1.1$ and all the displacement amplitudes real. The significant effect of over-constraining the triflate during the structure solution in this particular case is illustrated in Fig. 6.

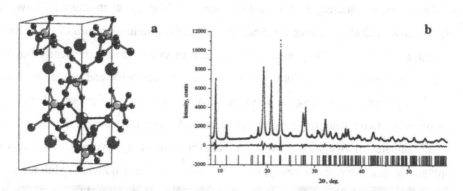

Figure 5. The crystal structure(a) and fit(b) for poly(ethylene oxide):NaSO₃CF₃.

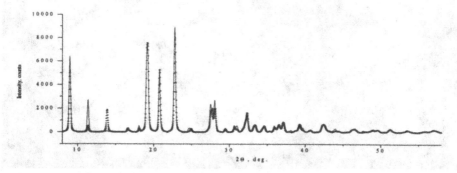

Figure 6. Solid line - calculated pattern of the true structure of poly(ethylene oxide): NaSO₃CF₃. Crosses -calculated pattern of the poly(ethylene oxide):NaSO₃CF₃ structure modified by averaging all like bond lengths and bond angles in the triflate anion.

Note the significant change in the appearance of the calculated diffraction pattern after averaging the bond lengths and bond angles of the triflate. Further computational details and discussion of the structure may be found in [11].

4. The structure of polymer electrolytes

The crystal structure of poly(ethylene oxide)$_3$:LiCF$_3$SO$_3$ (3 ether oxygens per Li$^+$ cation) has been established by profile refinement using the Rietveld method [12]. A total of 113 independent parameters with 88 soft constraints were used in the refinement. The final χ^2 was 1.8. The already known crystal structure of PEO$_3$:NaI was used to generate a starting model for the refinement. The crystal structure is shown in Fig. 7. Each PEO chain adopts a helical conformation. Within each turn of the helix is located a lithium ion. The analogy with a stack of crown ethers covalently bonded together is obvious. The lithium ions fit neatly into the helix and are coordinated directly by three ether oxygens from the chain. The coordination environment is completed by two further oxygens, one from each of two triflate anions, thus forming a five-coordinate arrangement around the lithium ions in approximately a trigonal bipyramidal geometry. Interestingly, the triflate anions bridge neighbouring lithium ions

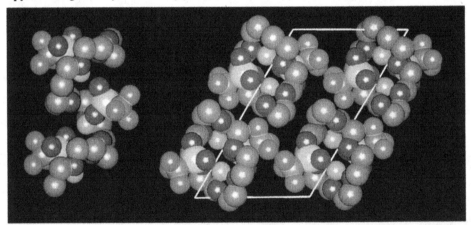

Figure 7. Part of the crystal structure of poly(ethylene oxide)$_3$:LiCF$_3$SO$_3$: one PEO chain with its associated ions (left); projection down the b-axis (fibre axis) of the unit cell showing four helical chains (right): green, carbon; red, oxygen; blue, lithium; yellow, sulphur; and purple, fluorine.

along the chain by donating one of their oxygens to each of two lithiums. The remaining oxygen of each triflate group is uncoordinated. It is evident that the $-CF_3$ groups of the triflate anions sit outside the dimensions of the helix. Figure 7(right) shows a projection down the *b* axis of the unit cell, which coincides with the fibre axis and is also coincident with the 2_1 screw axis of the helix. It is evident that every chain is associated with a dedicated set of cations and anions which do not bond to any of the other chains. In this sense, each chain forms an isolated infinitely long one-dimensional coordination complex.

We have studied a systematic series of polymer electrolytes of increasing cation radius (which doubles from Li^+ to Rb^+) and the gross structural features are summarised in Table 1 [12-15]. Included in these data are 4:1 complexes. These were the first structures of 4:1 complexes to be obtained by *ab initio* methods from powder diffraction data applied to polymer-salt complexes. Direct methods were employed. The first point to note is that in all cases the PEO chain adopts a helical conformation and the cation remains inside the chain. This is in complete contrast to the long held view that cations larger than Na^+ could not be located within the chain but were located instead in the interchain space. The chain conformation is also summarised in the Table 1. Whereas all C-O bonds are *trans* the C-C bonds can be *gauche* or *gauche minus*.

TABLE 1. Basic structural details of some polymer-salt complexes
(Ionic radii are taken from Shannon, R.D. and Prewitt, C.T. (1970) *Acta Crystallographica B* **26**, 1046 and are based on a six-coordinate oxygen environment)

$PEO_3LiCF_3SO_3$	helix	ttgttgttḡ	0.76	3	2	5
PEO_3:$NaClO_4$	helix	ttgttgttḡ	1.02	4	2	6
PEO_4: KSCN	helix	ttgttgttḡ ttḡ	1.38	5	2	7
PEO_4: NH_4SCN	helix	ttgttgttḡ ttḡ	1.61^a	5	2	7
PEO_4:RbSCN	helix	ttgttgttḡ ttḡ	1.52	5	2	7

^{a} The effective ionic radius of NH_4^+ can vary substantially from solid to solid. Our lattice parameter measurements place it between K^+ and Rb^+ in polyethers.

In order to accommodate the larger cations the PEO chain changes its conformation as shown in the Table, to produce a wider, fatter helix capable of accommodating the cations. This new conformation provides five ether oxygens in the coordination sphere of the cations. In all cases each chain is associated with a dedicated set of cations and anions and there is no evidence of ionic cross-linking between the chains. This is contrast to the 1:1 complexes such as $PEO:NaSO_3CF_3$. The chain in 1:1 complexes adopts a stretched zig-zag conformation with each cation coordinated by two neighbouring ether oxygens along the chain. Four anions coordinate each cation *via* the oxygens of the $-SO_3$ groups [11,16]. The 1:1 complexes exhibit extensive ionic cross-linking between polymer chains arising from the fact that a given cation is coordinated simultaneously by a polymer chain and anions, the latter also coordinates other cations which are attached to different chains. The 1:1 complexes form therefore a true three-dimensional structure as opposed to the one-dimensional structure of the 3:1 or 4:1 complexes. The 1:1 complexes melt at around 250 – 300° C whereas the melting temperature of the 3 or 4:1 complexes is typically 100 or 200° lower.

We have also solved the structure of $PEO_3:LiN(CF_3SO_2)_2$ in which the anion maybe thought of as two triflate ions fused together by removing an oxygen from each and replacing it by one bridging nitrogen [17]. This crystal structure could not be refined starting from the structure of $PEO_3:LiCF_3SO_3$ despite the expected similarity of the two structures. Instead it was necessary to employ the simulated annealing method as no other approach based on direct methods or Patterson synthesis could yield a satisfactory structure or even one that could be subjected to subsequent refinement. The simulated annealing method rapidly provided a crystal structure for this compound. The PEO chain is again helical with the lithium ions located in each turn of the helix and coordinated by five oxygens, 3 ether oxygens and 2 oxygens from one $-SO_2$ group. Although it would be possible for this compound to exhibit anionic bridging involving the oxygens of the two halves of the anion coordinating lithium ions on neighbouring chains, this does not happen. Instead the non-bonded side of the anion folds around towards its own chain.

5. From crystalline to amorphous polymer electrolytes

Having established a sound foundation in the structural chemistry of polymer electrolytes *via* crystallographic studies, our aim was to extend our understanding into the amorphous state. This we have achieved by variable temperature IR studies. A sample of $PEO_3:LiCF_3SO_3$ was heated in an infra red cell from room temperature through the melting transition at 178° C and into the amorphous state to while following any changes with infra red spectroscopy [5]. The infra red spectra probe, in particular, the vibrational states of the triflate anions. The technique is able to distinguish between triflate anions that are free and those that have lithium ions directly attached to the oxygens of the anion. Only one triflate species is evident in the spectrum of the crystalline state, and from the crystal structure data it is clear that this is a triflate ion with one lithium ion attached to each of two oxygens of the triflate. On melting no other triflate species are detected suggesting that the triflate ions remain attached to two lithium ions simultaneously and that, furthermore, short chains of $-Li^+-CF_3SO_3-Li^+-$ must be present in the amorphous state! In other words each chain remains associated with its dedicated set of cations and anions the main difference between the crystalline and amorphous states being the loss of register between the chains. By combining crystallography and spectroscopy, unprecedented insight has been gained into the structural chemistry of these materials which has not been possible by extensive studies over the last 20 years using spectroscopic techniques alone.

6. Structural implications for ion transport in materials design

It is clear from the diffraction and spectroscopic studies that the presence of cations within the PEO helix is common in complexes more dilute in salt than 1:1. It seems likely therefore, that ion transport occurs along the PEO helices facilitated by the chain motion. It is certainly true that the more difficult transition of ions between chains will be the rate limiting process in ion transport. Clearly a random arrangement of chains such as is anticipated in the amorphous polymer will not have an optimised geometry for ion transport. As well as the flexibility of the chains it will be important to organise these chains so that they are aligned, permitting rapid motion both along the chains and more particularly between chains. This is what is suggested by the diffraction and spectroscopic studies. Recently several groups have synthesised liquid crystal materials with poly(ethylene oxide) chains [18,19]. The liquid crystals organise the chains as well as retaining some chain flexibility. The indications are that conductivity associated with these organised systems is higher than the random (amorphous) materials. This appears to vindicate our view that a knowledge of structure could lead to design strategies that would improve ionic conductivity, possibly exceeding the threshold for practical applications. In the future, there is clearly much scope for design of new systems informed by the more "structured" view of polymer electrolytes.

Acknowledgement

PGB is indebted to the EPSRC and The Leverhulme Trust for financial support.

References

1. Bruce, P.G. (1996) Coordination chemistry in the solid state, *Philosophical Transactions of the Royal Society* **354**, 415-436.
2. Bruce, P. G. (1995) *Solid state electrochemistry*, Cambridge University Press.
3. *Polymer electrolyte reviews – 1*, J.R. MacCallum and C.A. Vincent (eds), Elsevier (1987).
4. Gray, F.M. (1991) *Solid polymer electrolytes*, VCH.
5. Frech, R., Chintapalli, S., Bruce, P.G., and Vincent, C.A. (1997) Structure of amorphous polymer electrolyte poly(ethylene oxide)$_3$:LiCF$_3$SO$_3$, *J. of the Chemical Society-Chemical Communications*, 157.
6. Newsam, J.M., Deem, M.W., and Freeman, C.M. (1992) Direct space methods of structure solution from powder diffraction data, in *Accuracy in Powder Diffraction II*. NIST Special Publications No. 846, pp. 80-91.
7. Rietveld, H.M. (1969) A profile refinement method for nuclear and magnetic structures, *J. of Applied Crystallography* **2**, 65-71.
8. Kirkpatrick, S., Gelatt, C.D., and Vecchi, M.P. (1983) Optimization by simulated

annealing, *Science* **220**, 671-680.

9. Baerlocher, C. (1993) Restraints and constraints in Rietveld refinement, in R.A. Young (ed.), *The Rietveld Method*, Oxford University Press, pp. 186-196.

10. Andreev, Y.G., Lightfoot, P., and Bruce, P.G. (1997) A general Monte Carlo Approach to structure solution from powder-diffraction data: application to poly(ethylene oxide)$_3$:LiN(SO$_2$CF$_3$)$_2$, *J. of Applied Crystallography* **30**, 294-305.

11. Andreev, Y.G., MacGlashan, G.S., and Bruce, P.G. (1997) *Ab initio* solution of a complex crystal structure from powder-diffraction data using simulated-annealing method and a high degree of molecular flexibility, *Physical Review B* **55**, 12011-12017.

12. Lightfoot, P., Mehta, M.A., and Bruce, P.G. (1993) Crystal structure of the polymer electrolyte poly(ethylene oxide)$_3$:LiCF$_3$SO$_3$, *Science* **262**, 883-885.

13. Lightfoot, P., Nowinski, J. L., and Bruce P.G. (1994) The crystal structures of the polymer electrolytes poly(ethylene oxide)$_4$:MSCN (M= NH$_4$, K)., *J. of American Chemical Society* **116**, 7469-7470.

14. Lightfoot, P., Mehta, M-A, and Bruce, P.G. (1992) Structures of the poly(ethylene oxide) : NaClO$_4$ complex PEO$_3$:NaClO$_4$ from powder X-Ray diffraction data., *J. of Material Chemistry* **2**, 379-381.

15. Thomson, J. B., Lightfoot, P., and Bruce, P.G. (1996) Structure of the polymer electrolyte poly(ethylene oxide)$_4$:RbSCN, *Solid State Ionics* **85**, 203-208.

16. MacGlashan, G.S., Andreev, Y.G., and Bruce, P.G. (1998) Structure of the polymer electrolyte poly(ethylene oxide):KCF$_3$SO$_3$ determined from powder diffraction data, *J. of the Chemical Society-Dalton Transactions*, in press.

17. Andreev, Y.G., Lightfoot, P., and Bruce, P.G. (1996) Structure of the polymer electrolyte poly(ethylene oxide)$_3$:LiN(SO$_2$CF$_3$)$_2$ determined by powder diffraction using a powerful Monte Carlo approach, *J. of the Chemical Society-Chemical Communications*, 2169-2170.

18. Dias, F.B., Batty, S.V., Voss, J.P., Ungar, G., and Wright, P.V. (1996) Ionic-conductivity of a novel smectic polymer electrolyte, *Solid State Ionics* **85**, 43-49.

19. Dias, F.B., Batty, S.V., Ungar, G., Voss, J.P., and Wright, P.V. (1996) Ionic-conduction of lithium and magnesium salts within laminar arrays in a smectic liquid-crystal polymer electrolyte, *J. of the Chemical Society-Faraday Transactions* **92**, 2599-2606.

QUANTUM MECHANICAL MODELING OF STRUCTURE EVOLUTION OF TRANSITION METAL CLUSTERS AND METALLOCARBOHEDRENES

HANSONG CHENG
Air Products and Chemicals, Inc.
7201 Hamilton Boulevard, Allentown, PA 18195-1501, USA

LAI-SHENG WANG
Department of Physics, Washington State University, Richland, WA 99352
and W. R. Wiley Environmental Molecular Sciences Laboratory,
MS K8-88, Pacific Northwest National Laboratory, Richland, WA 99352.

Abstract

Ab initio quantum-mechanical modeling based on density functional theory (DFT) was used to study transition metal clusters and metallo-carbohedrenes (MetCars). Combined with the state-of-the-art spectroscopic experiments, DFT calculations are capable of yielding much insight into the structures, chemical bonding and growth mechanisms of these clusters. Two specific cluster systems were investigated in detail: one involves small chromium clusters and another contains titanium carbides. Exhaustive structural search was performed by fully optimizing a variety of cluster geometries. For the chromium clusters, we found that a tightly-bound Cr_2 dimer plays a key role in determining the cluster structures. A dimer growth route is discovered for clusters up to Cr_{11}, at which a structural transition occurs from the dimer growth to a bulk-like body-centered-cubic structure. The uncovered structural evolution is consistent with the currently available experimental observations. For MetCars, we found that three factors, i.e., the C_2 dimer, cubic framework and layered structures, play an essential role in determining the structures and chemical bonding of the titanium carbide clusters. A growth pathway from Ti_3C_8 to $Ti_{13}C_{22}$ with Ti_4C_8, Ti_6C_{13}, Ti_7C_{13} and Ti_9C_{15} as intermediates is thus proposed. Both theory and experiments suggest that the cubic layered growth with C_2 dimers can lead to a new type of highly stable one-dimensional quantum wires.

1. Introduction

There have been active efforts recently to understand the structure and chemical bonding of transition metal and metallocarbohedrene clusters [1-4]. These clusters are important in many aspects. For instance, transition metal clusters can serve as models for studies in

135

J.A.K. Howard et al. (eds.),
Implications of Molecular and Materials Structure for New Technologies, 135–150.
© 1999 *Kluwer Academic Publishers. Printed in the Netherlands.*

surface phenomena and heterogeneous catalysis. The recently discovered MetCars represent a new class of clusters that possesses unusual stability. Understanding the cluster structures, chemical bonding and growth mechanisms is essential in order to develop new nano-materials with novel chemical and physical properties.

One of the fundamental issues concerning clusters is how their structures and chemical/physical properties evolve as they grow from an assembly of a few atoms to bulk [5]. It is rarely clear when a cluster should look and behave like a bulk. The situation is even more complex when the clusters are made of transition metals or metallocarbohedrenes since their electronic structures have not yet been fully understood. Despite extensive spectroscopic studies, structural characterisation and electronic state determination have been very difficult, partly because of lack of reliable fully quantum-mechanical based calculations. In particular, theoretical studies on clusters involving transition metals are highly challenging due to the presence of a large number of d-electronic states.

We have been undertaking systematic spectroscopic studies on transition metal clusters and MetCars recently [6-9]. Subsequently extensive quantum-mechanical calculations based on *ab initio* density functional theory have been carried out to model the cluster structures and to investigate their chemical and physical properties [10-14]. In this article, we will mainly focus on the theoretical aspect of the work. A comprehensive review of the studies is beyond the scope of this article. Therefore, we will specifically focus our discussions on two cluster systems: chromium and titanium carbohedrenes. We wish to gain insight into the growth mechanisms and properties of these clusters through modeling their structures.

2. Computational Methods

Quantum-mechanical calculations based on density functional theory were performed under both local spin density (LSD) and non-local spin density (NLSD) approximations. Many recent studies have demonstrated the value of DFT in providing accurate molecular geometries and energetics at a moderate computational cost [15-18]. By explicitly taking into account the electron correlation effect, DFT methods surpass the conventional Hartree-Fock *ab initio* methods (HF) in providing a description of molecular interactions which is often accurate [16]. The calculated results are, in general, superior to post-HF treatment [19-22]. Our DFT calculations utilized the double numerical basis set augmented by polarization functions. All core electrons were frozen to simplify the computation without significantly sacrificing the accuracy, as will be demonstrated by the calculated results. The LSD calculations employed the Vosko-Wilk-Nusair local correlation functional, while the NLSD calculations utilized Becke's gradient-corrected exchange and Perdew-Wang's gradient-corrected correlation functionals. The gradient-corrections were incorporated in the SCF cycles in an iterative manner. Spin-polarized Kohn-Sham scheme was utilized throughout.

Since no qualitative difference in relative energetics between the LSD and the NLSD calculations was found for Cr_n (n=2-4), we only carried out LSD calculations to study the chromium cluster system. NLSD calculations were performed for all the

MetCar clusters. The density of states (DOS) spectra were simulated by convoluting Gaussian functions to the occupied electronic states for comparison with the measured photoelectron spectra. The electron affinities were derived by evaluating the energy difference between the anionic and the neutral species.

3. Chromium Clusters

Of the first row transition metals, chromium is rather unique due to its half-filled valence shell. Its $3d^5 4s^1$ configuration gives rise to strong d-d bonding in the dimer, yielding an exceptionally short bond distance of 1.68Å [23-26], while bulk chromium is of a body-centered cubic (bcc) structure with a nearest neighbor distance of 2.50Å. A fundamental issue concerning the chromium clusters thus arises: what is the cluster growth pathway from a tightly-bonded dimer to the bulk structure? We attempted to address the issue by carrying out extensive DFT calculations on a variety of atomic configurations of chromium clusters up to Cr_{15}.

3.1. ELECTRONIC STRUCTURE OF THE DIMER

An important step to gain insight into the structure and chemical bonding of small chromium clusters is to understand the electronic configuration of the dimer [26], which nicely forms a closed-shell with a sextuple bond. The strong d-electron pairing in Cr_2 results in unusually short bond distance, forcing the $4s$-$4s$ interaction into a slightly repulsive regime due to Pauli exclusion. In general, this is a relatively stable electronic structure and it thus requires strong orbital interactions to "unlock" the d-electrons from the bonding orbitals. This special electronic configuration in Cr_2 is the basis on which small chromium clusters grow.

The optimized bond parameters and binding energies obtained at both LSD and NLSD levels are shown in Table 1. The calculated bond distance agrees with the experimental value very well. The LSD binding energy, however, is significantly overestimated as expected due to its nature of over-binding. The binding energy is largely and overly corrected in the NLSD calculations.

Table 1. Cr_2 bond distance and binding energy.

Method	$d_{Cr\text{-}Cr}$(Å)	ΔE (eV)
LSD	1.62	-2.15
NLSD	1.65	-0.01
exp.	1.68	-1.443

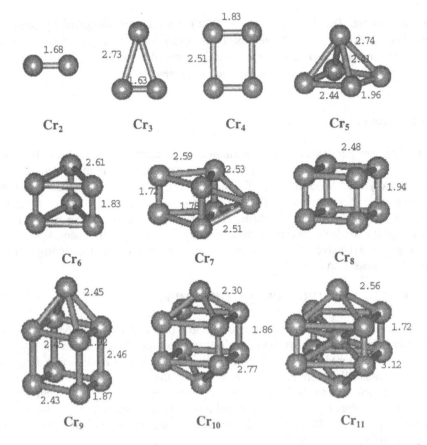

Figure 1. Fully optimized structures of chromium clusters Cr$_n$ for n=2-11. The bond distances are in Å.

3.2. STRUCTURE AND BONDING OF CHROMIUM CLUSTERS

Structures of small chromium clusters are largely determined by the closed-shell dimers due to their relative stability in the electronic configuration. Figure 1 displays the fully optimized cluster structures from Cr$_3$ to Cr$_{11}$ composed of dimers and isolated atoms only. Their symmetry groups and the calculated LSD binding energies are shown in Table 2. These clusters are selected not only because they are of the lowest energy structures among the numerous isomers we obtained, but also they collectively represent a growth path that seems to explain the existing experimental facts, as will be shown later.

As expected, Cr$_3$ is essentially composed of a dimer plus an atom. While the dimer electronic configuration is unperturbed, the third atom essentially remains in its atomic electronic state, leaving six unpaired electrons in the cluster. Cr$_4$ is formed by two weakly-bonded dimers whose electronic states are hardly changed. The weak bonding allows the dimers to "slide" against each other without costing too much energy. Indeed, we obtained a rhombus isomer of C$_{2v}$ symmetry only slightly higher in energy. It is clear that the relatively "inert" *d*-electrons in these dimers are responsible for the weak dimer-atom and dimer-dimer interactions. The dimers, together with the isolated atoms, constitute essential components in small chromium clusters.

Table 2. Cluster symmetry and the calculated LSD binding energy (eV) for Cr_n.

n	symmetry	BE	n	symmetry	BE
3	C_{2v}	-3.302	10	D_{4h}	-21.658
4	D_{2h}	-5.846	11	D_{4h}	-24.779
5	C_{2v}	-8.483	12	C_{2v}	-29.392
6	D_{3h}	-10.715	13	D_{4h}	-33.748
7	C_{2v}	-13.460	14	C_{4v}	-38.124
8	D_{4h}	-16.574	15	D_{4h}	-42.031
9	C_{2v}	-18.823			

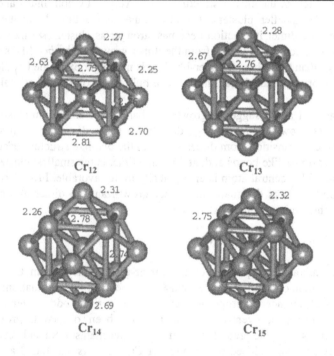

Figure 2. Fully optimized structures of chromium clusters Cr_n for n=12-15. The bond distances are in Å.

It is therefore not surprising that a slightly "slid" structure in Cr_5 is of the lowest energy. Up to Cr_9, the structures essentially are either dimers stacking together in even clusters or an assembly of dimers with an extra atom in odd clusters. At Cr_{10}, however, simply stacking five dimers together can no longer stabilize the cluster and the dimer at the center of the assembly is "squeezed" out, resulting in two isolated atoms. The structure of Cr_{11} shown in Fig. 1 is thus a natural choice. In general, Fig. 1 describes a dimer growth path for the chromium clusters to evolve up to Cr_{11}.

It is important to point out that the optimized geometry of Cr_{11} embodies both a dimer-packing structure with its characteristic short bond distances and a *bcc* framework and thus represents a natural structural transition from a dimer-stacking structure to *bcc*-like structure, leading to further evolution to chromium bulk. The optimized structures of larger clusters indeed resemble the bulk with slightly shorter bond distances. The fully optimized geometries are shown in Fig. 2.

For clusters smaller than Cr_{11}, the dimer-like bonds do not exceed 2.0Å. The relatively weak interactions among dimers or between dimers and the isolated atoms result in longer bonds. These weak interactions are also responsible for a range of dimer bondlengths. For example, in Cr_3, the dimer-like bond becomes shorter due to the σ_{4s}-$4s$ orbital overlap between the dimer and the third atom, resulting in a slight shift of the σ_{4s}-electron density towards the third atom. In Cr_4, however, the dimer-dimer interaction includes both inter-dimer σ_{4s}-σ_{4s} and $3d$-$3d$ overlaps, yielding a slight stretch to the dimer bonds. Clearly, the cluster structures are governed by both inter- and intra-dimer interactions. For smaller clusters, the intra-dimer interaction dominates the cluster growth. The inter-dimer interaction becomes stronger as cluster size increases, leading to the ultimate structural transition from the dimer growth to the *bcc*-like structures. For clusters larger than Cr_{11}, though dimer-like bonds no longer exist, the bondlengths are in general still more contracted than those in the bulk to maximize the orbital overlaps, as expected.

Figures 1 and 2 suggest a growth mechanism for chromium clusters: they first follow a dimer growth path. At Cr_{10}, the cluster starts deviating from simple dimer growth. A natural transition from dimer stacks to the *bcc*-like structure occurs at Cr_{11}: it possesses four dimer-like bonds and, at the same time, is the smallest cluster where *bcc*-like structure with a central atom is energetically more favorable. From Cr_{12} and up, the bulk-like *bcc* structures dominate the cluster growth and no dimer-like bonds can be identified in the minimum energy structures.

3.3. EVEN-ODD ALTERNATION

One of the natural consequences of the dimer-growth mechanism that governs the structural evolution of small chromium clusters is the even-odd alternation: there should be observable differences in properties between the even and odd clusters. Indeed, in at least two experiments, even-odd alternation has been observed, providing direct qualitative supports for the predicted dimer-growth mechanism. Su and Armentrout [27] measured the collision-induced dissociation of Cr_n^+ clusters and found a distinct even-odd alternation in dissociation energies for n<10. More importantly, they observed dimer loss for n up to 6. Both observations are consistent with the dimer-growth structures, even though the binding of the clusters may be altered by the positive charge. We recently measured photoelectron spectra of size-selected anion Cr clusters and found significant differences between the even and odd clusters for n<12. The spectra of the even ones show distinct peaks and gaps near the Fermi level while the odd clusters all show more congested spectra near the Fermi level due to the electronic states of the odd atom [14]. The experimental results are remarkably consistent with the dimer-growth mechanism.

4. MetCars

Metallocarbohedrenes (MetCars) were first discovered in 1992 by Castleman and his co-workers. They contain 8 early transition metal atoms and 12 C atoms proposed as a new class of stable molecular clusters [28]. They were observed as "magic" positive ion peaks in mass spectra from laser vaporization experiments of early transition metals with hydrocarbon-seeded carrier gases. These magic clusters, consisting of eight metal atoms and twelve carbon atoms, were proposed to be of a cage-like molecular shape similar to the fullerenes as shown in Fig. 3. If made in bulk quantity, these cluster materials are expected to exhibit novel and rich chemical and physical properties because of the presence of transition metals in their chemical compositions. However, little progress has been made in the bulk synthesis of these materials due to lack of information about their growth pathways and their definitive structures, despite extensive theoretical [29-31] and experimental [32-36] efforts following the original discovery.

Unlike fullerenes, which grow by expanding the cage, it was proposed that the MetCars follow a multicage growth path, based on the observation of magic cluster cations in mass spectra of zirconium carbide clusters at $Zr_{13}C_{22}^+$, $Zr_{18}C_{29}^+$, and $Zr_{22}C_{35}^+$ [37]. However, Pilgrim and Duncan subsequently showed that Ti and V carbide clusters form cubic nanocrystals with a composition of $M_{14}C_{13}^+$, a 3×3×3 cubic cluster much like the bulk crystal lattice [38]. The composition for the double-cage at $M_{13}C_{22}^+$ was not observed. In a very recent work, we combined anion photoelectron spectroscopy (PES) and density functional theory calculations to elucidate the structure of the $Ti_{13}C_{22}$ cluster [39]. We found that $Ti_{13}C_{22}$ has an unusual cubic structure with 8 C_2 dimers at the cube corners. This observation led to the proposal of a novel layer-by-layer growth model for the large carbide clusters that can account uniquely for the magic numbers in the multicage growth model. In this section, we present the results of both experimental and theoretical studies on the structure and bonding of a series of clusters with smaller magic numbers, and discuss the implication for the growth pathways of the metal carbide clusters (Ti_xC_y, x/y = 3/8, 4/8, 6/13, 7/13, 9/15, 13/22). We have obtained PES spectra of all these clusters and further performed DFT calculations. The calculated electron affinities (EAs) and single particle density of states (DOS) were compared with the experimental measurements. Structural information was obtained from the combined experimental and theoretical studies and was used to understand the growth pathways of the titanium carbide clusters.

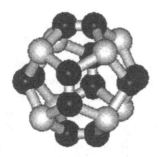

Figure 3. Cage-like dodecahedral structure with T_h symmetry.
The white balls denote Ti atoms and the black balls represent C atoms.

4.1. NEW MAGIC NUMBERS AND ANION PES SPECTRA

Fig. 4 shows a mass spectrum of $Ti_xC_y^-$ clusters, produced by laser vaporization of a Ti target with a 5% CH_4/He carrier gas. The anion spectrum differs significantly from the positive ion mass spectra, where the $Ti_8C_{12}^+$ MetCar peak and the $Ti_{14}C_{13}^+$ cubic cluster dominate the low and higher mass range, respectively. Several prominent anion clusters are observed, namely, $Ti_3C_8^-$, $Ti_4C_8^-$, $Ti_6C_{13}^-$, $Ti_7C_{13}^-$, $Ti_9C_{15}^-$, and $Ti_{13}C_{22}^-$. Most surprising in the anion mass distribution is the total absence of the anticipated Ti_8C_{12} peak, which is so prominent in the positive ion channel. The anion mass spectrum shows that clusters containing 8 to 12 Ti atoms have very low abundance, except for $Ti_9C_{15}^-$, which shows up as a prominent peak in an otherwise very low abundant mass range. The low abundance of the clusters containing 8 to 12 Ti atoms suggests that either these clusters are relatively unstable or the growth kinetics are such that they form the 13/22 cluster rapidly. In either case, the 9/15 cluster is unusually abundant and should provide clues for the growth from small clusters to the 13/22 cluster.

The $Ti_{13}C_{22}$ cluster has the same composition as the previously observed $Zr_{13}C_{22}^+$ and $Nb_{13}C_{22}^+$ positive ion clusters which were proposed to have the double-cage structure [37]. We recently found that the $Ti_{13}C_{22}$ cluster has an unusual cubic structure with 8 C_2 dimers occupying the 8 corners of the cube [11] rather than the previously proposed double cage structure. We also suggested that the Ti_9C_{15} cluster was an intermediate in the formation of the cubic $Ti_{13}C_{22}$ structure. However, the small magic clusters were not investigated in detail.

The properties of the new magic clusters observed in the anions may be studied by a variety of experimental techniques, including PES and ion mobility measurements. PES is particularly useful to probe the electronic structure of these clusters and allows the EAs and electronic density of states of the neutral clusters to be measured directly, which provides valuable electronic and spectroscopic information that can be used to verify theoretical calculations.

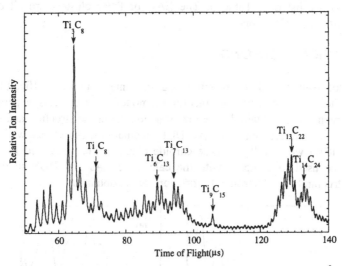

Figure 4. Mass spectrum of $Ti_xC_y^-$ anion from laser vaporization of a pure Ti target with a He carrier gas containing 5% CH_4.

Fig. 5 shows the PES spectra of $Ti_3C_8^-$, $Ti_4C_8^-$, $Ti_6C_{13}^-$, $Ti_7C_{13}^-$, $Ti_9C_{15}^-$ and $Ti_{13}C_{22}^-$ measured at 6.42 eV (193nm) photon energy. These spectra represent transitions from the ground state of the anions to the states of the neutrals. All the spectra exhibit broad features, suggesting that there are significant geometry changes between the anions and neutrals or high densities of low-lying states. We have also taken spectra at lower photon energies to improve the spectral resolution for the lower binding energy features, but no fine structures were further resolved, probably due to the high densities of electronic states accompanied by excitations of many vibrational modes in the detachment processes. The resolved spectral features were reproducible under different experimental conditions, such as different stagnation pressures or laser vaporization power, and identical spectra were obtained when ^{13}C-isotope labeled CH_4 was used, suggesting that the observed spectra are mainly due to a single isomer or a consistent set of isomers. We also took PES spectra of $Ti_3C_8^-$ and $Ti_4C_8^-$ produced using a TiC target with pure He carrier gas and obtained identical spectra.

Table 3. Estimated and calculated adiabatic electron affinities of the Ti_xC_y clusters (unit: eV).

Cluster	Ti_3C_8	Ti_4C_8	Ti_6C_{13}	Ti_7C_{13}	Ti_9C_{15}	$Ti_{13}C_{22}$
Cal.	2.33	2.12	2.30	2.34	1.80	3.00
Exp.[a]	2.5	1.8	2.2	2.1	1.8	3.0

a. The experimental uncertainties were estimated to be ± 0.2 eV due to lack of sharp threshold features in the PES spectra.

From the detachment thresholds, EAs for the neutral clusters can be determined. If sharp features or 0-0 vibrational transitions can be clearly resolved, the adiabatic EAs can be determined rather accurately from the anion PES spectra. However, due to the broad spectral features, the EAs in the present cases can only be estimated and are listed in Table 3. The EAs for these clusters are all quite high, reflecting their high carbon contents.

4.2. THEORETICAL MODELING

The broad PES features and lack of vibrational information make it difficult to interpret the spectra shown in Fig. 5 without theoretical modeling of the cluster geometric and electronic structures. To obtain detailed information about the structure and bonding of these clusters, we carried out extensive DFT calculations. As we showed previously [11], DFT should work well for these carbide clusters. The measured EAs and PES spectra can be used to compare with the calculated EAs and DOS so that reliable theoretical structural and electronic information can be obtained.

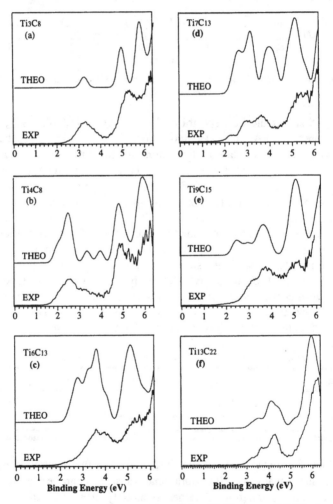

Figure 5. Photoelectron spectra and the calculated DOS spectra of $Ti_xC_y^-$ anions (x/y=3/8, 4/8, 6/13, 7/13, 9/15 and 13/22)

There may exist many possible isomers for the current Ti_xC_y clusters. Therefore, some knowledge on the bonding and structural motifs of the titanium carbide clusters is essential for judicially selecting initial structures for theoretical calculations. From previous investigations on MetCars and our own work on $Ti_{13}C_{22}$, structural features were noted: (1) the importance of C_2 dimers in the carbon-rich clusters, and (2) the cubic structural motif exemplified in the 3×3×3 cubic $Ti_{14}C_{13}^+$ nanocrystal and the C_2-decorated cubic $Ti_{13}C_{22}$ cluster. Hence, for Ti_3C_8 and Ti_4C_8 we considered various configurations between 3 and 4 Ti atoms and 4 C_2 dimers. For Ti_6C_{13}, Ti_7C_{13}, and Ti_9C_{15}, we calculated different configurations involving cubic structural motifs and various C_2 dimers. We optimized structures of both the anions and neutrals. The energy differences between the anions and the neutral species yields the adiabatic EAs. The DOS spectra of the anions were computed by convoluting Gaussians to occupied single particle energy levels and were used to compare qualitatively to the observed PES spectra.

Table 4. Calculated electronic binding energies for the lowest energy structures shown in Fig. 6.

Cluster	Ti_3C_8	Ti_4C_8	Ti_6C_{13}	Ti_7C_{13}	Ti_9C_{15}	$Ti_{13}C_{22}$
ΔE (eV)	-64.89	-70.65	-115.92	-120.84	-153.63	-231.06

We obtained several minimum energy structures for each cluster. Figure 6 displays only the lowest energy structures for the neutral clusters, Ti_3C_8, Ti_4C_8, Ti_6C_{13}, Ti_7C_{13}, Ti_9C_{15} and $Ti_{13}C_{22}$. The key optimized geometric parameters are labeled in the structures. The calculated electronic binding energies are shown in Table 4. The calculated EAs are shown in Table 3 together with the measured ones. In general, we found that the lowest energy structure of each cluster yields an EA in good agreement with the experimental value. Our calculations indicate that the calculated EAs are sensitive to the cluster structures. A larger geometric difference from the lowest energy isomer often results in significant deviation from the experimental EA. The optimized cluster structures exhibit significant bond relaxation from the anions to the neutrals, consistent with the observed broad PES features.

4.3. CLUSTER STRUCTURES AND BONDING

The optimized structures shown in Fig. 6 suggest that carbon dimers indeed play an important role in stabilizing the clusters. We found that in all cases the low-lying states of C_2 are actively involved in the chemical bonding. In particular, we found that there is a net electron transfer from the Ti to the C_2. Consequently, C_2 in the Ti_xC_y clusters can be viewed as acetylene-like C_2 anion with a similar electronic structure to that of CO or CN^-. Similar Ti-C_2 interactions were found in Ti_8C_{12}, which involves 6 C_2 dimers. Detailed insight into the chemical bonding in the Ti_xC_y clusters can be obtained by analyzing the calculated wavefunctions and comparing the calculated DOS with the PES spectra. Here we give detailed discussions on the electronic structures of Ti_3C_8 and Ti_4C_8; the electronic structures of larger Ti_xC_y clusters, while being much complex, exhibit similar features to these two smaller clusters.

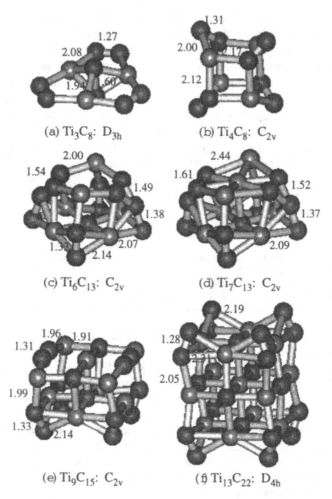

(a) Ti_3C_8: D_{3h}

(b) Ti_4C_8: C_{2v}

(c) Ti_6C_{13}: C_{2v}

(d) Ti_7C_{13}: C_{2v}

(e) Ti_9C_{15}: C_{2v}

(f) $Ti_{13}C_{22}$: D_{4h}

Figure 6. Optimized minimum energy structures of Ti_xC_y (x/y = 3/8, 4/8, 6/13, 7/13, 9/15 and 13/22).

For Ti_3C_8, the optimized structures involve a trigonal bipyramidal Ti_3C_2 unit with 3 C_2 dimers on the same plane. The 3 C_2 dimers are of very short C-C bondlengths, similar to a C-C triple bond while the distance between the two capping C atoms is considerably longer, facilitating the formation of more close-packed cluster structures. The bonding is primarily dominated by the nearly perfect orbital overlap between the π-orbitals of the 3 C_2 dimers and the well-oriented $3d$-orbitals of Ti. The extra-electron in the anion species goes to the π^*-orbitals of the C_2 dimers. The calculated Mulliken charges are shown in Table 5. As expected, the bonding is of covalent nature due to the efficient orbital overlap. The simulated DOS spectrum of structure resembles the PES spectrum very well as shown in Fig. 5(a). The DOS spectra obtained from other isomers deviate considerably from the PES spectra. This suggests that the electronic structure is also very sensitive to the structural change. The consistency among the calculated BE, EA and DOS of the lowest energy structure and their excellent agreement with the experiment nicely validate the optimized structure of the Ti_3C_8 magic cluster.

For Ti_4C_8, the lowest energy structure involves 4 C_2 bonded at the 4 faces of the tetrahedral Ti_4 clusters. The calculated EA is in good agreement with the experimental value. In particular, the DOS of the structure fits the PES spectrum remarkably well, as shown in Fig. 5(b). Therefore, we are confident that structure of Ti_4C_8 is of D_{2d} symmetry with 4 C_2 dimers bonded to the four faces of the tetrahedral Ti_4. Interestingly, this structure can be viewed as replacing the 4 corner C atoms of a 2×2×2 cubic Ti_4C_4 with 4 C_2 dimers. In this sense, Ti_4C_8 seems to be rather similar to the cubic $Ti_{13}C_{22}$ cluster, which can be viewed as replacing the 8 corner C atoms of a 3×3×3 cubic $Ti_{13}C_{14}$ nanocrystal with 8 C_2 dimers. This reinforces the prominent role of the C_2 dimers and the cubic framework in determining the structures of small titanium carbide clusters. Indeed, the driving force in the chemical bonding in Ti_4C_8 involves interactions between the π-orbitals of C_2 and the $3d$-orbitals of Ti, similar to the situation in Ti_3C_8. However, the orbital mixing is much enhanced in Ti_4C_8 and this can be seen clearly from the enhanced Mulliken charges on the C_2 dimers, as shown in Table 5.

Table 5. Calculated Mulliken charges in Ti_3C_8 and Ti_4C_8.

Cluster	C_2 dimer	Capping C atom	Ti
Ti_3C_8	-0.508	-0.662	0.949
Ti_4C_8	-0.744	-	0.744

To understand the structures of larger clusters, we first examine $Ti_{13}C_{22}$. While a detailed description of this cluster were reported recently [39], we would like to note here that it follows a layer-by-layer growth model: the 13/22 cluster can be viewed as a three layer A-B-A structure (Fig. 6), where A represents a Ti_4C_9 layer and B represents a Ti_5C_4 layer. The calculated EA is in excellent agreement with the experimental value. The simulated DOS spectrum resembles the PES spectrum extremely well, as shown in Fig. 5(f). The stability of this structure derives from the fact that all the Ti atoms are optimally bonded to the C atoms and C_2 dimers. We note that clusters with both capping A layers seems to make especially stable structures due to the optimized bonding of the metal atoms with the C_2 dimers.

The mass spectrum shown in Fig. 4 suggests that the growth from the 7/13 to the 13/22 clusters is rather rapid, resulting in very low abundance for clusters containing 8 to 12 Ti atoms. The prominent abundance of the 9/15 cluster in this size range indicates that it is particularly stable and may be an intermediate step along the growth path to $Ti_{13}C_{22}$. Therefore the structure of the Ti_9C_{15} cluster may embody features of $Ti_{13}C_{22}$. Based on the layer-by-layer growth model, a two-layer A-B structure would give a 9/13 cluster. Therefore, we considered the 9/15 cluster to be such a A-B structure with the two extra carbon atoms either being a C_2 dimer on top of the carbon-deficient B-layer or forming two additional C_2 dimers within the B-layer. There are two possible positions for the C_2 dimer on top of the B layer, a symmetric and an asymmetric one, which were found nearly degenerate. The structure shown for Ti_9C_{15} in Fig. 6 is of the highest binding energy, and the calculated EA of this structure is also in excellent agreement with the measured value (Table 3). The simulated DOS spectrum reproduces the key features of the PES spectrum as shown in Fig. 5(e). Again, the consistent agreement among the calculated BE, EA, and DOS of structure with the experiment supports this minimum energy structure of Ti_9C_{15}.

We considered the structures of both Ti_7C_{13} and Ti_6C_{13} on the basis of the two-layer Ti_9C_{13} cluster with incomplete B-layers: Ti_7C_{13} and Ti_6C_{13} can be viewed as removing two and three Ti atoms from the complete Ti_5C_4 B-layer, respectively. The minimum energy structures in both cases give EA values in good agreement with the experiment. The 4 C atoms in the partial B-layer are rearranged to two C_2 dimers. The two C_2 dimers from the A-layer are bonded to the two C_2 dimers in the partial B-layer. Once again, the calculated DOS spectra for the two clusters, shown in Figs. 5(c) and 5(d), respectively, are in good agreement with the PES spectra. While the calculated DOS curves for Ti_3C_8 and Ti_4C_8 give almost quantitative agreement with the experiment, the DOS curves of Ti_7C_{13} and Ti_6C_{13} can only be considered qualitatively or semi-quantitatively. It is likely that in Ti_7C_{13} and Ti_6C_{13} several isomers were also present in the PES experiments since some of them are nearly degenerate with the minimum energy structures and the overall features of the optimized structures are quite similar to the structures shown in Fig. 5. However, the better agreement between the calculated EAs and DOS and the experiments do lend credence to minimum energy structures for Ti_7C_{13} and Ti_6C_{13}.

4.4. GROWTH PATHWAY

The observed magic anion clusters and the elucidation of their structure and bonding provide important insight into the growth of the carbide anion clusters and the formation of the cubic $Ti_{13}C_{22}$. The structures of the magic clusters, $Ti_3C_8^-$, $Ti_4C_8^-$, $Ti_6C_{13}^-$, $Ti_7C_{13}^-$, and $Ti_9C_{15}^-$ strengthen the importance of the cubic framework and the C_2 dimers in the growth of the Ti_xC_y clusters and provide possible intermediates to the formation of the cubic $Ti_{13}C_{22}$. From the mass distributions (Fig. 4), there seem to be two rapid growth regions. The first one is from $Ti_3C_8^-$ to $Ti_6C_{13}^-$ and $Ti_7C_{13}^-$, where the clusters in between exhibit lower abundance. Many intermediate clusters are formed in this growth region with $Ti_4C_8^-$ being the more prominent one. The second growth region is from $Ti_7C_{13}^-$ to $Ti_{13}C_{22}^-$. The growth in this region must proceed extremely fast since almost no intermediate clusters are formed except the $Ti_9C_{15}^-$ cluster. The $Ti_3C_8^-$ cluster, being the most prominent in the small size range, may be the key to set out the growth sequence. Therefore, Fig. 6 probably represents the major growth steps leading to the formation of the $Ti_{13}C_{22}$ cluster starting from Ti_3C_8. The structural evolution of these clusters is highly systematic with the C_2 dimers, cubic framework, and layered structures being the essential components.

5. Conclusions

Quantum-mechanical calculations based on density functional theory have been used to model structures and chemical bonding of transition metal clusters and MetCars. DFT is particularly effective in dealing with molecular systems containing transition metal elements. We show much insight into the cluster structures, bonding nature and growth mechanisms can be obtained. The utility of full quantum-mechanical modeling is demonstrated through two important cluster systems.

The structures of a series of small chromium clusters were fully elucidated. We discovered a unique dimer growth route that gives rise to even-odd alternations of some key physical properties for clusters up to Cr_{11}. A major structural transition occurs at Cr_{11} from the dimer growth to *bcc*-like structures and the clusters start resembling the bulk. A tightly bound Cr_2 dimer was found to play a key role in determining the small cluster structures and bonding. The suggested growth mechanism is consistent with the known experimental results.

New "magic" numbers were found in the anion mass spectra of titanium carbide clusters, $Ti_xC_y^-$, where x/y=3/8, 4/8, 6/13, 7/13, 9/15, 13/22, that did not exist in the positive ion mass spectra. A combined photoelectron spectroscopy and density functional theoretical study was carried out to understand the structure and bonding of these clusters and obtain insight into their growth pathways. All Ti_xC_y clusters were found to be primarily composed of three major components: C_2 dimers, cubic framework (except Ti_3C_8), and layered structures. Substantial charge transfer was found from Ti to C_2, and consequently C_2 in the Ti_xC_y clusters can be viewed as acetylene-like C_2 anion. The favorable orbital overlaps between the π-orbitals of C_2 and the *d*-orbitals of Ti give rise to strong bonding in the Ti_xC_y clusters. The calculated EAs and the simulated DOS spectra for all the clusters are in excellent agreement with the experimental measurements, lending considerable credence to the validity of the optimized cluster structures. It is remarkable that even moderate variations in the cluster geometries among the different isomers can result in significantly different features in the DOS spectra and EA values and the lowest energy structures in all cases give the best agreement with experiments. A growth sequence is proposed starting from $Ti_3C_8^-$ to $Ti_{13}C_{22}^-$ with prominent intermediates at $Ti_4C_8^-$, $Ti_6C_{13}^-$, $Ti_7C_{13}^-$, and $Ti_9C_{15}^-$. Experimental and theoretical evidence indicates that the cubic layered growth with C_2 dimers can lead to a new type of highly stable one-dimensional quantum wires.

Acknowledgments

The support of this research by the National Science Foundation (to L.S.W) is gratefully acknowledged. The experimental work was performed at Pacific Northwest National Laboratory, operated for the U.S. Department of Energy by Battelle (contract DE-AC06-76RLO 1830). The theoretical work was conducted at Air Products and Chemicals, Inc. We gratefully acknowledge Drs. C. A. Valenzuela and J. B. Pfeiffer for supporting this collaborative research. L. S. W. is an Alfred P. Sloan Research Fellow.

150

References

1. A. G. Castleman, Jr., and K. H. Bowen, Jr., *J. Phys. Chem.* **100**, 12911 (1996).
2. I. M. L. Billas, A. Chatelain, and W. A. de Heer, *Science* **265**, 1682 (1994).
3. W.A. de Heer, P. Milani, and A. Chatelain, *Phys. Rev. Lett.* **65**, 488 (1990).
4. D. M. Cox, D. J. Trevor, R. L. Whetten, E. A. Rohlfing, and A. Kaldor, *Phys. Rev. B* **32**, 7290 (1985).
5. See, for example, *Physics and Chemistry of Finite Systems: From Clusters to Crystals*, edited by P. Jena, S. N. Khanna, and B. K. Rao (Kluwer Academic, Boston, 1992), Vols. I and II.
6. L. S. Wang, S. Li, and H. Wu, *J. Phys. Chem.* **100**, 19211 (1996).
7. H. Wu, S. R. Desai, and L. S. Wang, *Phys. Rev. Lett.* **76**, 212 (1996).
8. H. Wu, and L. S. Wang, *Phys. Rev. Lett.* **77**, 2436 (1996).
9. S. Li, H. Wu, and L. S. Wang, *J. Am. Chem. Soc.* **119**, 7417 (1997).
10. H. Cheng, and L. S. Wang, *Phys. Rev. Lett.* **77**, 51 (1996).
11. L. S. Wang, and H. Cheng, *Phys. Rev. Lett.* **119**, 7417 (1997).
12. L. S. Wang, X. B. Wang, H. Wu, and H. Cheng, *J. Am. Chem. Soc.* (submitted).
13. L. S. Wang, H. Cheng, and J. Fan, *J. Chem. Phys.* **102**, 9480 (1995).
14. L. S. Wang, H. Wu, and H. Cheng, *Phys. Rev. B* **55**, 12884 (1997).
15. J. Baker, M. Muir, and J. Andzelm, *J. Chem. Phys.* **102**, 2063 (1995).
16. T. Ziegler., *Chem. Rev.* **91**, 651 (1991), and references therein.
17. V. Barone, and L. Orlandini, *J. Phys. Chem.* **98**, 13185 (1994).
18. B. G. Johnson, P. M. W. Gill, and J. A. Pople, J. A., *J. Chem. Phys.* **98**, 5612 (1993).
19. B. Delley, *J. Chem. Phys.* **92**, 508 (1990).
20. G. J. Lamming, V. Termath, and N. C. Handy, *J. Chem. Phys.* **99**, 8765 (1993).
21. G. Fitzgerald, and J. Andzelm, *J. Phys. Chem.* **95**, 10531 (1991).
22. N. Godbout, D. R. Salahub, J. Andzelm, and E. Wimmer, *Can. J. Chem.* **70**, 560 (1992).
23. E. P. Kundig, M. Moskovits, G. Ozin, *Nature* **254**, 503 (1975).
24. L. Andersson, *Chem. Phys. Lett.* **237**, 212 (1995).
25. S. M. Casey and D. G. Leopold, *J. Chem. Phys.* **97**, 816 (1993).
26. M. M. Goodgame and W. A. Goddard, *Phys. Rev. Lett.* **54**, 661 (1985).
27. C.-X. Su and P. B. Armentrout, *J. Chem. Phys.* **99**, 6506 (1993).
28. B. C. Guo, K. P. Kerns, and A. W. Castleman, Jr., *Science*, **255**, 1411 (1992); B. C. Guo, S. Wei, J. Purnell, S. Buzza, and A. W. Castleman, Jr., *Science* **256**, 511 (1992).
29. J. S. Pilgrim, and M. A. Duncan, *J. Am. Chem. Soc.* **115**, 6958 (1993).
30. R. W. Grimes, and J. D. Gale, *J. Chem. Soc. Chem. Commun.*, 1222 (1992); T. T. Rantala, D. A. Jelski, J. R. Bowser, X. Xia, and T. F. George, *Z. Phys. D* **26**, S255 (1992); L. Pauling, *Proc.Natl. Acad. Sci. USA* **89**, 8175 (1992); Z. Lin, and M. B. Hall, *J. Am. Chem. Soc.* **114**, 10054 (1992); B. V. Reddy, S. N. Khanna, and P. Jena, *Science* **258**, 640 (1992); M. Methfessel, M. van Schilfgaarde, and M. Scheffler, *Phys. Rev. Lett.* **71**, 209 (1993); P. J. Hay, *J. Phys. Chem.* **97**, 3081 (1993; R. W. Grimes, and J. D. Gale, *ibid.* **97**, 4616 (1993); A. Khan, *ibid.* **99**, 4923 (1995); L. Lou, T. Guo, P. Nordlander, and R. E. Smalley, *J. Chem. Phys.* **99**, 5301 (1993).
31. A. Ceulemans, and P. W. Fowler, *J. Chem. Soc. Faraday Trans.* **88**, 2797 (1992); M. Rohmer, P. de Vaal, and M. Benard, *J. Am. Chem. Soc.* **114**, 9696 (1992); H. Chen, M. Feyereisen, X. P. Long, and G. Fitzgerald, *Phys. Rev. Lett.* **71**, 1732 (1993).
32. D. J. Dance, *Chem. Soc. Chem. Commun.* 1779 (1992); M. Rohmer, M. Benard, C. Henriet, C. Bo, and J. Poblet, *ibid.* 1182, (1993); I. Dance, *J. Am. Chem. Soc.* **118**, 2699, 6309 (1996); Z. Lin, and M. B. Hall, *ibid.* **115**, 11165 (1993); M. Rohmer, M. Benard, C. Bo, and J. Poblet, *ibid.* **117**, 508 (1995); *J. Phys. Chem.* **99**, 16913 (1995).
33. S. Wei, B. C. Guo, J. Purnell, S. Buzza, and A. W. Castleman, Jr., *J. Phys. Chem.* **96**, 4166 (1992); B. D. May, S. F. Cartier, and A. W. Castleman Jr., *Chem. Phys. Lett.* **242**, 265 (1995); S. F. Cartier, B. D. May, and A. W. Castleman Jr., *ibid.* **116**, 5295 (1994); *J. Chem. Phys.* **100**, 5384 (1994); K. P. Kerns, B. C. Guo, H. T. Deng, and A. W. Castleman Jr., *ibid.* **101**, 8529 (1994); S. F. Cartier, B. D. May, and A. W. Castleman Jr., *ibid.* **104**, 3423 (1996); *J. Phys. Chem.* **100**, 8175 (1996); K. P. Kerns, B. C. Guo, H. T. Deng, and A. W. Castleman, Jr., *J. Am. Chem. Soc.* **117**, 4026 (1995); H. T. Deng, K. P. Kerns, and A. W. Castleman, Jr., *ibid.* **118**, 446 (1996).
34. J. S. Pilgrim, and M. A. Duncan, *J. Am. Chem. Soc.* **115**, 4395 (1993); J. S. Pilgrim, L. R. Brock, and M. A. Duncan, *J. Phys. Chem.* **99**, 544 (1995); L. R. Brock, and M. A. Duncan, *ibid.* **100**, 5654 (1996).
35. Y. G. Byun, S. A. Lee, and B. S. Freiser, *ibid.* **100**, 14281 (1996); C. S. Yeh, S. Afzaal, S. A. Lee, Y. G. Byun, and B. S. Freiser, *J. Am. Chem. Soc.* **116**, 8806 (1994); Y. G. Byun, and B. S. Freiser, *ibid.* **118**, 3681 (1996).
36. S. Lee, N. G. Gotts, G. von Helden, and M. T. Bowers, *Science* **267**, 999 (1995).
37. S. Wei, B. C. Guo, J. Purnell, S. Buzza, and A. W. Castleman, Jr., *Science* **256**, 818 (1992).
38. J. S. Pilgrim, and M. A. Duncan, *J. Am. Chem. Soc.* **115**, 9724 (1993); J. S. Pilgrim, and M. A. Duncan, *Int. J. Mass Spectrom. Ion Processes* **138**, 283 (1994).
39. B. V. Reddy, and S. N. Khanna, *Chem. Phys. Lett.* **209**, 104 (1993); *J. Phys. Chem.* **98**, 9446 (1994).

SOLID-STATE REACTIVITY AND IMPLICATIONS FOR CATALYTIC PROCESSES

ELENA V. BOLDYREVA

Institute of Solid-State Chemistry Russian Academy of Sciences,

Kutateladze 18, Novosibirsk 128, 630128 RUSSIA

1. Introduction

The attitude of the scientific community to the problem of solid-state reactivity has changed noticeably with time. The early skepticism related to the possibility of solid-state reactions can be illustrated by the quotation "Corpora non agunt nisi fluida" ascribed to Aristotle. Later, in the 18th century Lavoisier wrote: "Most of the earths, stones and crystallizations are things absolutely new to the chemist, the examination of which could furnish an inexhaustible source of experiments and discoveries". In the 20th century solid-state chemistry was eventually recognized as one of the most fascinating fields of chemistry contributing to our knowledge of the fundamental laws of chemistry and leading to important practical applications.

Solid-state chemistry deals with the *synthesis* and *characterization* of solid substances (inorganic, organic, organometallic, or bioorganic); it also concerns their physical and chemical *properties* and considers various aspects of *reactions* involving solid reagents or / and products (Fig. 1) [1].

J.A.K. Howard et al. (eds.),
Implications of Molecular and Materials Structure for New Technologies, 151–174.
© 1999 *Kluwer Academic Publishers. Printed in the Netherlands.*

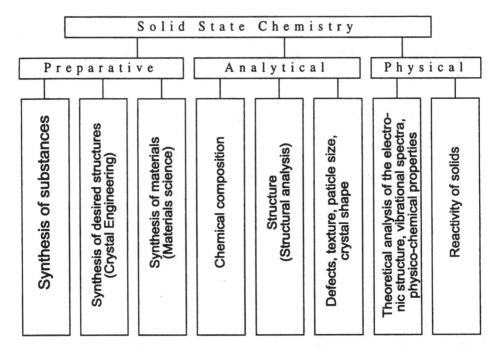

Figure 1. Problems considered by solid-state chemistry [1].

A chemist dealing with *molecules* aims at obtaining the product with the desired *chemical composition* and *molecular structure*. In *solid-state* chemistry, in addition to this, one may be interested in a particular *crystal structure*. Besides, the solid product may be needed as a *single crystal* (of a particular *shape* and *size*), as a *crystalline* or an *amorphous powder* (with controlled *surface area*), as a *film*, as a *wire*, *etc*. The type and the concentrations of *defects* (deviations from the ideal chemical composition and/or crystal structure) should be, in general, controlled. All these additional requirements are of special importance for practical applications of solid products as various materials, drugs, catalysts or sensors. Knowing the crystal structure, the size and the shape of crystals, the type and the concentration of the defects is also necessary for understanding the factors affecting reactivity of solids.

Application of structural analysis for identifying the products of solid-state reactions and for characterizing their molecular and crystal structures is probably the most obvious application of crystallographic techniques in solid-state chemistry. At the same time, crystallography can do much more than this (especially when combined with other techniques). It can be used as a valuable tool for studying the *mechanisms* of the solid-state processes and for finding the ways of controlling and optimizing them. These aspects of applying crystallography to studying *reactivity* of solids will be the subject of this chapter.

It is widely recognized that materials science and catalysis should benefit from the scientific knowledge developed in the field of solid-state chemistry [2-5]. Many of the solid-state reactions which are discussed here have industrial applications. In particular, they can be used for the preparation of new materials and catalysts. I shall aim to show that we need to study the *process, reactivity of solids,* even if we are interested primarily in the *material.* In addition, of course, control of the reactivity of solids is of primary importance for developing new *technologies.*

2. What is a Structure?

Since we are interested in *structural aspects* of solid-state reactivity, let us first define what a *structure* is. The Oxford Dictionary gives the following definitions of the term, which are probably close to an intuitive feeling of a structural chemist:

1. Structure is a way in which something is put together, organized.
2. Structure is a framework or essential parts of any complex whole.

The third definition of structure was given by a materials scientist [6].

3. Structure is any assemblage of materials which is intended to sustain loads.

In this chapter I shall discuss the possible role of structure in three aspects of solid-state reactions, basing the discussion on each of the three definitions.

3. Topochemical Reactions, or Structure as a "way in which something is put together"

For at least a century, it has been known in the theory of chemical reactions what can be termed as *"the rule of least molecular deformation"*[7], *"the principle of minimal structural change"* [8], or *"the principle of least motion"* [9]. These three principles can be summarized as follows: "Those elementary reactions will be favored that involve the least change in atomic positions and electronic configuration" [10]. It is, essentially, the same principle, which, when applied to solid-state reactions, is commonly referred to as a *"topochemical principle"*.

3.1. INORGANIC REACTIONS

V. Kohlschütter [11-13] and, later, W. Feitknecht [14], introduced the concept of "topochemical reactions" (from Greek "τοποζ- site) at the beginning of this century. According to Kohlschütter and Feitknecht, reactions in the solid state differ noticeably from the reactions of "free molecules" since reacting species have a pre-determined location in the crystal structure or at the crystal surface. The reactions studied by Kohlschütter, Feitknecht and their successors were thermal decomposition of inorganic salts and hydroxides, exchange reactions "solid + liquid", reactions "solid + gas", and reactions at solid electrodes.

Let us take as an example the way in which the topochemical principle may be applied to the solid-state dehydration of $Mg(OH)_2$. The reaction follows a very simple chemical equation $(Mg(OH)_2 \rightarrow MgO + H_2O)$, but it allows one to demonstrate many of the peculiar problems arising when studying solid-state reactivity, and we shall refer to this reaction throughout the lecture. The reaction is also of practical importance since its solid product - MgO - is used as a catalyst and as a catalyst support, as a flame retardant, as an insulator in electrical devices, as a neutralizing agent in medicine, *etc.* [15-20].

The initial stages of the dehydration of brucite, $Mg(OH)_2$, were carefully studied in the 1960s-1970s by Freund and co-workers [21]. To form a water molecule, two hydroxyl ions must come into contact, so that the proton of one of them becomes linked to the other oxygen atom (Fig. 2a).

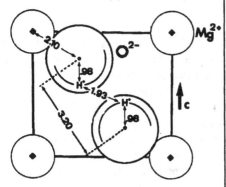
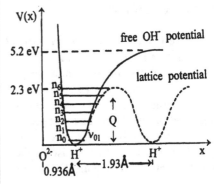

Figure 2a. Section along (110) through the unit cell of $Mg(OH)_2$ with the equilibrium distances at room temperature [21].

Figure 2b. Potential of the proton in a free OH^--ion and in $Mg(OH)_2$ [21].

The very first step of the solid state decomposition involves proton tunneling assisted by lattice phonons:

$$OH^- + OH^- \rightarrow O^{2-} + HOH$$

The probability of this elementary act is dependent on the crystal structure of $Mg(OH)_2$, that is on the way in which OH^--ions are put together. $Mg(OH)_2$ crystallizes in a hexagonal close-packed layer structure of the CdI_2-type. Oxygens are coordinated tetrahedrally to three Mg^{2+}-ions (only one of which is shown for each oxygen in Fig. 2a) and to one hydrogen. The height and the thickness of the potential barrier for a proton transfer can be estimated from X-ray structural data (interatomic distances) and vibrational spectra (the overtones of the O-H stretching vibrations up to the optical dissociation edge) (Fig. 2b). The equilibrium O-O distance is 3.20 Å, and the equilibrium H-H distance is slightly less than 1.93 Å at room temperature.

The estimates with this potential have shown the probability for a proton either to overcome the barrier, or to tunnel through it, to be vanishingly low. The reaction should not take place at all. It does, however. There is a way out of this dilemma by taking into account the thermal motions of the OH^--ions in the hydroxide lattice. There are two lattice modes which are important for the reaction. One of them brings the two hydroxyl ions into closer contact, and another provides the steric condition necessary for the proton transfer (tunneling distance ≈ 0.42 Å, H-O-H angle close to 110°) (Fig. 2c). As a result, the probability of proton tunneling increases sufficiently, to make the reaction possible.

So far only two adjacent OH^--ions have been considered. However, in the $Mg(OH)_2$ structure each OH^--ion is surrounded by OH^--neighbours within the same and in the adjacent (0001) planes (Fig. 3) [22]. The proton of each OH^- will have the choice of tunnelling into any of the neighbouring OH⁻-ions. Once transferred onto a neighbouring OH⁻, the proton will preferentially carry on tunneling to other OH⁻- ions [21].

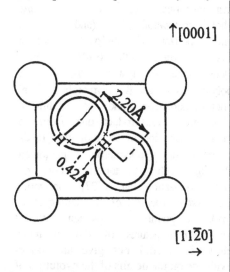

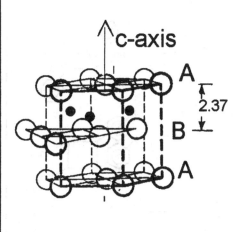

Figure 2c. Section along (110) through the unit cell of Mg(OH)₂ (the shortest distances assume twice the mean amplitudes at 500K) [21]

Figure 3. Mg(OH)₂ (open circles - oxygen and hydroxyl ions; black circles - Mg⁺²) [22]

One may talk of a "water molecule" only when the residence time of the proton on a OH⁻-ion exceeds, for example, the time an H₂O-molecule would need to execute rotational motion (10^{-10} sec) [21]. At ambient temperatures, the residence time of the proton at each OH⁻-ion is very short, and therefore no water molecules are formed, although protons migrate through the structure of Mg(OH)₂. With increasing temperature, however, the mean free path of a proton decreases due to scattering processes. The very same lattice phonons which have initially assisted bringing the proton to the tunneling level now disturb its subsequent long-range tunneling by upsetting the symmetry of the multi-minimum potential. The higher the temperature, the shorter the mean free path of a proton. Eventually, the free path between two collisions will be reduced to unit length. Then and only then may one consider the formation of a molecular H₂O-species at an OH⁻-site. If this H₂O molecule happens to be at or near the surface, it may leave the hydroxide lattice by a classical diffusion and desorption process. At this stage only will gaseous water be evolved from the sample. From then on one may talk of actual "dehydration" [21].

3.2. ORGANIC REACTIONS

Organic solid state chemists usually associate the "topochemical principle" with the names of Schmidt and Cohen [23-26], who have formulated it as follows: "The formation of the product of a solid-state reaction is pre-determined by mutual juxtaposition of the molecules in the parent crystal".

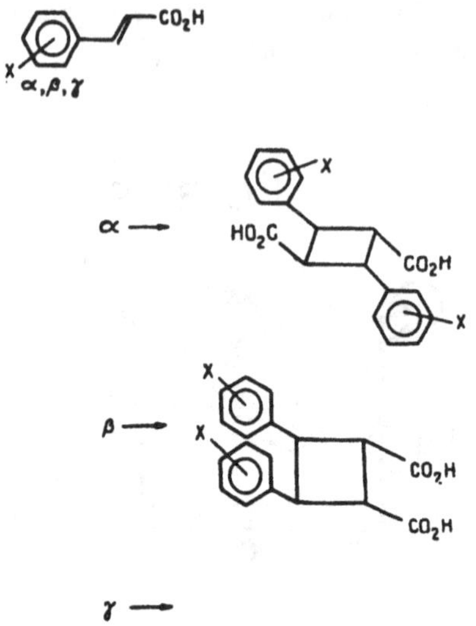

156

A common example is photochemical dimerization of *o*-ethoxy-*trans*-cinnamic acid. *Cis – trans*-photoisomerization (and no photodimerization) is observed in the melt. Crystallization from ether, benzene and ethyl alcohol gives the crystals of α-, β- and γ-polymorphs; α- and β- polymorphs show no photoisomerization and give different photodimerization products due to different mutual orientation of the molecules in the parent crystal, γ-polymorph is photoinert, showing neither photoisomerization, nor photodimerization (Fig. 4) [27]. Considering the *crystal* structures of the starting solids made it possible to explain different *molecular* structures of the products. However, it did not give any direct evidence on the details of the evolution of crystal structure.

Figure 4. Different behavior of three polymorphs of *o*-ethoxy-*trans*-cinnamic acid on irradiation

Only relatively recently, the solid-state photodimerization was followed by single-crystal X-ray diffraction at different degrees of the transformation [28-31], and some of these details became available.

Applying the topochemical principle to the thermal dehydration of *inorganic* brucite and to the photodimerization of the *organic* cinnamic acids is similar in many respects. In both examples the starting crystal structure pre-determines the mutual *juxtaposition* of the species participating in the reaction. The crystal structure should affect also the *migration* of protons (decomposition of $Mg(OH)_2$), or of the photo-excitation (photodimerization of cinnamic acids) prior to them being *trapped* at a structural imperfection. In order to describe the elementary steps of the reactions not only *static* but also *dynamic* characteristics of the crystal structures are required. To get a better insight into the mechanism of the dehydration of brucite, it was necessary to combine the X-ray crystallography with other techniques, for example, with vibrational spectroscopy and measurements of the proton conductivity [21]. Similarly, optical and Raman spectroscopy could be helpful for studying the photodimerization of the cinnamic acids.

3.3. MODIFIED TOPOCHEMICAL PRINCIPLE (DEFECTS IN CRYSTALS)

Usually, *regular* structural sites are considered when the "topochemical principle" is applied in describing solid-state reactions. However, examples are known, both in inorganic and in organic solid-state chemistry, when the same principle should be applied in a modified way, taking into consideration the deviations from the regular structure.

As the first example I shall consider thermal decomposition of silver oxalate:

$$Ag_2C_2O_4 \rightarrow 2Ag + 2CO_2.$$

This solid-state reaction was studied as a model, in order to understand the mechanism of the latent image formation in photographic systems. Besides, the solid reaction product - silver - is an important catalyst.

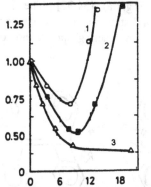

Figure 5. Relative diffusion coefficient (1), ionic conductivity (2) and decomposition rate constant (3) of $Ag_2C_2O_4$ as a function of Cu^{+2} - impurity concentration $(10^{-4}$ mol %$)$ [32]

In the crystals of silver oxalate there are always a few Ag-ions which are located not at regular lattice sites, but in the interstitials. The concentration of these structural defects, interstitial silver-ions, $Ag_i^{\bullet}$, can be decreased by doping the crystals of silver oxalate with small amounts of Cu^{2+}. A decrease in the concentration of $Ag_i^{\bullet}$ can be followed by measurement of Ag-self-diffusion coefficient, or by measurements in ionic conductivity (Fig. 5).

Isomorphous substitution of silver cations at regular lattice sites results not only in the decrease in the concentration of more mobile interstitial silver-ions, but, also, simultaneously, in the increase in the concentration of less mobile silver vacancies, V_{Ag}':

$$Ag_{Ag}^* + \underline{V_i^* \quad \Longleftrightarrow \quad V_{Ag}' + Ag_i^{\bullet}}$$
$$Cu_{Ag}^{\bullet} + V_{Ag}'$$

$$\sigma = \sigma(Ag_i^{\bullet})\big| + \sigma(V_{Ag}')\big|$$
$$\sigma = c(Ag_i^{\bullet})\,\mu(Ag_i^{\bullet}) + c(V_{Ag}')\,\mu(V_{Ag}')$$
$$\mu(Ag_i^{\bullet}) > \mu(V_{Ag}')$$

Therefore, the minimum at the Koch-Wagner self-diffusion and ionic conductivity isotherms is observed. The rate of thermal decomposition was shown to decrease with the decrease in the concentration of $Ag_i^{\bullet}$ [32].

Although the studies of the effect of interstitial silver cations on the thermal decompositions are already old and date back to the 1960s, it was not until very recently that reliable data on the crystal structure of silver oxalate were obtained [33]. The interstitial positions were localized in the structure (Fig. 6a) [34], and a structural explanation of the effect of $Ag_i^{\bullet}$ on the solid-state decomposition was proposed [35]. If

$Ag_i^{\bullet}$ occupies an interstitial position, the C-C bond in the adjacent oxalate-anions weakens, and electron transfer from $(C_2O_4)^{2-}$ to $Ag_i^{\bullet}$ is facilitated. Besides, a silver cluster serving as a pre-nucleus of the solid product is formed by the interstitial silver cation and the silver ions at regular positions (Fig. 6a) [34,35]. The modified topochemical principle can be applied also to this solid-state reaction, if one considers the structurally pre-determined mutual juxtapositions of reacting species not only at *regular*, but also at the *interstitial* sites.

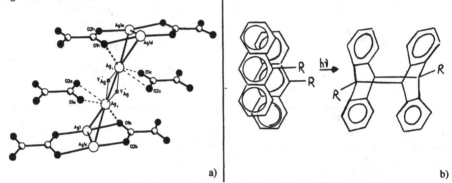

Figure 6. Applying the topochemical principle to defect regions in crystals.
(a) decomposition of $Ag_2C_2O_4$ [34,35], (b) photodimerization of substituted anthracenes [36].

The second example of extending the application of the topochemical principle to structurally imperfect crystal sites is from the field of organic solid-state reactions. Some compounds photodimerize, although from their packing geometry they would not be expected to do so. The photodimerization of the substituted anthracenes often gives products of different symmetry to that expected from packing geometry in the starting crystal structure. Thus, 9-cyanoanthracene has a crystal structure in which adjacent molecules are arranged head-to-head (Fig. 6b). Consequently, the *cis* dimer would be expected to form. In fact, the *trans* dimer is obtained. The topochemical principle, however, is not violated: rather, it should be applied to this system in a modified way, since the reaction takes place at defect sites of the crystal structure. The displacement of molecules at stacking faults of the crystals effectively brings molecules on either side of the fault into a head-to-tail overlap. The formation of dimers at stacking faults was observed directly in the electron microscope [36].

4. Topotactic Reactions, or Structure as a "Framework"

Thus far we have considered solid-state reactions as reactions between two chemical species located somewhere in the structure. A crystal, however, can hardly be reduced to two molecules or ions. A solid-state reaction is not only a sequence of chemical steps at particular sites of the crystal. It is also a complicated process of the formation of the crystal structure of the solid product from the structure of the parent crystal. Hence, when discussing the formation of the product *phase*, we have to consider the crystal structure as a framework, as a complex whole (*i.e.,* as in the second definition given by the Oxford Dictionary). Let us consider some examples as illustrations.

4.1. AN EXAMPLE

First, let us return to our example from the previous section - the dehydration of $Mg(OH)_2$. The close-packed anion layers of hydroxide undergo interlayer rearrangement to form the cubic MgO structure. The parent structure consists of sheets of closest-packed hydroxyl ions where Mg^{+2} ions occupy octahedral interstices between alternate layers, so that the stacking sequence is AmB,AmB (m - ionic Mg^{+2} layers). The product MgO has a face-centered cubic (NaCl) crystal structure, devoid of the parent's anisotropy. It can be viewed in terms of stacking sequence AmBmCm,AmBmCm of close-packed O^{2-} (A, B, and C) and Mg^{+2} (m) layers (Fig. 7). At the same time, a well-defined reproducible orientation relationship between the parent and the product lattice is preserved.

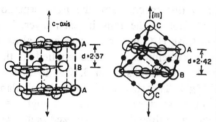

$$Mg(OH)_2 \rightarrow MgO+H_2O$$

Figure 7. Orientation relationship between brucite, $Mg(OH)_2$, and its dehydration product, MgO [22]

Despite the substantial atomic rearrangement needed to produce MgO from $Mg(OH)_2$, there is a well-defined accord between the crystal lattices of the two solids: $\{0001\} \parallel \{111\}$; $\{00\bar{1}0\} \parallel \{011\}$, $\{10\bar{1}1\} \parallel \{001\}$, or $\{1\bar{1}01\} \parallel \{001\}$[17,37]. The close-packed layers of anions, perpendicular to the c axis of the hydroxide and the [111] direction of the oxide, are parallel to each other. The a axis of the reactant coincides with the $[1\bar{1}0]$ direction of the product.

4.2. SOME DEFINITIONS

The existence of reproducible orientations between the crystallographic directions of the parent and product structures makes it possible to define the reaction as *topotactic*.

The first detailed studies of reactions preserving at least part of the structural network of the parent phase were carried out by V. Kohlschütter, W. Feitknecht, and their co-workers, although a special term for such processes - *topotactic reactions* - is usually associated with the name of Lotgering [38-40].

There is no gap between *topotactic* and *topochemical* reactions. A topotactic reaction is at the same time a topochemical one. The reverse is not always true. Most of the inorganic topochemical reactions studied by V. Kohlschütter, Feitknecht and their successors are topotactic. Many of the organic solid-state topochemical reactions studied by Schmidt, Cohen and their successors are not topotactic; moreover, they give non-crystalline, amorphous solid products.

4.3. DISPLACIVE AND RECONSTRUCTIVE TOPOTACTIC TRANSFORMATIONS

It is more difficult to understand a solid-state *process*, than to describe the final *result* - the structure and the orientation of the product, the size of particles and the morphology. Without understanding the process, however, it is hardly possible to *control* the result - the crystal structure, the orientation and the size of the particles, the porosity, the specific surface area of the product - and this is what is required for practical applications. Let us illustrate this on the same example - topotactic dehydration of $Mg(OH)_2$.

The relative orientations of the crystallographic directions of the parent structure of $Mg(OH)_2$ and the product structure of MgO alone do not give information on the mechanism of the structural transformation of $Mg(OH)_2$ to MgO. Several models were proposed in the literature with arguments in favor of each of them.

A so-called "homogeneous" mechanism was proposed [41] based on the transmission electron microscopy. According to this mechanism, the transformation changes the relative positions of the layers but not of the ions in each layer. The hydroxyl groups between the Mg sheets react to form H_2O molecules (we have already discussed this process when considering the topochemical principle applied to this reaction). The interatomic distance inside the layers shrinks from 3.12 Å to 2.99 Å, equal to that required in the {111} plane of MgO. The shrinkage in the basal plane produces cracks. The water molecules escape to the exterior of the crystal through these cracks or the space between the layers, and at the same time the close-packed planes are gradually reshuffled. After the collapse of the original structure is complete, the oxide consists of small crystallites separated by pores. A homogeneous model was supported later by another group, which described the dehydration of $Mg(OH)_2$ as a diffusionless shear transformation proceeding without nucleation and growth [42].

Assuming the homogeneous model to be correct, the thermal dehydration of $Mg(OH)_2$ can be qualified as a *displacive topotactic transformation* [43,44]. During a displacive transformation most atoms in the structure move slightly if at all, a three- or two-dimensional structural skeleton is maintained during the course of the reaction, the transformation produces distortions, which lead to a variation of the lattice parameters and possibly a modification of the symmetry. As examples of other displacive transformations one can mention cationic exchange reactions or intercalation-deintercalation reactions.

Ball and Taylor, who studied dehydration of $Mg(OH)_2$ by X-ray diffraction [45], have proposed another mechanism, which they have called "inhomogeneous", proceeding *via* oriented nucleation and nuclei growth of the product phase. According to this second model, a spinel-like intermediate is formed when $Mg(OH)_2$ is transformed into MgO.

If one assumes this inhomogeneous model to be correct, the dehydration of $Mg(OH)_2$ should be called a *reconstructive transformation*. During a reconstructive transformation the entire structural network must be unlinked, then relinked to form a new structure. Despite this, a definite and reproducible crystallographic orientation relationship between the parent and the product phases can be observed. This becomes possible because of the oriented nucleation and oriented growth with the displacement of the mother/daughter interface [43,44,46]. Examples of other reconstructive topotactic transformations are dehydration and thermal decomposition of many other compounds, redox 'gas + solid' reactions and condensation reactions.

Bernal pointed out, almost 40 years ago, that in solid-state transformations "the nucleus of a new form appearing in the lattice of an old one could hardly escape the influence of the lattice in its formation" [47]. Since that time this idea was developed further [46,48], and it has been shown that oriented nucleation and growth can be considered as the basic processes in topotactic reconstructive transformations. Oriented nucleation can be, for example, of epitactic origin. Epitactic nucleation can occur either at an external or internal reaction interface due to a coherency between certain crystal planes of the otherwise unrelated parent and product lattices. If the transformation spreads through the volume of the crystal as a result of a chemical reaction, it becomes topotactic, although the mechanism is epitactic. This possibility was mentioned already in the first publication of Lotgering on topotaxy [38].

4.4. FORMATION OF METASTABLE PHASES IN TOPOTACTIC TRANSFORMATIONS

Topotactic reactions often give new *metastable* phases, which cannot be obtained from any of the known forms at thermodynamic equilibrium. Structural, thermodynamic and kinetic aspects of considering topotactic solid-state reactions are most closely interrelated. The structure of solid products is often determined by purely *kinetic* factors, that is by the ease of transforming one structure into another with the maximum amount of the original structural framework being preserved, and not by the minimum in the free energy of the final phase. It is remarkable that the main ideas which are of primary importance for the preparation of metastable phases by topotactic reactions were formulated by W. Ostwald as early as 1897 [49]. According to Ostwald's rule, "in all the processes it is not the most stable state with the least amount of free energy which is initially obtained, but the least stable lying nearest to the original state in free energy".

"Lying nearest to the original state in free energy" - for two solid phases this usually means the existence of structural filiation. As well as the "topochemical principle", this is also (but in another aspect) a continuation of the principles "of least motion" [9], "of least molecular deformation" [7], "of minimal structural change" [8]. The preservation of structural motifs reduces the number of bonds to be broken and decreases the lengths of diffusion paths. It minimizes the amount of atomic rearrangement, providing a special energetically favorable reaction path.

4.5 FORMATION OF PSEUDOMORPHS IN TOPOTACTIC REACTIONS

Topotactic reactions form the basis of many processes of practical interest. Metastable phases formed in the course of topotactic reactions are often characterized by unusual and interesting physical and chemical properties. They can be used as catalysts or as advanced materials. Many of the implications of topotactic reactions for the synthesis of practically important materials stem from the large surface area the solid products of these reactions have. Topotaxy often creates *pseudomorphs*. A pseudomorph is a crystal that has become converted into another substance, or a mixture of substances, without change in its external form.

162

Pseudomorphs often show diffraction patterns close to those of the single crystals due to the nearly identical orientation of the numerous crystallites that comprise them, but differ noticeably from single crystals by high specific surface area and, often, by high catalytic activity.

If in a reaction, by virtue of pseudomorphism, the shape of the particle does not change, but at the same time the product has a substantially greater density than the parent, the solid inevitably develops voids. The voids of a pseudomorph can be of various kinds. In iron prepared by topotactic reduction of magnetite, they take the appearance of large cracks [50,51]. More commonly, the pores are in the form of very small cavities spread rather uniformly throughout the crystal, giving it the texture of a sponge. This makes it possible to apply the products of topotactic reactions as support materials in catalysts, or as catalysts themselves. The product of decomposition of $Mg(OH)_2$ again can serve as an example. The "active" form of MgO obtained by the topotactic dehydration at 623K has a large surface area and is used as a catalyst support [16,20], and the presence of paramagnetic defects at its exposed {111} planes makes it an unusually active catalyst for H_2-D_2 equilibration at 78K [15]. Even when used as a support, MgO is not merely an inert carrier for the real catalyst (metallic Fe), but instead interacts with the Fe in a manner essential for the ultimate production and stabilization of small metallic particles [16,20]. It is also worth mentioning that the catalytic activity of MgO may be more than 5 orders of magnitude different, depending on the way in which the solid sample was produced [15,17,18]. It was shown in the early studies [52] that the lower the temperature and the shorter the time of synthesis of MgO are, the higher is its catalytic activity.

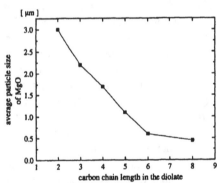

Figure 8. The decrease of the particle size of MgO with increasing carbon chain length in the magnesium diolate precursors [18]

A recent study [18] has shown the possibility to control the particle size of MgO obtained by thermal decomposition of magnesium alcoholates and diolates (and, hence, its activity) by modifying the length of the carbon chain of the precursor (Fig. 8). For another oxide, CdO, the size and the shape of the crystallites, and hence the catalytic properties, are strongly dependent on whether the precursor of this compound was hydroxide or carbonate [53]. A strong dependence of properties from the method of preparation and from the preliminary treatment is typical for

solid materials. This explains why one does need to study *processes* even if one is interested primarily in the *materials*.

4.6. SIZE EFFECTS IN TOPOTACTIC REACTIONS

The control of the size of starting precursor crystals is of primary importance. Numerous examples are known when crystals of different size not only show different reaction rates, but also give different polymorphs of the solid products [54-56]. For example, γ-

Fe_2O_3 characterized by high catalytic activity can be obtained from Fe_3O_4 only if the size of starting particles does not exceed 3000Å. From larger crystallites another polymorph α-Fe_2O_3 (with no catalytic properties) is formed [54,55].

The size effect may account also for the frequent observation that different methods of studying topotactic reactions (transmission electron microscopy (TEM) and X-ray diffraction) may reveal very different amounts of orientation between the parent and the product phase for the same reaction. TEM may show perfect topotaxy whereas X-ray diffraction finds only a low degree of orientation between the phases if any orientation at all [17]. This was demonstrated, for example, by Mackay [57] for the transformation of lepidocrocite (γ-FeOOH) to maghaemite (γ-Fe_2O_3). It may well be that the different conclusions on the homogeneous or the inhomogeneous mechanism for the decomposition of $Mg(OH)_2$ mentioned above also result from the difference in the methods of investigation used. Topotaxy is much more often observed by TEM than by X-ray diffraction since the samples used in TEM are very small, thinner than 0.1μm, and topotaxy usually requires small particle size in the precursors.

4.7. APPLICATIONS OF CRYSTALLOGRAPHY TO THE STUDY OF TOPOTACTIC REACTIONS

Topotaxy is obviously a very fruitful field for applying the knowledge of crystallography and the experimental techniques of structural analysis (X-ray and neutron diffraction, high-resolution electron microscopy). It is important to be able to find the elements of structural similarity in two solid phases, and all the variety of crystallographic tricks aimed at solving the problem can be useful for solid-state chemists. Among these tricks are finding the supergroup-subgroup relationships [58,59], selecting similar units, structure-forming-units, or synthons [60-63], the analysis of the bond-networks using graph theory [64-68], finding pseudo-symmetry in the structures [59,69] and analysis of structural classes and their subclasses [69,70].

Until very recently, considering crystal structures as networks was more common in inorganic crystallography, and most topotactic reactions studied and described in the literature involve inorganic compounds. However, in recent years, organic crystallography has also moved in the direction of presenting structures as networks, mainly in relation to the problems of crystal engineering and materials design, or to the problem of the polymorphism [60-63,66,67]. Various methods of searching for structural similarities were suggested [59,69]. All this could also form the basis of finding and studying many topotactic solid-state reactions in the world of organic solids. The main point is to make the next step after describing an organic solid-state reaction as a topochemical one, having correlated the composition of the products and the mutual juxtaposition of the reacting groups or species. One would need to consider the transformation of the structure as a whole. Unfortunately, today it remains unclear from many of the publications on reactions in organic crystals if the reaction gave crystalline or amorphous solid products, and/or if there was any orientation relationship between the parent and the product structures.

4.8. TIME-RESOLVED CRYSTALLOGRAPHY AND TOPOTACTIC TRANSFORMATIONS. NON-CRYSTALLOGRAPHIC TECHNIQUES

Direct observations of the spatial propagation of topotactic reactions (formation and growth of the product phase), as well as thermal analysis and kinetic studies are no less important for evaluating and understanding topotaxy than the structural studies [44,45,71-74]. Sometimes structural changes during displacive topotactic transformations are so small that a routine X-ray structural analysis does not allow one to distinguish between the two phases (Fig. 9). In this case a combination of other methods, such as neutron diffraction, IR-spectroscopy or differential scanning calorimetry can help, as was shown in the study of the supermetastable WO_3-phase formed using $WO_3 \bullet 1/3 H_2O$ as the precursor [75-77] (WO_3 is a compound of large practical importance and can be used in catalysis or in electronic devices) [44,75-79].

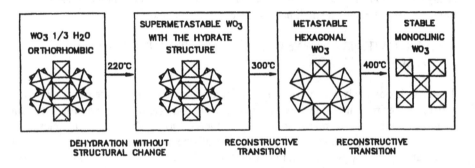

Figure 9. ($WO_3 \bullet 1/3 H_2O$) as precursor and different WO_3 phases [75-77]

Modern experimental techniques of time-resolved X-ray diffraction make it possible not only to characterize the starting and the final stages of a topotactic transformation (the structures of the parent and the product phases and their mutual orientation and the crystallite size in the solid product), but also to monitor the structural transformation (Fig. 10) [80]. For other examples of time-resolved crystallographic studies of topotactic solid-state reactions see [81-87].

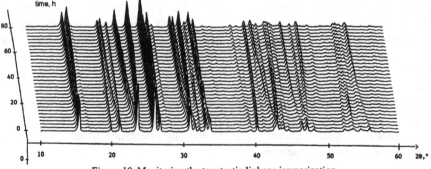

Figure 10. Monitoring the topotactic linkage isomerization in [Co(NH_3)_5ONO]Br_2 by X-ray powder diffraction [80]

5. Stress and Strain in Solid-State Reactions, or Structure as "an Assemblage of Materials Intended to Sustain Load"

5.1. GENERAL INTRODUCTION TO THE PROBLEM

When discussing solid-state reactivity, it is always important to take into account the role of stress and strain. I say "always" in order to emphasize that mechanical stress and strain arise in solids not only when external mechanical load is applied, but during any solid-state transformation (phase transition or chemical reaction). The main reasons for this are the changes in molar volume and structural reconstruction. Manifestations of stress and strain generation can be observed with a naked eye: the nuclei of the product phase formed in solid-state reactions are usually cracked and the crystals themselves are often fragmented or deformed in the course of solid-state reactions (Fig. 11, [88]).

 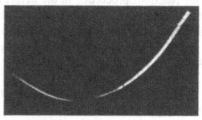

Figure 11. Fragmentation (left) and elastic bending (right) of single crystals of $[Co(NH_3)_5NO_2]Cl(NO_3)$ in the course of photoisomerization [88].

Sometimes, what we could take for a "nucleus of the product phase", is in reality nothing other than a system of intersecting and interpenetrating cracks [89-92].

The role of stresses and strains in solid-state reactions is very extensive. "The ability of the structure to sustain loads" in many cases determines the rate of a solid-state reaction, its spatial propagation, the morphology and the properties of the products formed (important for materials!), the possibility of carrying out the same chemically reversible reaction repeatedly in the same crystal (important for devices!), and, finally, the very possibility of the reaction occurring, the composition and the structure of products [93-109]. It is necessary to consider the generation and the relaxation of the mechanical stresses during a solid-state transformation, in order to understand why some of the topochemical reactions give non-oriented crystalline or even amorphous solid products, whereas the others are topotactic, and why some of the solid-state reactions proceed in a single-crystal-to-single-crystal mode without fragmentation of the parent crystals or a deterioration in their quality, whereas others do not. Taking the role of mechanical stresses into account, one can carry out solid-state reactions in a single-crystal-to-single-crystal mode [95]. Moreover, one can vary the experimental conditions and/or the crystal size and habit in such a way that the *same* solid-state reaction proceeds either with violent fragmentation of the parent crystals or in a single-crystal-to-single-crystal mode [28-31,88,108,109]. The ability to carry out chemically reversible solid-state reactions in a single-crystal-to-single-crystal mode could be successfully applied in photometric devices [110] or in holographic materials [111].

5.2. EXTERNAL MECHANICAL ACTION AND SOLID-STATE REACTIVITY: CHANGES IN THE MUTUAL JUXTAPOSITIONS OF THE REACTING SPECIES

Stresses and structural strain induced by external action can affect a solid-state reaction at all the stages and in various ways. I shall mention here only two of many possibilities. The molecular structure of the product may be affected by a *distorted juxtaposition of the species* involved in the reaction in a strained part of the crystal. For example, a "violation of the topochemical principle" (the formation of products other than predicted from the ideal crystal structure) was observed for some photodimerization reactions. We have already discussed this phenomenon before. It is commonly explained by the fact that the reactions proceeded in the defect regions of the crystals only (in the vicinity of dislocations), where the juxtaposition of the molecules was different from that in the rest of the structure [112-118]. It was recently shown by experiments in the diamond anvil cell that shear stresses alone can induce dimerization in the same crystals even in the dark [119].

5.3. EFFECT ON THE DISTORTION OR/AND RECONSTRUCTION OF STRUCTURAL NETWORKS

Mechanical stress and structural strain may facilitate or inhibit the *transformation of* one *structural network* into another, thus affecting the transition from the parent phase to the product phase. For example, shear stresses may induce many transformations in inorganic and organic solids, in particular, displacive transformations or oriented nucleation in reconstructive topotactic reactions [120-124]. Even isotropic mechanical loading (for example, applying hydrostatic pressure) can result in a highly anisotropic response of the structure, which can lead to chemical reactions or, at least, can facilitate or inhibit reactions induced by temperature or light. Sometimes, the anisotropy of a structural distortion plays such a predominant role, that compressing the structure under load inhibits reactions with the overall decrease in molar volume and, on the contrary, accelerates reactions with an increase in molar volume. Linkage isomerization in nitro- and nitrito-ammine Co^{III}-complexes may serve as an example.

The quantum yield of nitro-nitrito photoisomerization in the elastically compressed parts of the crystal of $[Co(NH_3)_5NO_2]Cl(NO_3)$ was shown to decrease [100] despite the overall negative volume change in the course of this reaction ($\Delta V/V = -0.7\%$) [125]. The reason for this seemingly unusual effect is the anisotropy of the compression of $[Co(NH_3)_5NO_2]Cl(NO_3)$ induced by isomerization [125]: in some of the directions the structure expands. The direction of maximum expansion coincides with the direction of the crystallographic c axis ($\Delta c/c = +3.5\%$), and it is this direction in which maximum compression is observed under external mechanical load (Fig. 12) [126].

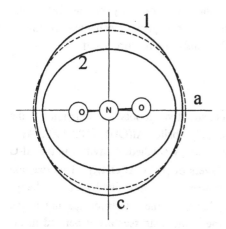

Figure 12. Principal sections of strain ellipsoids for the distortion of crystal structure of [Co(NH₃)₅NO₂]Cl(NO₃):
1 - nitro-nitrito isomerization;
2 - hydrostatic pressure.
Dashed line - reference sphere.
a,c - crystallographic axes.
Orientation of the NO₂-ligand is shown. Strain is exaggerated [126].

Similarly, hydrostatic pressure was shown to accelerate the thermal nitrito-nitro isomerization in [Co(NH₃)₅ONO]Br₂ [108,127,128] despite the overall increase in the molar volume ($\Delta V/V = +0.84\%$) in the course of this reaction [125]. The explanation of the effect is again to be sought in the anisotropy of structural distortion in the course of the isomerization: the structure contracts noticeably in particular directions, maximum linear contraction being equal to -3.4% [108,125,127].

5.4. OTHER MECHANISMS FOR THE EFFECT OF STRESS AND STRAIN ON SOLID-STATE REACTIVITY

Numerous other mechanisms for the effect of mechanical stress and strain on solid-state reactions can be imagined. In the section on topochemical reactions we have briefly discussed the importance of the migration of chemical species through the structure for solid-state reactions. The mobility of the species and the diffusion coefficients in the strain field may be very different as compared with the unstrained crystal [129]. We have also mentioned that many inorganic and organic solid-state reactions take place mainly at dislocations. For these reactions, the effect of stresses on the solid-state reactivity can be related to the generation of new dislocations in the stressed crystal. These dislocations can serve as new nucleation sites. This phenomenon was termed as a "dislocation-chain [96].

5.5. MECHANICAL STRESS AND STRAIN INDUCED BY THE SOLID-STATE REACTION ITSELF: THE FEEDBACK PHENOMENON"

The ability of the crystal structure to sustain mechanical stress determines the strain induced in the structure by the solid-state transformation (phase transition or chemical reaction) itself and the further course of the same transformation in this strain field. The effect of the mechanical stresses and strains induced in the crystal by the reaction on the reaction rate can be either positive (positive feedback), or negative (negative feedback) [105,107]. The generation and the relaxation of mechanical stresses can affect also the spatial propagation of the solid-state processes [93,94,96,101]. The reasons why the mechanical stress and strain can affect the solid-state reactivity remain the same, independently of whether the solid was subjected to some special mechanical treatment, or the stresses were generated in the course of the solid-state reaction itself.

5.6 AN EXAMPLE OF POSITIVE FEEDBACK

The first example is a displacive, non-reconstructive topotactic reaction, - the intercalation of LiCl from aqueous solution into gibbsite, $Al(OH)_3$ [102,130]. As a recent structural study [131] has shown, Li^+-ions occupy octahedral cavities in the Al-O layers, and the anions are located between the layers of $[Li^+ \bullet Al(OH)_3]_n$. This reaction with a conservation of structural framework is, however, accompanied by a large change in molar volume ($\Delta V/V = 35\%$) and a large change in the interlayer spacing in the structure (from 9.7 Å to 14.8 Å). As a result, mechanical stresses are generated in the crystal, and the structure of $Al(OH)_3$ becomes distorted already after the first 'intercalation-deintercalation' cycle. This distortion can be clearly seen in a Laue diffraction pattern (specific diffuse "tailings" of the diffraction spots are observed) [102]. This structural distortion has important chemical consequences. The re-intercalation of LiCl into the strained structure in the second cycle is much quicker than in the first one, that is a positive feedback 'reaction-strain' arises (Fig. 13a) [102]. The pre-activation of the $Al(OH)_3$ host by the intercalation of LiCl also enables other guests, often quite large in size (such as $C_6H_5(CH=CH)COO^-$), to intercalate into the same matrix [132,133]. This was successfully used for the synthesis of various host-guest systems based on $Al(OH)_3$, with the practical applications of the materials ranging from nanoscale-composites and nanoparticles with controlled size to new efficient drugs, pigments, sensors and catalysts [132-135].

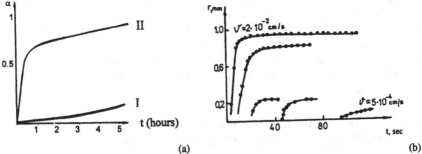

Figure 13. Examples of positive (a) and negative (b) feed-back in solid-state transformations.
(a) Kinetics of the intercalation of LiCl into $Al(OH)_3$ in two successive cycles [102];
(b) Successive growth of product nuclei during polymorphic transition in NH_4Cl [98].

5.7. AN EXAMPLE OF NEGATIVE FEEDBACK

A polymorphic transformation in NH_4Cl from the CsCl structural type to the NaCl structural type at 460K is also a non-reconstructive displacive transformation. It is accompanied by a large increase in the molar volume ($\Delta V/V = 19\%$). If one uses this value to estimate the 'internal pressure' that should be generated in the parent crystal by a growing nucleus of the new phase and then calculates the transition temperature corresponding to this pressure from the Clausius-Clapeyron equation, one gets an increase in temperature as large as 600K [98]. The phase transition should not be observed. It is observed, however, and the reason for this is the relaxation of the mechanical stresses decreasing the 'internal pressure' via plastic deformation of the parent crystal by punching and slip of dislocation semi-loops [98]. The strain around the growing nuclei could be visualized in the polarized light, and the punching and slip of the dislocation semi-loops was also followed experimentally using optical microscopy and etching techniques [98].

A feedback loop arises during the polymorphic transition in NH_4Cl: the formation of the product nucleus induces mechanical stresses in the parent crystal, the relaxation of these stresses results in punching of dislocation semi-loops near the growing nucleus and these semi-loops form a pile-up which gives a back-stress on the nucleus, thus inhibiting its further growth. The existence of this loop has important consequences for the transformation. Each nucleus starts growing faster than it continues. The growth of a nucleus stops when the nucleus reaches some final size. Each subsequent nucleus grows slower than the preceding ones did (Fig. 13b) [98].

The transformation could be controlled by modifying mechanical properties of NH_4Cl. This modification is achieved in practice by introducing impurities (doping), in the same way as mechanical properties of steel are modified in industry. Cu^{+2} can be used as such an impurity, since its concentration is easy to control by optical spectroscopy. Two NH_3 molecules instead of two NH_4^+-ions are formed per one introduced Cu^{+2} ion to compensate charge, and each Cu^{+2} ion forms a complex with two NH_3 molecules. This affects the hydrogen bond network in the crystal and may be one of the reasons for the change in the mechanical properties. The change in the mechanical properties with increasing concentration of the impurity was followed by direct experiments [98].

Introduction of small amounts of the Cu^{+2} impurity (≤ 0.13 mol%) affects both the spatial pattern of the transformation and the transformation temperature. Many small nuclei are formed during the transformation in the doped crystals, instead of a few large nuclei in pure NH_4Cl. The temperature of the polymorphic transformation increases with the increase in the concentration of the impurity. Special experiments were also carried out, in which only the surface layer of the NH_4Cl crystals was doped by Cu^{+2}. The transformation in crystals prepared in this manner started in the undoped bulk (and not at the surface, as would happen in an undoped crystal). The transformation temperature (as compared with a reference transition temperature in a pure crystal) was about 12K higher even for the undoped central part of the crystal, whereas for the surface layer doped with the impurity it was even 70K higher [98].

5.8. SOME CONSEQUENCES OF PRACTICAL IMPORTANCE

The polymorphic transition in NH_4Cl provides an example of the effect of the relaxation of mechanical stresses on the size and spatial distribution of the nuclei of the product phase.

Mechanical stresses arising during the intercalation of Li-salts into $Al(OH)_3$ account for the high surface area and catalytic properties of the de-intercalation product after several intercalation-deintercalation cycles. In general, the texture of the product, the size and the shape of the crystallites formed during solid-state reactions is often determined by the conditions of the generation and the relaxation of the mechanical stresses. This may be one of the reasons of the effect of the precursor choice and of the reaction conditions on the texture of the product. For example, it is to a large extent the strain field induced in the course of the dehydration of $Mg(OH)_2$ that determines the texture of MgO. The size effect mentioned previously in the section on topotactic transformations (the formation of different polymorphs of the product from the starting particles of different size) is also explained by various conditions of the relaxation of the mechanical stresses.

6. Conclusions

We have discussed briefly some of the structural aspects of solid-state reactions, employing as the basis of our discussion three different definitions of structure. Of course, it is rather artificial to consider the effect of the mutual juxtaposition of molecular fragments on the solid-state reaction separately from the role of extended networks of structural fragments in the same crystal or from the ability of the same structure to sustain a load. In reality, all three aspects of the effect of crystal structure on a solid-state reaction are closely interrelated.

When considering a solid-state reaction we may start with the simplest analysis of the mutual juxtaposition of the neighbouring species in the structure. However, then we must inevitably make the next step and proceed to a consideration of the whole structure as a network and take into account the generation and the relaxation of the mechanical stresses in the crystal. The structure is not something "fixed forever", it distorts as the reaction proceeds *via* the cooperative movements of the whole structural network. Feedback arises and must be taken into account when treating the effects observed [105,107].

In 1979 Dunitz wrote: *"Time to ponder the significance of the results is a luxury that few present-day crystallographers can afford. One might have hoped that the increased ease and rapidity of crystal structure analysis would have left more time for thinking, but the contrary seems to be true. The structural facts may be trying to tell us something, but we have no time to listen. Another long-overdue paper is waiting to be written"* [136]. I think that these words sound today no less true than in 1979. It is quite often the case that structural studies are not related to chemical processes in the same solids. I have shown examples of relatively old chemical studies, but the structural aspects of the same processes often remained unstudied until very recently. The structural data do have something to say not only about the *properties* of crystals and materials, but also on the *processes* leading to these crystals and these materials; they can contribute noticeably to our understanding of solid-state *reactivity*. We must just care to listen. If this chapter has shown that the problems of solid-state reactivity are worthy of careful and varied crystallographic studies, I shall consider my task fulfilled.

7. References

1. Boldyreva, E. V. (1993) *J. Chem. Educ.* **70**, N.7, 551-556.
2. Hüttig, G. (1943) in: G. M. Schwab (ed.), *Handbuch der Katalyse*, Springer Verlag, Wien, Bd.6, 322-357.
3. Boldyrev, V. V., Bulens, M., Delmon, B. (1979) *Control of the Reactivity of Solids*, Elsevier, Amsterdam.
4. Delmon, B. (1982) in Dyrek K, Haber J., Nowotny J. (eds.), *Reactivity of Solids*, Materials Science Monographs, 10, Vol.1, Elsevier, Amsterdam, 327-369.
5. Delmon, B. (1997) *Solid State Ionics* **101-103**, 655-660.
6. Gordon, J. E. (1991) *Structures, or Why Things Don't Fall Down*. Penguin Books, London.
7. Muller, A. (1886) *Bull. Soc. Chim., Paris* **45**, 438-440.
8. Hückel, W. (1934) *Theoretische Grundlagen der Organischen Chemie*, Leipzig.
9. Rice, F. and Teller, E. (1938) *J. Chem. Phys.* **6**, 489-496.
10. Hoffmann, R., Minkin, V. I. and Carpenter, B. K. (1996) *Bull. Soc. Chim. Fr., Paris* **133**, 117-130.
11. Kohlschütter, V. (1919) *Z. Anorg. Allgem. Chemie* **105**, 1-25.
12. Kohlschütter, V. (1927) *Kolloid-Zeitschrift* **42**, 254-268.
13. Kohlschütter, V. (1929) *Helv. chim. Acta* **12**, 512-529.
14. Feitknecht, W. (1934) *Kolloid Zeitschrift* **68**(2), 128-132.
15. Boudart, M., Delbouille, Derouane, E. G., Indovina, V. and Walters, A. B. (1972) *J. Am. Chem. Soc.* **94**, 6622-6630.
16. Boudart, M., Delbouille, A., Dumesic, J. A., Khammouma, S. and Topsoe, H. (1975) *J. Catal.* **37**, 486-502.
17. Volpe, L., Boudart, M. (1985) *Catal. Rev. Sci. Eng.* **27**, N.4, 515-538.
18. Thoms, H., Epple, M. and Reller, A. (1997) *Solid State Ionics* **101-103**, 79-84.
19. Jost, H., Braun, M. and Carius, Ch. (1997) *Solid State Ionics* **101-103**, 221-228.
20. Bond, G., Molloy, K. C. and Stone, F. S. (1997) *Solid State Ionics* **101-103**, 697 - 705.
21. Freund F, Gieseke, W. and Nägerl, H. (1975) in P. Barret (ed.), *Reaction kinetics in heterogeneous chemical system*, Elsevier, 258-277.
22. Brindley, G. W. (1963) In: *Progress in Ceramic Science*, V. 3, 1-55.
23. Schmidt, G. (1967) *Photochemistry of the Solid State*, Interscience, New York.
24. Cohen, M. (1975) *Angew. Chem. Int. Ed. Engl.* **14**, 386-393.
25. Cohen, M. (1979) *Mol. Cryst. Liq. Cryst.* **50**, 1-10.
26. Cohen, M. (1987) *Tetrahedron* **43**, 1211-1225.
27. Cohen, M.D. and Green, B. S. (1973) *Chem. Britain*, **9**, 490-497.
28. Enkelmann, V., Wegner, G., Novak, K. and Wagener, K. B. (1993) *J. Amer. Chem. Soc.* **115**, 10390-10391.
29. Enkelmann, V., Wegner, G., Novak, K. and Wagener, K. B. (1994) *Mol. Cryst. Liq. Cryst. Inc. Nonlin. Opt.* **242**, 121-127.
30. Novak, K., Enkelmann, V., Wegner, G. and Wagener, K. B. (1993) *Angew. Chem. Int. Ed. Engl.* **32**(11), 1614-1616.
31. Novak, K., Enkelmann, V., Köhler, W. Wegner, G. and Wagener, K. B. (1994) *Mol. Cryst. Liq. Cryst. Inc. Nonlin. Opt.* **242**, 1-8.
32. Boldyrev, V. V., Eroshkin, V. I., Pis'menko, O. T., Ryzhak, I. A., Medvinskii, A. A., Shmidt, I. V., Kefeli, L. M. (1968) *Kin. Katal. (Russian Journal of Kinetics and Catalysis)* **9**, N.2, 260-268.
33. Naumov, D. Yu., Virovets, A. V., Podberezskaya, N. V. and Boldyreva, E. V. (1995) *Acta Cryst.* **C51**, 60-62.
34. Naumov, D. Yu., Podberezskaya, N. V., Virovets, A. V. and Boldyreva, E. V. (1994) *Russ. J. Struct. Chem.* **35**, N.6, 158-168.
35. Naumov, D. Yu., Boldyreva, E. V., Podberezskaya, N. V. and Howard, J. A. K. (1997) *Solid State Ionics* **101-103**, 1315-1320.
36. Thomas, J. M. (1974) *Phil. Trans. Roy. Soc.* **277**, 251-286.
37. Carrido, J. (1951) *Am. Mineral.* **36**, 773-776.
38. Lotgering, F. K. (1959) *J. Inorg. Nucl. Chem.* **9**, 113-123.
39. Lotgering, F. K. (1960) *J. Inorg. Nucl. Chem.* **16**, 100-108.
40. Lotgering, F. K. (1961) in J.H. de Boer (ed.), *Reactivity of Solids*, Proceed. 4th Intern. Symp. Reactivity of Solids, (Amsterdam, 1960), Elsevier: Amsterdam, 584-586.

172

41. Goodman, J. F. (1958) *Proc. R. Soc. London* **A247**, 346-352.
42. Niepce, J. C., Watelle, G. and Brett, N. H. (1978) *J. Chem. Soc. Faraday Trans. I.* **74**, 1530-1537.
43. Figlarz, M. (1992) A Plenary Lecture at the 12th Int. Symp. React. Solids, Madrid.
44. Figlarz, M. (1995) in K. J. Rao (ed.), *Perspectives in Solid State Chemistry,* Naroda, New Dehli, 1-21.
45. Ball, M. C. and Taylor, H. F. W. (1961) *Mineral. Mag.* **32**, 754-766.
46. Oswald, H. R. and Günter J. R. (1977) In: *Crystal |Growth and Materials,* Ed. by Kladis E. and Scheel H. J., North-Holland Publishing Company, 416-433.
47. Bernal, J. D. (1960) *Schweiz. Arch.* **26**, 69-75.
48. Figlarz, M., Gerand, B., Delahaye-Vidal, A., Dumont, B., Harb, F., Coucou, A. and Fievet, F. (1990) *Solid State Ionics* **43**, 143-70.
49. Ostwald, W. (1897) *Z. Phys. Chem.,* **22**, 289-330.
50. Westrik, R. and Zwietering, P. (1953) *Proc. K. Ned. Akad. Wet.* **B56**, 492-497.
51. Pluschkell W. and Sarma B. V. S. (1973) *Arch. Eisenhüttenwes.,* **44**, N. 3, 161-166.
52. Fricke, R. and Lücke, J. (1935) *Z. Elektrochem.* **41**, 174-183.
53. Niepce, J. C., Mesnier, M. and Louer, Th. (1977) *J. of Solid State Chem.,* **2**, 341-351.
54. Egger, K. and Feitknecht, W. (1962) *Helv. Chim. Acta* **45**, 2042-2057
55. Feitknecht, W. (1964) *Pure applied Chemistry* **9**, 423-440.
56. Giovanoli, R. and Leuenberger, U. (1969) *Helv. Chim. Acta* **52**(8), 2333-2347.
57. Mackay, A. L. (1961) in J.H. de Boer (ed.), *Reactivity of Solids,* Proceed. 4th Intern. Symp. Reactivity of Solids, (Amsterdam, 1960), Elsevier: Amsterdam, 571-583.
58. Burzlaff, H. and Rothammel, W. (1992) *Acta Cryst.* **A48**, 483-490.
59. Dzyabchenko, A. V. (1994) *Acta Cryst.* **B50**, 414-425.
60. Desiraju, G. R. (1995) Angew. Chem. **107**, 2541-2558; *Angew. Chem. Int. Ed. Engl.* **34**(21), 2311-27.
61. Desiraju, G. R. (Ed.) (1996a) *The Crystal as a Supramolecular Entity,* Perspectives in Supramolecular Chemistry, Vol.2, Wiley, New York.
62. Desiraju, G. R. (1996b) *J. Mol. Struct.* **374**, 191-8.
63. Desiraju, G. R. (1997a) *Chem. Com.* N.16, 1475-1482.
64. Zorky, P. M. and Kuleshova, L. N. (1981) *Russian Journal of Structural Chemistry* **22**(6), 153-156.
65. Kuleshova, L. N. (1982) *Crystal chemistry of organic solids with hydrogen bonds,* Ph. D. Thesis, Moscow State University.
66. Etter, M. C. (1990) *Acc. Chem. Res.* **23**, 120-126.
67. Bernstein, J., Etter, M. C. and Leiserovitz, L. (1994) in H.-B. Bürgi, J. D. Dunitz (eds.), *Structure Correlation,* VCH, Weinheim, V.2, Chapter 11, 431-507.
68. Videnova-Adrabinska, V. (1996) *J. Mol. Struct.* **374**, 199-222.
69. Zorky, P. M. (1996) *J. Mol. Struct.* **374**, 9-28.
70. Belsky, V. K., Zorkaya, O. N. and Zorky, P. M. (1995) *Acta Cryst.* **A51**(4), 473-81.
71. Günter J. R. and H. R. Oswald (1975) *Bull. Inst. Chem. Res. Kyoto Univ.* **53**, 249-255.
72. Oswald, H. and Reller, A. (1986) *Ber. Bunsenges. Phys. Chem.* **90**, 671-676.
73. Oswald, H. and Reller, A. (1989) *Pure Appl. Chem.* **61**, N.8, 1323-1330.
74. Stoch, L. (1989) *Thermochim. Acta* **148**, 149-164.
75. Gerand, B., Nowogrocki, G., Guenot, J. and Figlarz, M. (1979) *J. Solid State Chem.* **29**, N.3, 429-434.
76. Gerand, B., Nowogrocki, G. and Figlarz, M. (1981) *J. Solid State Chem.* **38**, N.3, 312-320.
77. Figlarz, M. (1988) *Chem. Scr.* **28**(1), 3-7.
78. Gomez-Romero (1997) *Solid State Ionics* **101-103**, 23-248.
79. Najbar, M and Camra, J. (1997) *Solid State Ionics* **101-103**, 707-711.
80. Masciocchi, N., Kolyshev, N., Dulepov, V., Boldyreva, E. and Sironi, E. (1994) *Inorg. Chem.* **33**, 2579-2585.
81. Weng, J., Larsen, E. M., Holt, J. B., Waide, P. A., Rupp, B. and Frahm, R. (1990) *Science* **249**, 1406-1409.
82. Ohashi, Y. (ed.) (1993) *Reactivity in Molecular Crystals,* VCH, Tokyo.
83. Christensen, A. N., Norby, P. and Hanson, J. C. (1997) *Acta Chem. Scand.* **51**(3), 249-258.
84. Muller, H., Svensson, S. O., Birch, J. and Kvick, A. (1997) *Inorg. Chem.* **36**(7), 1488-1494.
85. Nielsen, K., Ståhl, K., Hanson, J. C., Norby, P., Jiang, J. Z. and van Lanschot, J. (1997) *Proceed. ECM-17, Lisboa, Portugal*, 163.
86. Norby, P. (1997) *J. Am. Chem. Soc.* **119**(22), 5215-5221.
87. Svensson, S. O., Birch, J., Muller, H. and Kvick, A. (1997) *J. Synchr. Rad.* **4**, 83-94.

88. Boldyreva, E. V., Sidel'nikov, A. A., Chupakhin, A. P., Lyakhov, N. Z. and Boldyrev, V. V. (1984) *Dokl. Akad. Nauk SSSR* **277**, 893-896.

89. Galwey, A. K. and Mohamed, M. A. (1987) *Thermochim. Acta* **121**, 97-107.

90. Galwey, A. K. and Laverty, G. M. (1991) *Siberian Journal of Chemistry* N.1, 51-59.

91. Galwey, A. K., Laverty, G. M., Baranov, N. A. and Okhotnikov, V. B. (1994) *Phil. Trans. Roy. Soc. London* **A347**, 139-156, 157-184.

92. Sidel'nikov, A. A., Mitrofanova, R. P. and Boldyrev, V. V. (1994) *Thermochem. Acta* **234**, 269-274.

93. Morrison, J. A. and Nakayama, K. (1963) *Trans. Faraday Soc.* **59**, 2560-2568.

94. Boldyrev, V. V. (1973) *Russ. Chem. Rev. (Uspekhi Khimii)* **42**, 1161-1183.

95. Baughman, R. H. (1974) *J. Polymer Sci. Part A-2, Polymer Phys.* **12**, 1511-1535.

96. Raevsky, A. V. (1981) in G. Manelis (ed.) *Mechanism of Thermal Decomposition of Ammonium Perchlorate*, Institute of Chemical Physics, Chernogolovka, 30-100.

97. McBride, M. (1983) *Acc. Chem. Res.* **16**, 304-312.

98. Sidel'nikov, A. A., Chupakhin, A. P. and Boldyrev, V. V. (1985) *Izv. Sib. Otd. Akad. Nauk SSSR, Ser. Khim. Nauk*, N.17, Issue 6, 39-49.

99. McBride, J. M., Segmuller, B., Hollingsworth, M., Mills, D. and Weber, B. (1986) *Science* **234**, 830-835.

100. Boldyreva, E. V. and Sidel'nikov, A. A. (1987) *Izv. Sib. Otd. Akad. Nauk SSSR Ser. Khim Nauk* **5**, 139-145.

101. Chupakhin, A. P., Sidel'nikov, A. A. and Boldyrev, V. V. (1987) *React. Solids* **3**, 1-19.

102. Nemudry, A. P. (1987) *React. Solids* **3**, 317-327.

103. Ramamurthy, V. and Venkatesan, K. (1987) *Chem. Rev.* **87**(2), 433-481.

104. Hollingsworth, M. D. and McBride, J. M. (1988) *Mol. Cryst. Liq. Cryst. Inc. Nonlin. Optics* **161**, 25-41.

105. Boldyreva, E. V. (1990) *React. Solids* **8**, 269-282.

106. Hollingsworth, M. D. and McBride, J. M. (1990) In: *Advances in Photochemistry* (Ed. D. H. Volman, G. S. Hammond and K. Gollnick), **V.15**, Wiley, New York, 279-379.

107. Boldyreva, E. V. (1992) *J. Thermal Anal.* **38**, 89-97.

108. Boldyreva, E. V. (1996) in V. V. Boldyrev (ed.), *Reactivity of Solids. Past, Present, Future*, Blackwells Science Publ., Chapter 7, 141-184.

109. Boldyreva, E. V. (1997) *Solid State Ionics* **101-103**, 843-849.

110. Boldyreva, E. V., Sidel'nikov, A. A., Rukosuev, N. I., Chupakhin, A. P. and Lyakhov, N. Z. (1988) *Photometer*. Russian patent N.1368654 A1.

111. Köhler, W., Novak, K. and Enkelmann, V. (1994) *J. Chem. Phys.* **101**, N.12, 10474-10480.

112. Williams, J. O. and Thomas, J. M. (1967) *Trans. Faraday Soc.* **63**, 1720-1729.

113. Cohen, M., Ludmer, Z., Thomas, J. M. and Williams, J. O. (1971) *Proc. R. Soc. London* **324A**, 459-468.

114. Thomas, J. M. (1972) *Isr. J. Chem.* **10**, 563-580.

115. Desvergne, J. P., Bouas-Laurent, H., Lapouyade, R., Gaultier, J., Hauw, C., Dupuy, F. (1972) *Mol. Cryst. Liq. Cryst.* **19**, 63-85.

116. Desvergne, J. P., Thomas, J. M., Williams, J. O. and Bouas-Laurent, H. (1974) *J. Chem. Soc. Perkin Trans.* **2**, 362-368.

117. Thomas, J. M., Williams, J. O., Desvergne, J. P., Guarini, G. and Bouas-Laurent, H. (1975) *J. Chem. Soc. Perkin Trans.* **2**, 84-88.

118. Ramdas, S. and Thomas, J. M. and Goringe, M. J. (1977) *J. Chem. Soc. Faraday Trans.* **2**, **73**, 551-561.

119. Politov, A. A., Fursenko, B. A., Prosanov, I. Yu. and Boldyrev, V. V. (1994) *Int. J. Mechanochem. Mech. Alloying*, N.3, 172-176.

120. Bridgman, P. W. (1935) *Phys. Rev.* **48**, 825.

121. Bridgman, P. W. (1937) *Proc. Am. Acad. Arts Sci.* **71**, 387-460.

122. Dachille and Roy (1961) *Proceed. 4th Intern. Symp. React. Solids*, Elsevier, Amsterdam, 502-511.

123. Zharov, A. A. (1994) in A. L. Kovarskii (ed.) *High Pressure Chemistry and Physics of Polymers*, CRC Press, Roca Raton, FL, Chapter 7.

124. Gilman, J. I. (1996) *Science* **274**, 65.

125. Boldyreva, E. V., Virovets, A. V., Podberezskaya, N. V., Burleva, L. P. and Dulepov, V. E. (1993) *Russian Journal of Structural Chemistry, Russian Edition* **34**, N. 4, 128 - 140.

126. Boldyreva, E. V., Ahsbahs, H. and Uchtmann, H. (1994) *Ber. Bunsenges. Phys. Chem.* **98**(5), 738-745.

127. Boldyreva, E. V. (1994) *Mol. Cryst. Liq. Cryst. Inc. Nonlin. Opt.* **242**(2), 17-52.

174

128. Boldyreva, E. V., Kuz'mina, S. L. and Ahsbahs, H. (1998) *Russ. J. Struct. Chem.*, 343-349; 350-361.
129. Geguzin, Ya. E. (1979) *Diffusion zone*, Nauka, Moscow.
130. Nemudry, A. P., Isupov, V. P., Kotsupalo, N. P. and Boldyrev, V. V. (1986) *React. Solids* 1, 221-7.
131. Besserguenev, A. V., Fogg, A. M., Francis, R. J., Price, S. J., O'Hare, D., Isupov, V. P. and Tolochko, B. P. (1997) *Chem. Materials* 9, 241-247.
132. Isupov, V. P., Mitrofanova, R. P., Chupakhina, L. E. and Rogachev, A. Y. (1996) *Chem. Sustainable Development* 4, 213-217.
133. Isupov, V. P., Chupakhina, Mitrofanova, R. P., Tarasov, K. A., Rogachev, A. Yu. and Boldyrev, V. V. (1997) *Solid State Ionics* **101-103**, 265-270.
134. Uvarov, N. F., Isupov, V. P., Sharma, V. and Shukla, K. (1992) *Solid State Ionics* **51**, 41-52.
135. Uvarov, N. F., Bokhonov, B. B., Isupov, E. F. and Khairetdinov, E. F. (1994) *Solid State Ionics* **74**, 15-27.
136. Dunitz, J. D. (1979) *X-ray Analysis and the Structure of Organic Molecules*, Cornell University Press: Ithaca and London, 302-303.

WEAK INTERACTIONS IN MOLECULAR CRYSTALS

JACK D. DUNITZ
Organic Chemistry Laboratory,
Swiss Federal Institute of Technology,
ETH-Zentrum,
CH-8092 Zurich, Switzerland.

1. Weak Interactions

The earth's crust may be held together mainly by ionic forces, molecules by covalent bonds, but it is weak intermolecular interactions that hold us, along with the rest of the organic world, together. The hydrogen bond is the best known example, and hydrogen bonding has fufilled Pauling's sixty-year old expectation that its "significance for physiology would turn out to be greater than that of any other structural feature" [1]. Indeed, it has emerged as the most important organizing principle not only for the structures and properties of biologically important molecules but also for crystal engineering. The hydrogen bond can be classified as a weak interaction because its stabilization energy, typically 5 to 10 kcal mol^{-1}, is much less than that of a typical covalent bond (~100 kcal mol^{-1}), but it really occupies an intermediate position in the hierarchy of strong and weak bonding. It is weaker than a covalent bond but stronger than the other interactions we shall consider here, which fall mostly into the class of what may loosely be described as van der Waals or dispersion attractions (the net attractive interactions between the permanent and induced charge distributions of different closed-shell electrically neutral molecules), plus the Pauli repulsions that occur when closed shell orbitals overlap.

1.1. MOLECULAR PACKING

One of our main sources of knowledge about intermolecular "nonbonded" interactions is the systematic study of molecular packing patterns in crystals. Conversely, such patterns ask to be interpreted in terms of empirical rules, the simpler the better. Pauling's set of van der Waals radii [1], designed to reproduce contact distances between atoms in crystals, was one of the early successes of this kind of approach. Kitaigorodskii's close packing principle [2] was another. Since those early days, however, we do not seem to have come much further in

J.A.K. Howard et al. (eds.),
Implications of Molecular and Materials Structure for New Technologies, 175–184.
© 1999 *Kluwer Academic Publishers. Printed in the Netherlands.*

conceptual understanding or predictive capability. It is true that by quantum mechanical methods we can, in principle, calculate the energy of any given arrangement of molecules, but in practice it is all much too complicated, and the details elude us. A basic problem in our study of intermolecular interactions is that we do not really know what to look for. How should we classify the complex molecular arrangements found in crystals? According to symmetry? But long range periodicity is produced by directionally specific short-range interactions, nothing more. According to the geometric details of the crystal packing of common functional groups? How are we to reconcile these two points of view? In any case, it is clear that one cannot learn very much from the packing in any one particular crystal. One has to look at many structures to identify the recurring patterns, to notice exceptions from these patterns, and to enquire about the reasons for these exceptions.

What we have gained in the meantime to satisfy these needs is the Cambridge Structural Database (CSD), developed over the years as the main product of the Cambridge Crystallographic Data Centre (CCDC), which needs no introduction to this audience. Whatever its original objectives may have been, the CSD has developed into a scientific instrument that is quite indispensable for the study of intermolecular interactions. In its early days, following the trend of most organic chemical crystallographers, it was mainly concerned with the checking of published bond distances and angles. The principal interest then was more in the elucidation of molecular structures than in studying details of interactions between molecules. In recent years, however, stimulated by the recognition of the possibilities of crystal engineering — of designing crystal structures with particular features and desired properties — and, in general, by the emergence of supramolecular chemistry — the chemistry of non-covalent bonding — the focus of interest has shifted towards the analysis of intermolecular interactions. The CSD now provides computerized access to about 175,000 organic and organometallic crystal structures and makes it possible to scrutinize the packing in selected structures, to obtain statistics for any desired geometric parameters in selected groups of structures, and, in general, to obtain information about molecular packing arrangements in a relatively swift and convenient way. All the user has to do is to ask the right questions.

2. Atom-atom Potentials

Nowadays, much money, time, and effort are being invested in computer modeling, mostly using atom-atom pair potentials, as pioneered by Kitaigorodskii [2]. Various types of potential are in use, some based on attempts to imitate the underlying physical interactions (electrostatic terms, induction, dispersion, exchange repulsion, charge transfer, etc.), others on parametrization of standard potential types, such as $E(r) = A\exp(-Br) - Cr^{-6}$ or $E(r) = Ar^{-12} - Br^{-6}$, where A, B, C are simply treated as empirical constants for each type of atom-atom interaction and r is the internuclear distance. Coulombic terms may also be added, if regarded as necessary. Since there is no rigorous way of factorizing the total interaction energy into separate terms, and since we are interested in functions that are readily transferable from one system to another and are characteristic of the atom-atom pairs, regardless of their chemical environment, we admit to a preference for the empirical approach. Gavezzotti and Filippini have fitted "exp-6" potentials to observed geometries and heats of sublimation, where available, to produce reasonably satisfactory agreement for a data set comprising several hundred organic crystal structures [3]. A selection of these pair potentials is listed in Table 1.

TABLE 1 Atom-atom Potential Parameters: $E = A \exp(-BR_{ij}) - CR_{ij}^{-6}$

interaction	A	B	C
H...H	5774	4.01	26
H...C	28870	4.10	113
H...O	70610	4.82	105
C...C	54050	3.47	578
N...N	87300	3.65	691
O...O	46680	3.74	319
F...F	40850	4.22	135
S...S	259960	3.52	2571
Cl...Cl	140050	3.52	1385
OH...O	4509750	7.78	298

A in kcal mol^{-1} , B in Å^{-1}, C in kcal mol^{-1} Å^6 , R_{ij} in Å

With the help of these or other sets of atom-atom potentials, energy minimization ($\partial E = 0$) methods are applied to test trial structures for mechanical stability; a lattice dynamics calculation requires $\partial^2 E$ values as well. Increasingly, molecular dynamics simulations and Monte Carlo methods are also being applied. Thus, today's visitor to a typical organic chemistry laboratory will see not only people working at the bench but also people huddled over computer keyboards, peering at vivid molecular graphics displays on a viewing screen. The molecules on the screen are often so "real" that one can easily forget that one is dealing with models, not with reality. Indeed, for many users, the quality of the computer graphics and the ease of implementation of the programs are far more important than the quality of the potential functions or the meaningfulness of what is portrayed.

It must be emphasized that the potentials given in Table 1, like most others, are derived by fitting many variables to a large amount of experimental data and can be regarded at best as average or "typical values. Depending on details of the molecular structures involved, they may be of better or worse applicability. Nevertheless, some general features of such potential curves can be discussed with reference to the two extreme examples shown in Figure 1. The OH...O potential corresponds to a relatively strong hydrogen-bonded interaction as in an alcohol with $E(min) \sim -5$ kcal mol^{-1}, the (C)H...O potential to a much weaker interaction with $E(min) \sim -0.1$ kcal mol^{-1}. Note that the deeper the energy minimum, the sharper the curvature at the turning point. Indeed, for almost any "reasonable" set of atom-atom potentials, the force constant is linearly proportional to the depth of the potential well. For a shallow well, the force constant is therefore smaller, the vibrational levels are more closely spaced, and the vibrational entropy is larger than for a deep well. While the contribution of the vibrational entropy to the overall stabilization (free enthalpy) can be more or less neglected for strong covalent bonds, it becomes increasingly important as bonding interactions becomes weaker. This is one of the main factors behind the phenomenon of entropy-energy compensation [4]. In the limit, for a completely flat potential corresponding to zero bonding energy, the vibrational motion goes over into free translation, with appropriate change from vibrational to translational entropy.

Note also that the anharmonicity of the potential curve is such that the cubic term in a power series expansion about the minimum is always negative: that is, when the interatomic distance is increased, the energy rises less sharply from its minimum value than when it is decreased. Just as with covalent and other types of bonding, it is always easier to stretch a bond than to compress it. There is a difference, however. Stretching a C-C covalent bond by 0.2 Å would be an energetically expensive process, costing about 60 kcal mol^{-1}. A corresponding increase in a hydrogen bond distance (Figure 1) only costs only about 1 kcal mol^{-1}, and, for the weak H...O interaction, an increase of 0.2 Å in the interatomic distance makes practically no difference to the energy. For weak interactions, the stabilization energy may be small, but the field extends over quite large distances.

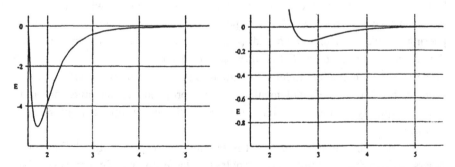

Figure 1. Potential curves for a typical hydrogen-bond interaction (left) and for a weak H...H interaction (right); vertical axis, energy (E) in kcal mol^{-1}, horizontal axis, interatomic distance in Å.

Atom-atom potentials used by different authors are derived from different sources and aimed at different targets. We may demand of them at least that they should yield zero or only small restoring forces when applied to known crystal structures, as well as "reasonable" lattice energies. What are "reasonable" lattice energies? Ideally, these quantities should be scaled to sublimation enthalpies at 0 K, but, since estimates of these quantities from experimental data are hard to come by and available only for a handful of substances, they are more usually scaled to sublimation enthalpies at 300 K. The experimental values can be reproduced to within a few kcal mol^{-1}, a range that corresponds roughly to the experimental uncertainties. There are problems about how to handle the heat capacity and the resulting enthalpic and entropic contributions to the free enthalpy, problems that are often dealt with simply by ignoring them, so that no conclusions about the temperature dependence of the free enthalpy can be drawn. As far as this type of calculation is concerned, phase transitions do not exist.

2.1. NON-BONDED CONTACT DISTANCES

Nevertheless, atom-atom potentials used by different authors do resemble one another in their general characteristics. Some remarks about the significance of the equilibrium distance R(equ) seem to be called for. This distance, where attractive and repulsive forces exactly balance for any given atom pair, is NOT the same as the van der Waals distance, defined as the sum of the van der Waals radii of the atoms concerned. Indeed, there appears to be no general agreement about the exact meaning of the so-called van der Waals radii, apart from their standing as a basis for estimating contact distances between atoms in crystals [1]. (Note that Bondi's set of radii [5] were actually selected for the calculation of molecular volumes and may not always be the most suitable for the calculation of contact distances in crystals.) The additivity assumption would imply that atoms in crystals are regarded as hard, incompressible spheres with characteristic radii, and it is obvious that an atom-atom potential curve for the mutual interaction of such spheres would not at all resemble the curves shown in Figure 1; it would be perfectly flat along the distance axis until the assumed van der Waals radius, when it would show a sharp discontinuity to infinite slope. In any case, it cannot be expected that van der Waals distances, intended to reproduce contact distances in a crystal, should correspond to equilibrium distances R(equ) of individual atom-atom potential curves. In a crystal it is not a matter of balancing forces from individual atom-atom interactions; the force on a given atom does not arise solely from its interaction with its closest non-bonded neighbour but is the resultant of forces arising from its interactions with many other atoms. Most of these will be attractive, for attractive forces extend over longer distances, while repulsive ones operate only at short distances and are therefore limited to interactions between pairs of atoms on the molecular peripheries — it is only those that come into actual contact with atoms in other molecules. For each such contact pair interaction, there are many non-contact pairs at longer distances in the intermediate attractive range. If the crystal is to be at equilibrium, the repulsive forces between contact pairs must be balanced by the attractive intermolecular forces between all pairs of atoms that are not in contact. Thus, insofar as the pair-pair potential model is applicable, contact distances between atoms in crystals must be shorter than R(equ) of the corresponding atom-atom potential curve. Table 2 shows that, for a number of common types, this is indeed the case.

TABLE 2 Atom-atom potentials and atomic radii

type	depth[a]	R(equ)[b]	R(0)[b]	R(vdW)[b]
H...H	0.010	3.36	2.98	2.4
H...C	0.049	3.29	2.92	2.9
H...O	0.121	2.80	2.48	2.6
C...C	0.093	3.89	3.45	3.4
N...N	0.150	3.70	3.28	3.0
O...O	0.080	3.61	3.20	2.8
F...F	0.070	3.20	2.84	2.7
S...S	0.445	3.83	3.39	3.7
Cl..Cl	0.240	3.83	3.39	3.6

a kcal mol^{-1} b Å

We can go further. Everyone knows that as molecules become larger, the corresponding crystals tend to have lower vapor pressures and higher melting points, and the liquids tend to have higher boiling points. The cohesive energy between the molecules tends to increase. But the greater the mutual intermolecular attractive forces, the greater must also be the repulsive forces necessary to preserve equilibrium in the crystal. So, in the crystalline state, as molecules become larger, the outer atoms in contact with other molecules will tend to be subject to stronger repulsive forces to maintain the balance in the equilibrium structure. This implies that, other things being equal, atom-atom contact distances should tend to be shorter in crystals of large molecules than in crystals of small ones. Of course, other things seldom are equal! For molecules, the balance of forces on each atom also involves the stiffness of covalent bonds and valency angles. For a rigid molecule in a crystal there is no net force or torque on the molecule as a whole.

These conclusions seem to be confirmed by inspection of actual crystal structures [6]. In crystals of aromatic hydrocarbons, the shortest intermolecular H...H (and H...C) contact distances tend to decrease with molecular size; the stronger repulsions are required to balance the increased net attractions. For the three largest, kekulene, $C_{46}H_{24}$, dibenzonaphthopyranthen), $C_{42}H_{20}$, and quaterrylene, $C_{40}H_{20}$, the shortest intermolecular H...H distances are 2.11, 2.15 and 2.16 Å, respectively, far less then the van der Waals diameter of around 2.5 Å and among the shortest such distances known; for comparison, the shortest intermolecular H...H distance in crystalline benzene is about 2.60 Å. Thus, a short intermolecular contact is not necessarily an indication of a strong specific attraction between the atoms concerned but may be merely an expression of strong overall attraction between the molecules. As far as the forces are concerned, a short intermolecular H...H contact is always repulsive.

3. Attractions and Repulsions.

As suggested by the previous discussion, we need to be careful in distinguishing between attraction meaning attractive force and attraction meaning stabilizing energy contribution. I propose that we use the term *attraction* to refer to the force and the term *stabilization* to refer to the energy. Similarly, we distinguish between *repulsion* (force) and *destabilization* (energy). In interactions involving contacts between pairs of atoms on the molecular peripheries of different molecules, the energy associated with a given interaction may be stabilizing or slightly destabilizing while the force acting between these atoms will invariably be repulsive. In terms of the atom-atom potential curve, we are climbing a little from the energy minimum up the repulsive slope, which is usually very steep, so we can not climb very far. Table 2 shows the interatomic distances R(0) at which the interaction energy is zero. Longer distances correspond to stabilization, shorter ones to destabilization. For the deeper atom-atom potentials (e.g., Cl...Cl), the van der Waals distance R(vdW), although shorter than R(equ), is longer than R(0), i.e., the normal contact distance corresponds to a repulsive force but to an energy stabilization. For the shallow ones (e.g., H...H), R(vdW) is shorter than both R(equ) and R(0); here the normal contact distance is repulsive and destabilizing.

The crystal structure represents not only a balance between attractive and repulsive forces (mechanical equilibrium) but also a balance between enthalpic and entropic contributions to the free energy. At low temperature, enthalpy wins, at sufficiently high temperature the entropic contribution dominates; at a transition temperature between different phases we have thermodynamical equilibrium, $\Delta H = T\Delta S$. In polymorphic systems it is usually the densest (most tightly packed) polymorph that is thermodynamically stable at low temperature. This underlies the prime importance of close packing, the avoidance of free space.

A rough indicator of the amount of free space is the packing coefficient k, the ratio of the molecular volume V_{mol} to the volume available in the particular condensed phase [2]. For crystals $k = ZV_{mol}/V_{cell}$. For cubic close packing of spheres $k = 0.74$, for organic crystals k lies in the range 0.65 to 0.80. Organic compounds expand on melting, typically by 5 to 10%, just enough to enable the molecules to slide past one another to change their relative positions. Only a fraction of the lattice energy is lost on melting. Despite the loss of long-range order, the attractions that hold the molecules together in the solid are still present to a large extent in the liquid. Around $k < 0.5$ compounds become gaseous.

182

4. Crystal Structure Prediction and Design

To a good approximation, the energy stabilization due to intermolecular attractions is already taken care of when we put the molecules into a box with the correct volume to give the correct density. Such an arrangement is unlikely to be optimal. It will usually have some regions where there is too much empty space and other regions where the molecules are too tightly compressed together. Both situations are energetically unfavorable. The total interaction energy of the starting arrangement can then be lowered by adjusting the molecular positions and orientations in such a way as to fill the available space as uniformly as possible, that is, essentially, by avoiding bumps against bumps and hollows against hollows, in favour of bumps against hollows — following Pauling's principle of complementariness [7] or Kitaigorodskii's principle of close packing [2]. However, there may be several almost equi-energetic ways to achieve this. Thus the frequent occurrence of polymorphism may come as no surprise.

The strength of "bump into hollow" as a guiding principle can be illustrated by an example involving space group statistics. In a 1994 scan through the Cambridge Crystallographic Database, we counted 190 structures in the enantiomorphic pair of space groups $P3_121$ and $P3_221$, and only 1 in the pair $P3_112$ and $P3_212$ [8]. The two pairs have the same symmetry elements, threefold screw axes and perpendicular twofold axes, so the reason for the difference in popularity may not be immediately obvious. In the former pair, the twofold axes are parallel to the translations, in the latter pair they are perpendicular to them. The corresponding plane groups of the projected patterns are $P31m$ and $P3m1$. Two-dimensional (projected) packing patterns produced by reproduction of an arbitrary shape according to the symmetry operations of these plane groups are shown in Figure 2. For the preferred pair of space groups, with the twofold axes parallel to the translations, we have bump into hollow; for the other pair, with twofold axes perpendicular, we have bump into bump, and hollow into hollow.

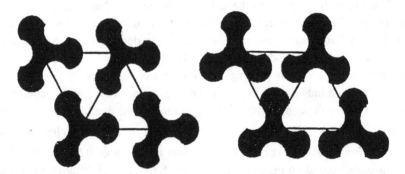

Figure 2. An arbitrary shape repeated in the plane groups $P31m$ (projection of $P321$) and $P3m1$ (projection of $P312$). The former packing has bump against hollow, the latter has bump against bump and hollow against hollow.

While it is not too difficult to predict possible crystal structures for a given molecule and it is sometimes possible even to hit on the correct one, a reliable prediction of the structure(s) actually adopted still seems to belong to the realm of wishful thinking. In fact, even if we could calculate the lattice energies of possible structures with confidence, it is by no means sure that the thermodynamically stable structure will actually be formed, for the crystallization process is under kinetic control and therefore does not necessarily lead to the thermodynamically stable product.

As an example, take the crystal structure of benzene. The orthorhombic structure in space group *Pbca* is approximately cubic close packed, and has provided material for many authors to discourse on the reasons for its stability. There is also a high-pressure (25 kbar) monoclinic $P2_1/c$ structure, approximately hexagonal close packed, the first member of the isostructural series of the linear condensed aromatic hydrocarbons (naphthalene, anthracene, etc.). On the computational side, there have been several papers during the last decade or so describing results of energy minimization of trial structures generated by various methods. They lead not to a single structure that is significantly more stable than any competitor but rather to a group of 10 to 20 structures that lie within a range of only 2 - 3 kcal mol^{-1}.[9, 10] The experimental *Pbca* structure is always represented in the group, but it is not always the one of lowest calculated energy. In a comparable study of monosaccharide structures, the authors comment: "...we are impressed by the astonishingly large number of structures within a few kcal/mol. This is equivalent to predicting a large tendency towards polymorphism..." [11]. Or it could be an indication that other factors, besides thermodynamic ones, are involved in the crystallization process.

The role of specific structure-directing interactions in crystal design has received considerable attention in the recent literature. Hydrogen-bonding interactions have a strong directional dependence, and packing patterns associated with particularly stable hydrogen-bonded systems (usually involving multiple hydrogen bonds) have been recognized for decades. New ones are still being discovered and applied with success in the design of crystal structures and molecular recognition systems. However, when it comes to much weaker intermolecular interations, one can find examples and counter examples for almost any proposed structure-directing interaction. Parallel stacking of benzene rings is not necessarily a result of π–π interactions; it is enough that it leads to a locally close packed arrangement. Similarly, the T-interaction between pairs of benzene rings in the crystal structures of aromatic hydrocarbons is not necessarily a result of interaction between a C-H bond of one ring and the π cloud of another; repeated, it leads to layers with the same close packed herring-bone arrangement that one can observe in a brick wall. There are certainly many examples of intermolecular contacts with interesting interpretations in terms of plausible chemical concepts. One thinks of orientationally directed C-H...O, Hal...Hal, O...C=O interactions, but none of these is strong enough to oppose the drive towards maximally close packing. Molecules tend to crystallize in packing arrangements that fill space as closely as possible;

other things equal, the higher the density (the packing coefficient) the lower is the lattice energy. Even for molecules containing good hydrogen bond donors and acceptors, close packed crystals lacking hydrogen bonds may be preferred to more open structures with hydrogen bonds (alloxan, for example, tightly packed with density 1.93 $g.cm^{-3}$ but lacking N-H...O hydrogen bonds [12]). The weaker interactions in question, with packing energy contributions of the order of a kcal mol^{-1} or less, can be structure directing only insofar as they can be realized within the demands of close packing.

References

1. Pauling, L. (1940) *The Nature of the Chemical Bond*, Cornell University Press, Ithaca, N.Y.
2. Kitaigorodsky, A. I. (1973) *Molecular Crystals and Molecules*, Academic Press, New York, N.Y.
3. Gavezzotti, A. and Filippini, G. (1997) Energetic Aspects of Crystal Packing: Experiment and Computer Simulation, in A. Gavezzotti (ed.), *Theoretical Aspects and Computer Modeling of the Molecular Solid State: The Molecular Solid State Volume 1*, Wiley, Chichester, England.
4. Dunitz, J. D. (1995) Win some, lose some: enthalpy-entropy compensation in weak intermolecular interactions, *Chemistry & Biology* **2**, 709-712.
5. Bondi, A. (1964) van der Waals Volumes and Radii, *J. Phys. Chem.* **68**, 441-451.
6. Gavezzotti, A. and Dunitz, J. D. (1998) Unpublished work.
7. Pauling, L. and Delbrück, M. (1940) The nature of the intermolecular forces operative in biological processes, *Science* **92**, 77-79.
8. Brock, C. P. and Dunitz, J. D. (1994) Towards a Grammar of Crystal Packing, *Chem. Mater.* **6**, 1118-1127.
9. Gibson, K. D. and Scheraga, H. A. (1995) Crystal Packing without Symmetry Constraints. 2. Possible Crystal Packings of Benzene Obtained by Energy Minimization from Multiple Starts, *J. Phys. Chem.* **99**, 3765-3773.
10. van Eijck, B. P., Spek, A. L., Mooij, W. T. M. and Kroon, J. (1998) Hypothetical Crystal Structures of Benzene at 0 and 30 kbar. *Acta Cryst.*, **B54**, 291-299.
11. van Eijck, B. P., Mooij, W. T. M. and Kroon, J. (1995) Attempted Prediction of the Crystal Structures of Six Monosaccharides, *Acta Cryst.* **B51**, 99-103.
12. Bolton, W. (1964) Crystal Structure of Alloxan, *Acta Cryst.* **17**, 147-152.

WEAK HYDROGEN BONDS

THOMAS STEINER
Department of Structural Biology,
Weizmann Institute of Science,
Rehovot 76100, Israel

Abstract

An overview is given on general properties and structural characteristics of weak hydrogen bond interactions. The focus is on C–H$\cdots$O and X–H$\cdots\pi$ hydrogen bonds (X = O, N, C), which are currently under intense investigation. Hydrogen bonds involving metal centers and the 'dihydrogen bond' X–H$^{\delta+}\cdots$H$^{\delta-}$–Y are briefly mentioned.

1. Introduction

The role of hydrogen bonds, *i.e.* directional cohesive interactions of the type X–H$\cdots$Y, in determining the architecture of supramolecular assemblies can hardly be overestimated. In hydrogen bonds, the donor can be any kind of acidic X–H group (O–H, N–H, S–H, Cl–H, C–H, *etc.*) and Y can be any base (O, N, S, halide ions, π-electron clouds, etc.). Because there are so many different types of donors and acceptors, and each donor/acceptor combination represents a different kind of hydrogen bond with different structural and energetic properties, combinatorics leads to a chemical variety of hydrogen bonds that is very hard to overlook. The acidities of different X–H donors and the basicities of different Y acceptors cover extremely wide ranges and, therefore, the strengths of different X–H$\cdots$Y hydrogen bonds cover a range that meets covalent bonds at one extreme, and van der Waals interactions at the other. The energies of hydrogen bonds range from *ca* 0.5 kcal/mol for the weakest to 30 kcal/mol for the strongest kinds.

Only for the most frequently occurring 'conventional' hydrogen bonds, like O–H$\cdots$O, N–H$\cdots$O, and related systems, are chemical and structural characteristics well-documented [1]. Although it is long known that numerous weaker hydrogen bond types (often called 'non-conventional' or 'unconventional' hydrogen bonds) occur in a vast variety of chemical and biological systems, their importance has been generally recognized only in recent years. In this contribution, a short overview will be given on the most relevant kinds of weak hydrogen bonds, with a focus on the structural characteristics as observed in crystallographic studies.

J.A.K. Howard et al. (eds.),
Implications of Molecular and Materials Structure for New Technologies, 185–196.
© 1999 *Kluwer Academic Publishers. Printed in the Netherlands.*

2. C–H···O Hydrogen Bonds

C–H···O interactions are the best known of all weak hydrogen bond types. Although structural evidence for the hydrogen bond nature of short C–H···O contacts was early given by Sutor from X-ray [2] and later by Taylor and Kennard from neutron diffraction data [3], the phenomenon was widely neglected for a surprisingly long period. This has now changed, and C–H···O hydrogen bonding has been investigated by a considerable number of authors. Several reviews describe recent developments and also the historical background of the research on C–H···O hydrogen bonds in the fields of general structural chemistry [4–7] and of structural biology [8].

2.1. ROLE OF C–H ACIDITY

In classical vibrational spectroscopic experiments, Allerhand and Schleyer have shown that the 'donor strength' of C–H groups depends on the carbon hybridization as $C(sp)–H > C(sp^2)–H > C(sp^3)–H$, and increases if C is bonded to electronegative substituents [9]. $C(sp)–H$ is a relatively strong donor even if it is bonded 'only' to carbon, $C\equiv C–H$, and becomes a classical carbon acid if bonded to nitrogen, $N\equiv C–H$. Substituent effects are most important for $C(sp^3)–H$: in this hybridization state, C–H can be very weakly polarized and hence a very weak hydrogen bond donor, such as in methyl groups bonded to carbon, and it can be an even stronger donor than $C\equiv C–H$, such as in chloroform or trinitromethane.

Structurally, donor strengths reflect in average H···O (or C···O) hydrogen bond distances in crystals, which are shorter for the stronger donors [4]. This circumstance was generally described by Pedireddi and Desiraju, who have shown a correlation of mean C···O distances in crystals with the conventional C–H acidity [10]. In Figure 1, this correlation is drawn for sterically unhindered donors. Sterically hindered donors form on the average longer hydrogen bonds than unhindered C–H groups of the same acidity. In addition to the data shown in the original publication [10], a data point for the donor $CH(NO_2)_3$ is included in Figure 1; $CH(NO_2)_3$ is an extremely strong carbon acid ($pK_a = 0.14$ compared to 4.8 for acetic acid), and in the crystalline adduct with dioxane, it forms the very short C–H···O hydrogen bond shown in Figure 2 [11]. A selection of further C–H groups that can donate short hydrogen bonds with H···O distances ≤ 2.1 Å is shown in Figure 3.

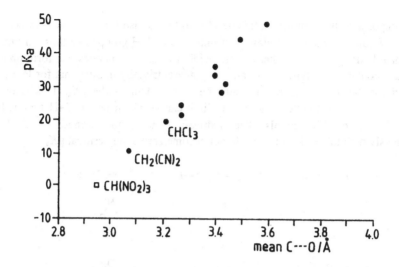

Figure 1. Correlation of the mean C···O distance with the C–H acidity (for sterically unhindered C–H···O hydrogen bonds); adapted from a publication of Pedireddi and Desiraju [10], who show in addition data for sterically hindered C–H groups. The square data point was included for the present paper, and represents a single crystal structure with a trinitromethane donor (crystal structure shown in Figure 2).

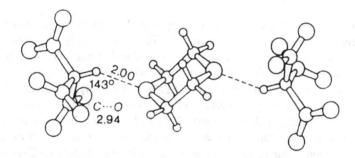

Figure 2. A highly acidic C–H group that forms a very short C–H···O hydrogen bond: the crystalline 2:1 adduct of trinitromethane to dioxane; crystal structure published by Bock and coworkers [11], Figure adapted from Ref. [7].

C–H acidities encompass an extremely broad range of about 60 orders of magnitude, which overlaps with acidities of organic and even of mineral acids. Unfortunately, no crystal structure is available as yet of the strongest carbon acids known, such as tricyanomethane, $CH(CN)_3$. Chemically different C–H groups cover almost the entire range of acidic behavior. Figure 1 shows that average C···O distances (*i.e.* hydrogen bond donor strengths) continuously increase with falling C–H acidity. It is a most important observation that this trend is followed also for very low acidities. This disproves frequently published views that only acidic C–H groups like those in Figure 3 can donate hydrogen bonds, whereas less acidic (or less polar) ones can form only van der Waals contacts to oxygen. The latter hypothesis trivially implies that for less acidic C–H groups, the C···O separations must be independent of the pK_a. This is not the behavior observed in Figure 1: no acidity limit can be given where C–H groups loose their hydrogen bond potentials. The hydrogen bond donor strength of C–H falls continuously with falling acidity, but do not become zero at any critical pK_a.

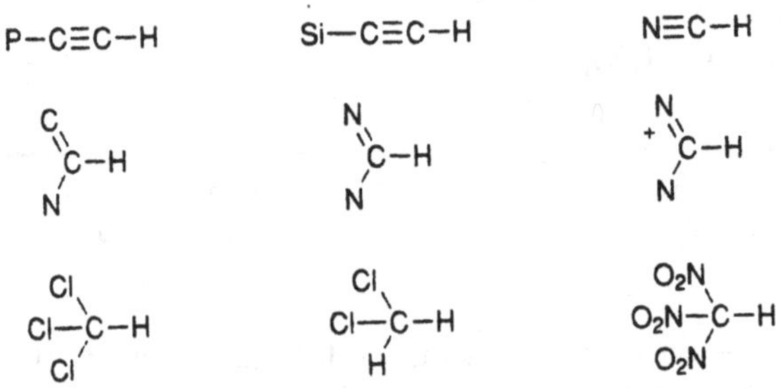

Figure 3. Selection of C–H groups that can form short C–H···O hydrogen bonds with H···O distances ≤ 2.1 Å.

2.2. ROLE OF ACCEPTOR BASICITY

The influence of acceptor basicity on hydrogen bond strengths is generally smaller than the influence of donor acidity, and more difficult to quantify [1]. Nevertheless, an influence of acceptor nature on mean intermolecular distances can be shown also for C–H···O hydrogen bonds. For example, mean H···O separations of CH_2Cl_2 donors to the acceptors C=O and C–O–C in crystals are 2.27(4) and 2.50(4) Å, respectively [12].

It is obvious that good hydrogen bond acceptors like C=O, P=O, H_2O *etc.* are also good in accepting hydrogen bonds from C–H donors. In the low region of the basicity scale, it is of importance that even the poor oxygen acceptors of carbonyl ligands in organometallic complexes can accept hydrogen bonds from C–H donors, with many examples documented for cyclopentadienyl donors and other ligands which are frequently used in organometallic chemistry [13]. An interesting example with an ethynyl donor and a very long H···O separation is shown in Figure 4: even for this very long contact, weakly hydrogen bond nature could be shown by infrared spectroscopy [14]. Figure 4 together with the spectroscopic data represents one of the examples where it is shown experimentally that the distance range of the hydrogen bond exceeds the van der Waals separation of donor and acceptor.

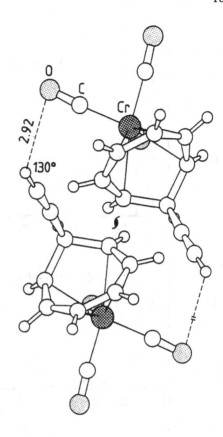

Figure 4. A very long C–H⋯O hydrogen bond formed by a C≡C–H group to a carbonyl ligand of an organometallic complex. Despite the very long H⋯O distance, weakly hydrogen bonding nature was shown by vibrational spectroscopy [14].

2.3. DIRECTIONALITY

An important and inherent property of all kinds of hydrogen bonds is their *directionality*. Ideally, hydrogen bonds are linear with the donor pointing exactly at the acceptor. Because of the relative weakness of the interaction, however, hydrogen bonds can be easily bent from linearity, leading to broad angular frequency distributions observed in the solid state [1]. As hydrogen bonds become weaker, they can be more easily bent from linearity, and angular frequency distributions become broader and broader. For C–H⋯O contacts formed by ethynyl, vinyl and methyl groups, this is shown in comparison to hydroxyl donors in Figure 5 [15]. The angular distribution of C≡C–H⋯O interactions is similar to that of conventional O–H⋯O hydrogen bonds, showing that the C≡C–H group is a relatively strong donor. For vinyl groups C=CH$_2$, the distribution is significantly broader, reflecting the lower donor strength. Finally, for methyl donors, the distribution is broadest but still shows a preference for linear contact geometries. This means that even for the very weakly polarized methyl groups C–CH$_3$, the interaction with oxygen atoms is directional, and essentially unlike the van der Waals contacts for which the related distribution is shown at the bottom of Figure 5.

190

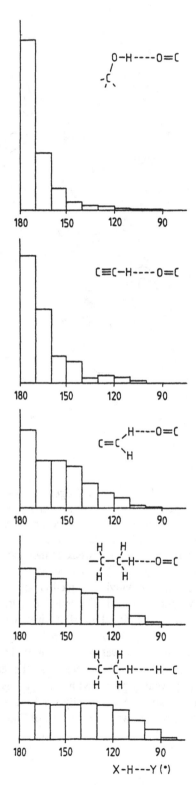

Figure 5. Angular distribution of C–H⋯O angles in contacts to O=C< formed by (a) hydroxyl groups, (b) ethynyl groups, (c) vinyl groups, (d) methyl groups of the type CH₂–CH₃. (e) Angular distribution of C–H⋯H angles in C–H⋯H–C van der Waals contacts of methyl groups [15].

Crystal statistics based on normalized H-atom positions, and the relatively permissive distance cutoff criteria H⋯O < 3.0 Å for (a) to (d), and H⋯H < 2.7 Å for (e). The distributions are weighted by the cone-correction factor $1/\sin\theta$ (J. Kroon and J. A. Kanters, *Nature*, 1974, **248**, 667), and scaled in such a way that they cover the same areas.

2.4. ENERGIES

Energies of C–H···O hydrogen bonds in the solid state cannot be measured directly. Theoretical calculations on C–H···O hydrogen bond energies have been performed by a large number of authors, who used different computational methods [7]. It appears that hydrogen bonds formed by the relatively strong ethynyl donor have energies around 2 kcal/mol, in some cases up to 3 kcal/mol, whereas those formed by methyl groups have energies ≤ 1 kcal/mol. This means that the energy range overlaps at one end with that of conventional O–H···O hydrogen bonds, and fades away to energies of pure van der Waals interactions at the other end. Similar to the geometric properties of C–H···O contacts, calculated interaction energies indicate that there are no defined borderlines between hydrogen bonds and van der Waals interactions.

2.5. COOPERATIVITY

C–H···O interactions experience cooperativity effects which are completely analogous to those in conventional hydrogen bonding [5—7]. This means that in arrays of interwoven hydrogen bonds, the individual hydrogen bonds can enhance each others energies by mutual polarization of the involved groups. This effect is very pronounced for functional groups that can donate and accept hydrogen bonds simultaneously [1], such as hydroxyl or ethynyl groups. The latter can accept hydrogen bonds directed at the C≡C bond, so that arrays of the type X–H···C≡C–H···Y can be formed. Within such chains, cooperativity enhancement has been theoretically calculated for the system C≡C–H···C≡C–H···C≡C [16], and was experimentally observed (by infrared spectroscopy) for O–H···C≡C–H···O–H chains in the crystalline sex steroid mestranol [17].

Some cooperatvity enhancement must be expected for chains of the type C–H···O–H···O, but for this assumption, experimental evidence is still missing. Slight cooperativity enhancement might also occur in motifs as the one shown in Scheme 1, where mutual polarization can in principle occur through the conjugated double bond systems [7].

(1)

2.6. PATTERNS

Within a supramolecular assembly, C–H⋯O hydrogen bonds may occur isolated or as part of small or extended hydrogen bond arrays. Numerous examples are shown in the cited literature [4—8, 17]. Such arrays can be composed of only C–H⋯O hydrogen bonds (e.g. that in Scheme 1), or can be formed of interconnected C–H⋯O and conventional hydrogen bonds. The role of C–H⋯O hydrogen bonds in interconnected arrays depends on the particular situation. There are examples where the weaker interactions appear to be bystanders in a frame of stronger and structure-determining hydrogen bonds, and there are other examples where the weak hydrogen bonds play decisive roles in a subtle interplay of strong and weak directional interactions [7].

Of greater interest are well-defined and robust motifs which occur frequently in crystals and can be classified as 'supramolecular synthons' in the sense introduced by Desiraju [18]. An example for such a synthon is motif (1), which occurs in many crystal structures containing C=CH–C=O moieties. Other supramolecular synthons are composed of conventional and of C–H⋯O hydrogen bonds; a typical example composed of a strong O=C–O–H⋯N and a much weaker C–H⋯O=C hydrogen bond is shown in Scheme 2 [19].

(2)

In biological macromolecules, C–H⋯O hydrogen bonds rarely occur isolated, but mostly in connection with stronger hydrogen bonds. A very important example is synthon (3), which occurs systematically in proteins with the donors Cα–H and peptide N–H.

(3)

Regular sequences of this synthon determine the structure of α-helices and β-sheets [20]. As a matter of fact, the conventional N–H⋯O=C hydrogen bond is the much stronger of the two.

3. X–H···π Hydrogen Bonds

In the so-called X–H···π hydrogen bonds, an X–H donor interacts with the electron cloud of π-bonded systems like phenyl rings, heteroarenes, C≡C and C=C bonds, and so on. Even for the formal single bonds in the highly strained cyclopropanes, which have some π-bond character, hydrogen bond acceptor potential could be shown [21]. Because there are many different π-acceptors which can form hydrogen bonds with all X–H donor types, a large number of X–H/π donor/acceptor combinations is possible. Many of these possible combinations have actually been observed in crystal structures, such as O–H···Ph, N–H···Ph, Cl–H···Ph, C≡C–H···Ph, Cl$_3$C–H···Ph, O–H···C≡C, N–H···C≡C, C–H···C≡C, O–H···C=C, and so on. Particularly strong hydrogen bonds can be formed to acceptor molecules which carry a negative charge like the tetraphenylborate ion, N$^+$–H···Ph–B$^-$. It is here not possible to show more than one of the many examples: in Figure 6, an O–H···Ph hydrogen bond is shown, which is donated from a water molecule to the phenyl ring of the Phe residue in the crystalline tripeptide Tyr-Tyr-Phe dihydrate [22].

In the optimal geometry of an X–H···π hydrogen bond, the donor points perpendicularly to the midpoint of the π-system. Corresponding theoretical calculations have been published for the system N–H···Ph [23, 24], and for a number of gas-phase dimers, perpendicular approach of the donor to the π-acceptor has been observed.

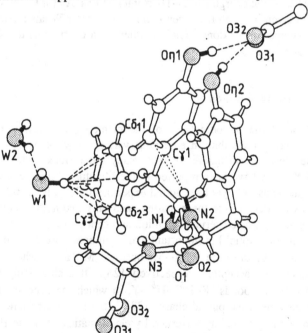

Figure 6. O$_W$–H···Ph hydrogen bond in the crystal structure of Tyr-Tyr-Phe dihydrate. The structure contains also an intramolecular aromatic-(i+1)-amine interaction [22].

However, the geometry of X–H···π hydrogen bonds is extremely soft, and it can be more easily distorted than with conventional hydrogen bonds. In crystals, X–H···π hydrogen bonds have been observed, where the donor points almost exactly at one of the individual C-atoms of the acceptor. With Ph acceptors, the donor can also point at the midpoint of one of the individual aromatic C–C bonds [22]. There is theoretical and also spectroscopic evidence that X–H···π hydrogen bonds directed at individual C-atoms are considerably weaker than those directed at the π-system centroids [25]. In optimal geometry, N–H···π hydrogen bonds have energies of about 3 to 4 kcal/mol.

In X–H···π hydrogen bonds, the donor/acceptor separations cover wide ranges. For water-to-phenyl hydrogen bonds like that shown in Figure 6, the distances from O_W to the aromatic centers M are typically between 3.2 and 3.7 Å (3.26 Å for the example shown). In N^+–H···Ph–B^- hydrogen bonds, they can be shorter, down to 3.0 Å. The distance of the H-atom to individual C-atoms is typically $\geq$ 2.4 Å, but somewhat shorter distances have also been observed [25].

Even C–H groups can donate hydrogen bonds to π-systems. Structural, spectroscopic and quantum chemical evidence could be given for the hydrogen bond nature of C≡C–H···C≡C and C≡C–H···Ph contacts [16]. These interactions can also be part of cooperative arrays like C≡C–H···C≡C–H···C≡C–H, O–H···C≡C–H···Ph, and so on. Also chloroform molecules have been found donating hydrogen bonds to C≡C and Ph acceptors. The energies of C≡C–H···π hydrogen bonds are around 1 to 2 kcal/mol, *i.e.* significantly below the energies of N–H···π interactions. As for C–H···O contacts, no borderline can be given between hydrogen bonds and van der Waals interactions. When varying contact geometries or donor acidities, there is a continuous transition between the two interaction types.

4. Other Weak Hydrogen Bonds

After a short look at the periodic table of the elements, and using some chemical imagination, it is easy to see that an enormous variety of weak hydrogen bond types must be expected to occur in nature. In fact, many other weak hydrogen bonds than C–H···O/N and X–H···π have been reported in the literature. Furthermore, many kinds that can be imagined have not yet been found experimentally. It is far beyond the scope of this article to view deeply into these matters, and only two hydrogen bond types are mentioned which have attracted serious attention in recent years.

One family of 'exotic' but important hydrogen bonds are those involving metal centers. They can be of the type X–H···M, but also M–H···Y, where X and Y are conventional donors and acceptors, respectively [26]. Of a chemically very different kind are the 'dihydrogen bonds' $X–H^{\delta+}···H^{\delta-}–Y$, in which the acceptor is a hydrogen atom which carries a negative partial charge [27]. From the electrostatic point of view (weak hydrogen bonding is mainly governed by electrostatics) a dipole $H^{\delta-}–Y^{\delta+}$ is just as good as acceptor of hydrogen bonds as, for example, $O^{\delta-}–H^{\delta+}$.

5. Acknowledgements

The author is on leave from the Institut für Kristallographie, Freie Universität Berlin, Takustraße 6, D-14195 Berlin, Germany. He wishes to thank the Minerva Foundation, Munich, for granting a fellowship to stay at the Weizmann Institute of Science in the research group of Professor Joel L. Sussman.

6. References

1. Jeffrey, G. A and Saenger, W. (1991). *Hydrogen Bonding in Biological Structures*, Springer, Berlin.
2. Sutor, D. J. (1962). The C–H···O Hydrogen Bond in Crystals. *Nature* **195**, 68–69.
3. Taylor, R. and Kennard, O. (1982). Crystallographic Evidence for the Existence of C–H···O, C–H···N and C–H···Cl Hydrogen Bonds. *J. Am. Chem. Soc.* **104**, 5063–5070.
4. Desiraju, G. R. (1991). The C–H···O Hydrogen Bond in Crystals: What Is It? *Acc. Chem. Res.* **24**, 290–296.
5. Desiraju, G. R. (1996). The C–H···O hydrogen bond: structural implications and supramolecular design. *Acc. Chem. Res.* **29**, 441–448.
6. Steiner, T. (1996). C–H···O hydrogen bonding in crystals. *Cryst. Rev.* **6**, 1–57.
7. Steiner, T. (1997). Unrolling the hydrogen bond properties of C–H···O interactions. *Chem. Commun.* 727–734 (Feature Article).
8. Wahl, M. C. and Sundaralingam, M. (1997). C–H···O hydrogen bonding in biology. *Trends in Biol. Sci.* **22**, 97–102.
9. Allerhand, A. and Schleyer, P. von R. (1963). A Survey of C–H Groups as Proton Donors in Hydrogen Bonding. *J. Am. Chem. Soc.* **85**, 1715–1723.
10. Pedireddi, V. R. and Desiraju, G. R. (1992). A Crystallographic Scale of Carbon Acidity. *J. Chem. Soc., Chem. Commun.*, 988–990.
11. Bock, H., Dienelt, R., Schödel, H. and Havlas, Z. (1993). The C–H···O Hydrogen Bond Adduct of two Trinitromethanes to Dioxane. *J. Chem. Soc., Chem. Commun.*, 1792–1793.
12. Steiner, T. (1994). Effect of Acceptor Strength on C–H···O Hydrogen Bond Lengths as Revealed by and Quantified from Crystallographic Data. *J. Chem. Soc., Chem. Commun.*, 2341–2342.
13. Braga, D., Grepioni, F., Biradha, K., Pedireddi, V. R. and Desiraju, G. R. (1995). Hydrogen Bonding in Organometallic Crystals. 2. C–H···O Hydrogen Bonds in Bridged and Terminal First-Row Metal Carbonyls. *J. Am. Chem. Soc.* **117**, 3156–3166.
14. Steiner, T., Lutz, B., van der Maas, J., Schreurs, A. M. M., Kroon, J. and Tamm, M. (1998). Very long C–H···O contacts can be hydrogen bonds: experimental evidence from crystalline $[Cr(CO)_3\{\eta^6\text{-}(7\text{-}exo\text{-}(C\equiv CH)C_7H_7)\}]$. *Chem. Commun.*, 171–172.
15. Steiner, T. and Desiraju, G. R. (1998). Distinction between the weak hydrogen bond and the van der Waals interaction. *Chem. Commun.*, 1998, 891–892.
16. Steiner, T., Starikov, E. B., Amado, A. M. and Teixeira-Dias, J. J. C. (1995). Weak Hydrogen Bonding. Part 2. The Hydrogen Bonding Nature of Short C–H···π Contacts: Crystallographic, Spectroscopic and Quantum Mechanical Studies of Some Terminal Alkynes. *J. Chem. Soc., Perkin Trans. 2*, 1321–1326.

196

17. Steiner, T., Lutz, B., van der Maas, J., Veldman, N., Schreurs, A. M. M., Kroon, J. and Kanters, J. A. (1997). Spectroscopic evidence for cooperativity effects involving C–H⋯O hydrogen bonds: crystalline mestranol. *Chem. Commun.*, 191–192.

18. Desiraju, G. R. (1995). Supramolecular Synthons in Crystal Engineering – A New Organic Synthesis. *Angew. Chem., Int. Ed. Engl.* **34**, 2311–2327.

19. Sharma, C. V. K. and Zaworotko, M. J. (1996). X-Ray crystal structure of $C_6H_3(CO_2H)_3$-1,3,5 · 1.5(4,4'-bipy): a 'super trimesic acid' chicken-wire grid. *Chem. Commun.*, 2655—2656.

20. Derewenda, Z. S., Lee, L. and Derewenda, U. (1995). The occurrence of C–H⋯O hydrogen bonds in proteins. *J. Mol. Biol.*, **252**, 248–262.

21. Allen, F. H., Lommerse, J. P. M., Hoy, V. J., Howard J. A. K. and Desiraju, G. R (1996). The Hydrogen Bond Donor and π-Acceptor Characteristics of Three-Membered Rings. *Acta Crystallogr., Sect. B.* **52**, 734–745.

22. Steiner, T., Schreurs, A. M. M., Kanters, J. A. and Kroon, J. (1998). Water Molecules Hydrogen Bonding to Aromatic Acceptors of Amino Acids: the Structure of Tyr-Tyr-Phe Dihydrate and a Crystallographic Database Study on Peptides. *Acta Crystallographica, Sect. D.*, **54**, 25–31.

23. Levitt M. and Perutz M. F. (1988). Aromatic Rings Act as Hydrogen Bond Acceptors. *J. Mol. Biol.* **201**, 751–754.

24. Worth, G. A. and Wade, R. C. (1995). The Aromatic (i+2) Amine Interaction in Peptides. *J. Phys. Chem.* **99**, 17473–17482.

25. Steiner, T., Mason, S. A. and Tamm, M. (1997). Neutron Diffraction Study of Aromatic Hydrogen Bonds: 5-Ethynyl-5*H*-dibenzo[*a,d*]cyclohepten-5-ol at 20 K. *Acta Crystallogr., Sect. B* **53**, 843–848.

26. Brammer, L., Zhao, D., Lapido, F. L. and Braddock-Wilkig (1995). Hydrogen Bonding Involving Transition Metal Centres – A Brief Review. *Acta Crystallogr., Sect. B*, **51**, 632–640.

27. Crabtree, R. H., Siegbahn, P. E. M., Eisenstein, O., Rheingold, A. L. and Koetzle, T. F. (1996). A New Intermolecular Interaction: Unconventional Hydrogen Bonds with Element-Hydride Bonds as Proton Acceptor. *Acc. Chem. Res.* **26**, 348–354.

DIRECT AND INDIRECT ROLES OF METAL CENTRES IN HYDROGEN BONDING

L. BRAMMER
Department of Chemistry, University of Missouri-St. Louis
8001 Natural Bridge Road, St. Louis, MO 63121-4499, USA

1. Introduction

Hydrogen bonding is perhaps the most important and most widely studied of all intermolecular interactions [1]. This is particularly recognized today in the fields of supramolecular chemistry [2] and molecular biology [1c]. Much of the emphasis to date has been on the involvement of hydrogen bonds in systems of an organic or biological nature, while comparatively less attention has been given until very recently to hydrogen bonds in inorganic chemistry [3]. This article is intended to introduce the reader to a pervasive aspect of the latter, namely the role that metal centres can play in hydrogen bonding.

Metal centres can be involved in hydrogen bonding in both a direct or an indirect manner. The distinction is that in the former the metal participates directly as either the hydrogen bond donor or acceptor, whereas in the latter the metal exerts either an electronic or spatial influence upon the hydrogen bond donor and/or acceptor groups. Our own work has evolved along a path that encompasses each of these roles for metals and in so doing has suggested applications from reaction chemistry to molecular conformation, and most recently to supramolecular chemistry and crystal engineering. This article will follow a similar evolution taking examples from our own work and also from the recent literature.

2. Direct Participation of Metals in Hydrogen Bonding

The hydrogen bond, D–H···A, is generally considered to comprise two components. A hydrogen atom is attached to an electronegative element, *e.g.* N, O, S, or even C, known as the donor (D), leading to a polar D–H bond in which the hydrogen carries a partial positive charge. The hydrogen atom can then interact with an acceptor (A) that possesses either lone-pair electrons or polarizable π-electrons. The interaction is usually predominantly electrostatic in nature. In recent years it has been shown that in certain cases metals can participate in either the hydrogen bond donor or acceptor role.

J.A.K. Howard et al. (eds.),
Implications of Molecular and Materials Structure for New Technologies, 197–210.
© 1999 *Kluwer Academic Publishers. Printed in the Netherlands.*

2.1. METALS AS THE DONOR GROUP (M–H···O)

The most common case of metals serving as hydrogen bond donors involves cationic metal "hydrides" in the formation of M–H···O hydrogen bonds [4]. Such hydrogen bonds have been recognized crystallographically and by IR spectroscopy, and there is also some evidence to suggest a characteristic downfield-shifted "hydride" signal in the ^{1}H NMR (Figure 1).

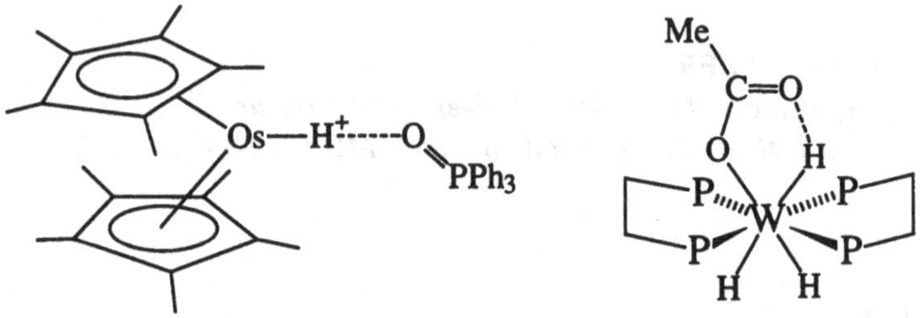

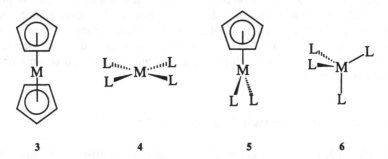

Figure 1. Examples of M–H···O hydrogen-bonded systems with characterization data (1, IR(CH$_2$Cl$_2$): Δν(Os–H) 20 cm^{-1}, Δν(P=O) 26 cm^{-1} [4b]; 2, X-ray: H···O 2.33(6) Å; ^{1}H NMR: W-H···O δ 2.92, W-H δ –2.78 [4c])

2.2. METALS AS THE ACCEPTOR GROUP (D–H···M, D = C, N, O)

Although sporadic reports appeared over previous decades [5], conclusive recognition of the fact that electron-rich transition metal centres can serve as hydrogen bond acceptors has come primarily in the past decade [6,7]. Such metal centres are typically late transition metals in low oxidation states, and require an accessible filled metal-bound orbital. The primary classes of compound studied to date are illustrated in Fig. 2.

Figure 2. Systems in which the metal centre is capable of acting as a hydrogen bond acceptor. (3, d^6 Cp$_2$M , M = Fe^{II}, Ru^{II}, Os^{II} [7h-j]; 4, d^8 square-planar ML$_4$, M = Rh^I, Ir^I Ni^{II}, Pd^{II}, Pt^{II} [6a,7c,d,f,g,l]; 5, d^8 CpML$_2$, M = Co^I, Rh^I, Ir^I [7e]; 6, d^{10} tetrahedral ML$_4$, Co^{-1}, Ni^0 [6b-d,7a,b]).

2.2.1. R$_3$NH$^+$Co(CO)$_3$L$^-$ Salts (L = CO, PAr$_3$; Ar = aryl group)
Building upon an early study by Calderazzo and coworkers [7a], we have over the past few years investigated in some detail the interaction of secondary and tertiary

ammonium cations with the Co(CO)$_3$L$^-$ anion (L = CO, PAr$_3$) [8]. The resulting salts have been studied both in the solid state and in solution [6b-d]. In many cases the primary cation-anion interaction is an N–H⋯Co hydrogen bond. A systematic study of steric and electronic effects on this hydrogen bond has been conducted. Broadly speaking, increased acidity of the cation, decreased steric hindrance at the cation, and increased basicity of the anion result in a decreased hydrogen bond separation in the solid state (and by inference an increase in the hydrogen bond strength). This is accompanied by a pyramidal distortion of the anion away from the tetrahedral

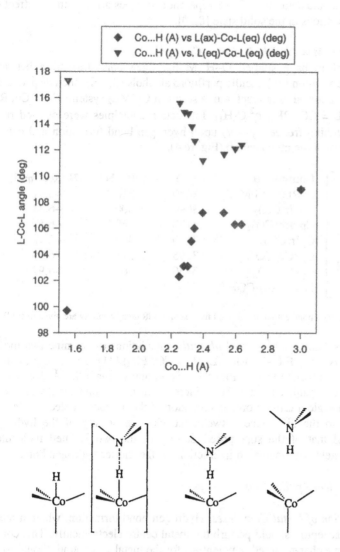

Figure 3. Model geometric reaction pathway (right to left) for protonation of Co(CO)$_3$L$^-$. (The points at Co⋯H 3.0 Å represent an estimated minimum separation for the undistorted tetrahedral anion).

geometry of the free anion, as quantified by monitoring the interligand angles. Focusing on these two geometric changes across the series of compounds studied permits mapping of an approximate geometric model of the reaction path for protonation of the anion (or deprotonation of the corresponding metal hydride) following the Structure Correlation approach of Bürgi and Dunitz [9] (see Figure 3). However, the variation of H⋯Co distance with L–Co–L angle based upon this set of crystal structures does not follow a monotonic curve. Some striking departures from the general trend described above have been noted, wherein H⋯Co separations have been observed that are some 0.1–0.15 Å longer than anticipated from comparison with structures of related salts and with solution spectroscopic data [10]. These apparent anomalies arise from the effect of other competing interactions in the solid state [8,10].

2.2.2. Cp'ML₂ Systems (M = Co, Rh, Ir; Cp' = Cp, Cp*)

Poliakoff and coworkers have studied O–H⋯M hydrogen bond formation between neutral species in lXe solution [7e]. Acidic perfluoro alcohols (R_fOH), such as perfluoro tertiary butanol, were allowed to interact with a series of Cp'ML₂ systems (M = Co, Rh, Ir; Cp' = Cp, Cp*; L = CO, PR_3, η^2-C_2H_4). Interaction enthalpies were derived from changes in O–H stretching frequency (Δv) upon hydrogen bond formation and can be correlated with basicity of the metal centre (Figure 4).

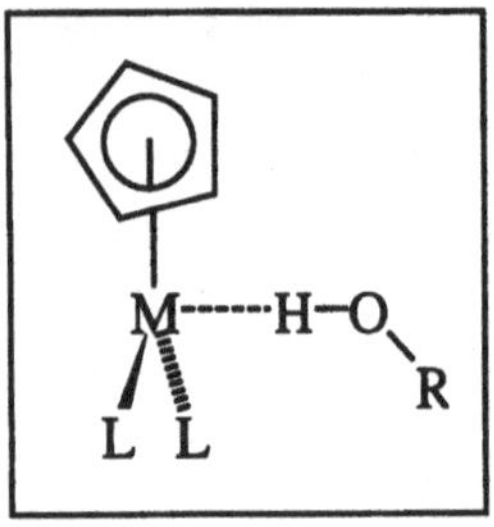

Compound	v(O–H)	Δv(O–H)	ΔH (kcal/mol)
CpRh(CO)PMe₃	3080	510	–6.91
Cp*Ir(CO)₂	3090	500	–6.84
Cp*Rh(CO)₂	3130	460	–6.57
CpIr(CO)₂	3190	400	–6.12
Cp*Co(CO)₂	3195	395	–6.09
CpCo(CO)₂	3330	260	–4.92
Free (CF₃)₃COH	3590		

Figure 4. O–H⋯M hydrogen bond enthalpies derived from solution IR data (v and Δv are given in cm⁻¹).

In a follow-up to this study we have used *ab initio* calculations to examine two model complexes for this system, $CF_3O–H⋯RhCpL_2$ (L = CO, PH₃) [11]. These studies show that on changing from L = CO to L = PH₃ there is an approximate 0.1 Å decrease in H⋯Rh separation accompanied by a 0.01 Å increase in O–H bond length and a 2° increase in O–H⋯Rh angle. Furthermore, examination of the interaction electron density [12], corresponding to the difference between the electron density of the hydrogen bonded complex and that of the superposed density of the two isolated molecules, shows features consistent with increased interaction for the shorter hydrogen bond.

2.2.3. Potential Chemical Applications

Electronic Perturbation of Metal Complexes. Hydrogen bond formation, where a metal centre serves as the acceptor, should perturb the metal centre electronically. This could, for example, result in a change in redox potentials for the metal compound, though such applications have not been exploited to our knowledge. However, in the *ab initio* study of the interaction electron density referred to in section 2.2.2 it is apparent that formation of the O–H···Rh hydrogen bond results in some rearrangement of electron density among the metal d-orbitals. This suggests that in the right system such hydrogen bond formation may even have applications in chemically perturbing a metal complex, for example activating ligands towards desired chemical reactions.

Oxidative Addition at Metal Centres. We have suggested previously [3a] that hydrogen bonds involving metals as the acceptor may in some cases serve as a step on the path towards (stepwise heterolytic) oxidative addition at the metal centre. At present we are aware of no definitive examples of hydrogen bond involvement in such reactions. However, in a few cases hydrogen-bonded compounds and the putative oxidative-addition products have been isolated (Figure 5).

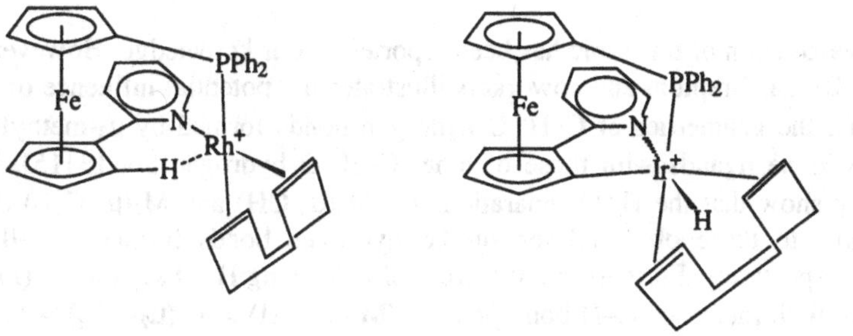

7 (16e Rh^I)   8 (18e Ir^{III})

Figure 5a. C–H oxidative addition, possibly *via* C–H···M hydrogen bonding shown in 7, can be inferred in the Ir system (8). 7.PF$_6$ and 8.PF$_6$ were isolated and characterized spectroscopically and crystallographically [13].

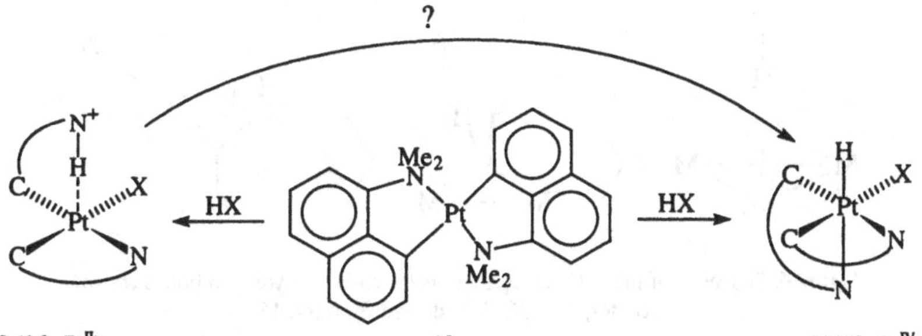

9 (16e Pt^{II})   10   11(18e Pt^{IV})

Figure 5b. Possible N–H oxidative addition via N–H···M hydrogen bonding (adapted from refs. 3d, 7c). Compounds 9 and 11 were prepared from 10 and characterized spectroscopically and crystallographically. Compound 9 suggests a possible pathway to the oxidative addition product 11 involving stepwise heterolytic cleavage of the N–H bond *via* proton transfer to the metal.

Organometallic Crystal Engineering. The occurrence and range of geometries of both M–H···O and D–H···M (D = C, N, O) hydrogen bonds in organometallic crystal structures has been explored [4e,7j] using the CSD [14]. However, while such hydrogen

202

bonds may have applications in organometallic crystal engineering it is likely that their role would be only in supplementing the more abundant inter-ligand interactions in the construction of extended hydrogen-bonded networks.

3. Indirect Participation of Metals in Hydrogen Bonding: Electronic Influence of the Metal

In this section we will discuss the electronic effect on hydrogen bond formation of coordination of hydrogen bond donors or acceptors to metal centres. Thus, *via* acceptance or donation of electrons from/to the group participating in hydrogen bonding, the metal can be regarded as participating indirectly in the hydrogen bond.

3.1. METAL INFLUENCE UPON HYDROGEN BOND DONOR GROUPS

Little investigation of this topic has been reported to our knowledge. However, a recent study by Braga, Grepioni and coworkers illustrates the potential influence of metals by comparing the geometries of C–H$\cdots$O hydrogen bonds formed by μ_3-methylidyne and μ_2-methylidene ligands with those of other C–H$\cdots$O hydrogen bonds [15]. Using the CSD they show that the H$\cdots$O separations for $M_3(\mu_3$-CH) and $M_2(\mu_2$-CH$_2$) donors are comparable to those observed for similar hydrogen bonds formed by alkynes and alkenes, respectively. Furthermore the order of increasing H$\cdots$O separation (Figure 6) is consistent with increased C–H bond polarity ($M_3(\mu_3$-CH) > $M_2(\mu_2$-CH$_2$) > MCH$_3$), and could possibly be attributed to a carbon hydridization model comparable to the organic analogues, *i.e.* bond polarity: C_{sp}–H > C_{sp^2}–H > C_{sp^3}–H. However, a simpler model would be one of diminishing C–H bond polarity arising from the reduced electron-withdrawing effect of coordination to fewer metal centres.

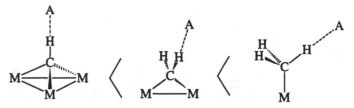

Figure 6. Sequence of mean H$\cdots$O separations for C–H$\cdots$A hydrogen bonds shown (acceptor, A = OC–M) adapted from ref. 15.

3.2. METAL INFLUENCE UPON HYDROGEN BOND ACCEPTORS

3.2.1. *Metal Halides as Hydrogen Bond Acceptors*
  In recent studies using the CSD we have shown that, in contrast to the very poor hydrogen bond acceptor capability of organochlorine [16a], metal-bound chlorine is an excellent hydrogen bond acceptor, comparable to the chloride ion [16b]. This effect can be attributed to the greater ionic character of the M–Cl bond *vs.* the C–Cl bond, hence the greater build-up of negative charge on the chlorine in the former. As a

result, coordination to a metal centre has a dramatic effect on the hydrogen bond acceptor capability of chlorine. The approach of hydrogen bond donors to the terminal chlorides is also markedly anisotropic, with a preferred angle of approach of *ca.* 110 °. This is consistent with the greater basicity of the chlorine p-orbital lone pairs than the sp-orbital lone pair [17] or can alternatively be described in terms of the quadrupolar nature of the electronic charge associated with the chlorine. Further work has shown that these conclusions hold true for the heavier halogens, but there are some marked differences in the case metal fluorides; hydrogen bond acceptor strength decreases upon descending the halogen group [18].

We have sought to use these observations in the design of crystalline solids and in doing so have identified new *inorganic* supramolecular synthons D–H$\cdots$X$_n$M (D = N, O, etc.; X = halogen; n = 2,3) in which the M–X bonds are mutually orthogonal (or approximately so). New crystal structures based upon these synthons (12-14) are shown in Figure 7, and exhibit zigzagged or helical chains due to the additional flexibility permitted for n = 2, and linear chains for n = 3 [19].

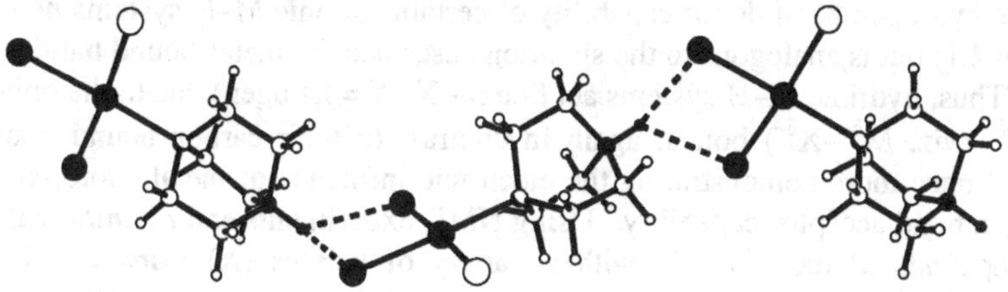

12

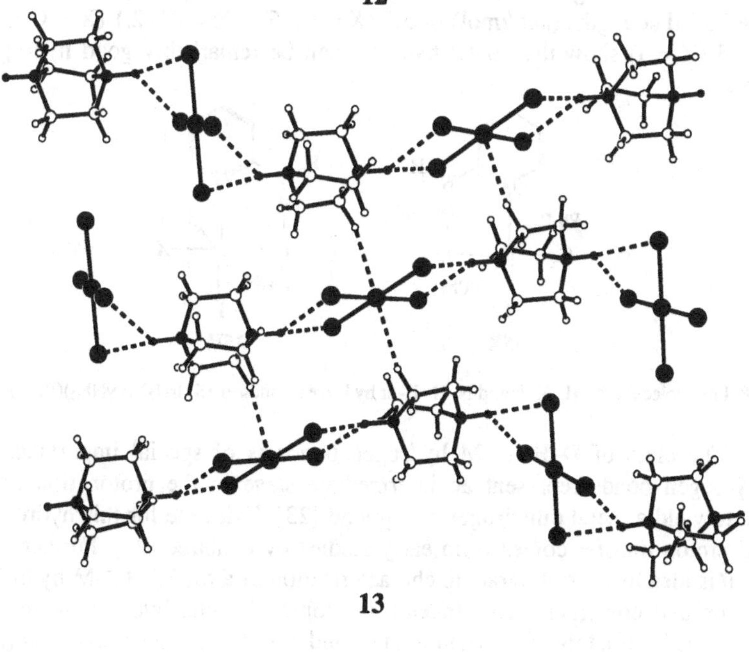

13

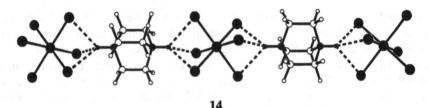

14

Figure 7. Crystal structures of PtCl₃(DABCO)H **12**, (DABCO)H₂[PtCl₄] **13**, and (DABCO)H₂[PtCl₆] **14**, illustrating hydrogen-bonded networks based upon N–H⋯Clₙ Pt supramolecular synthons (highlighted by shading), with n = 2, 2, and 3, respectively (DABCO = 1,4-diazabicyclooctane). In **12** and **13**, the zigzagged chains are further cross-linked by C–H⋯Pt hydrogen bonds, *i.e.* directly involving the metal as the acceptor.

3.2.2. Metal Hydrides as Hydrogen Bond Acceptors – A Special Case

Recently it has been recognized that hydride ligands in transition metal complexes can serve as hydrogen bond acceptors [20], *i.e.* D–H⋯H–M (D = C, N, O). This contrasts with the hydrogen bond donor capability of certain cationic M–H systems noted earlier (section 2.1) but is analogous to the situation described for metal-bound halides (section 3.2.1). Thus, hydridic M–H systems act like M–X (X = halogen) due to the quite polar $M^{\delta+}$–$H^{\delta-}$ (viz. $M^{\delta+}$–$X^{\delta-}$) bonds, again in contrast to their carbon-bound counterparts [21] and once more demonstrating the electronic influence of metal coordination upon hydrogen bond acceptor capability. Using NMR experiments and *ab initio* calculations on compounds of the type **15**, with a variety of halides (X) *trans* to the hydride, Crabtree, Eisenstein and coworkers have determined hydrogen bond strengths for the intramolecularly hydrogen bonded **15X** and **15H** isomers (Fig. 8) [22]. The N–H⋯X–Ir hydrogen bond strengths (kcal/mol) of 5.2 (X = F), 5.0 (X = H), 2.1 (X = Cl), 1.8 (X = Br), < 1.3 (X = I) show that metal hydrides can be remarkably good hydrogen bond acceptors.

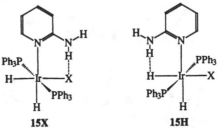

15X   **15H**

Figure 8. Intramolecular N–H⋯X–Ir and N–H⋯H–Ir hydrogen bonds in **15** (IrH₂X(pyNH₂)(PPh₃)₂) (ref. 22).

The class of D–H⋯H–M hydrogen bonds is of special importance because such hydrogen bonds represent an intermediate stage in the protonation of a metal hydride to yield a metal dihydrogen compound [23]. Evidence for this hydrogen bond-assisted proton transfer comes from early studies by Crabtree [24]. Further support is leant to this idea by crystallographic characterization of a M(H₂)⋯Cl–M hydrogen bond by Caulton and coworkers [25]. Indeed, the topic of metal hydride *vs.* metal halide protonation is itself a topic of current interest and has recently been reviewed [26].

3.2.3. *Metal Carbonyls as Hydrogen Bond Acceptors*

While a systematic comparison of the hydrogen bond acceptor capabilities of metal-bound carbonyl groups *vs.* organic carbonyls analogous to the discussion in section 3.2.1 for halogens has not been reported, the influence of metal coordination upon the acceptor capability of carbonyls has been investigated. Braga, Grepioni, Desiraju, and coworkers have shown using the CSD that the mean H⋯O separation for the abundance of C–H⋯O hydrogen bonds found in crystal structures of organometallic compounds is shorter for bridging than for terminal carbonyls [3f, 27] (Figure 9). This is consistent with enhanced basicity of the carbonyl oxygen as π-back-donation from the metal(s) is increased upon coordination to a greater number of metal centres.

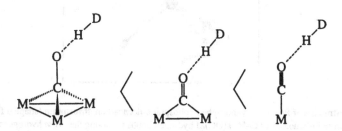

Figure 9. Decrease in mean H⋯O separations for C–H⋯O hydrogen bonds on going from terminal to μ_2 to μ_3 bridging carbonyls as the hydrogen bond acceptor (from refs 3f, 27).

In studies of *ca.* 10 salts of the form $R_3NH^+Co(CO)_4^-$ [8] we have found consistently short C–H⋯O hydrogen bonds between the cations and anions. Mean H⋯O separations are comparable with those of hydrogen bonds involving μ_2-CO ligands rather than terminal CO ligands bound to cobalt [27]. While it is clear that such hydrogen bonds will be "charge-assisted", *i.e.* have an enhanced electrostatic component compared with neutral hydrogen bonds, one can also invoke unusually high oxygen basicity. This results from the fact that the metal is very electron rich, consistent with the formal metal oxidation state of -I and d^{10} electron configuration. Thus, greater π-back-donation 9is expected than for typical terminal carbonyls, as confirmed by the low $v(CO)$ stretching frequency (1890 cm^{-1}) for the solvated $Co(CO)_4^-$ anion [6d, 8].

In a recent study based upon crystal structure geometries retrieved from the CSD we have suggested that the $\pi(C{\equiv}O)$ bonds of terminal metal carbonyls can serve as hydrogen bond acceptors [28] in a similar way to the π-bond of acetylenic groups [29]. D–H⋯$\pi(OC)M$ hydrogen bonds are found to be uncommon for D = N, O, but reasonably abundant for D = C. Indeed it is particularly interesting to note examples of intra-molecular C–H⋯$\pi(OC)M$ hydrogen bonds suggestive that such interactions might be useful in determining the conformation of some organometallic complexes (Figure 10). Indeed, one might envisage applications for such an approach in homogeneous catalysis, given the right system.

206

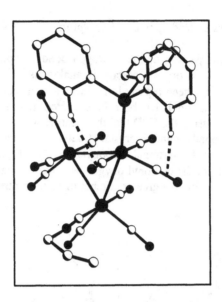

Figure 10. Structure of Propylisocyanotriphenylphosphino-decacarbonyltriosmium (adapted from ref. 30) showing intramolecular C–H···π(OC)M hydrogen bonds involving the *ortho* hydrogens of a triphenylphosphine ligand. (H···π 2.27, 2.42 Å; C–H···π 165, 156 °).

4. Indirect Participation of Metals in Hydrogen Bonding: Spatial Influence of the Metal

In this section we will discuss hydrogen bonds between ligand-based functional groups at the periphery of organometallic molecules. In such cases the distance (and number of intervening bonds) substantially mitigate the electronic influence of the metal on hydrogen-bonding interactions. However, the metal can play another valuable *indirect* role in the hydrogen bonds formed, in this case a spatial one. A suitable choice of ligands and a knowledge of preferred metal coordination geometries can in principle exert some control over the specific directions along which hydrogen-bonded coupling of molecules occurs.

An analogous approach has been applied quite successfully in what has been termed supramolecular coordination chemistry. In recent years there have been a growing number of elegant examples of the use of transition metals in preparing coordination polymers [31] or their smaller cousins molecular squares, cubes, hexagons, *etc.* [32]. These structures are based upon the combination of bi- (or tri-) functional organic ligands coordinated to metal centres that have predictable coordination geometry. The metal coordination geometry (tetrahedral, square planar, octahedral) is utilized in defining the relative orientation of the ligands, which themselves link two or more metal centres.

Using a similar approach we have recently prepared metal complexes containing ligands bearing hydrogen-bonding groups common in organic crystal engineering and supramolecular chemistry, such as carboxylic acid and amide groups.

In the two examples presented in Figure 11, homoleptic dications of Pt(II) have been prepared as their chloride salts using nicotinamide (**16**) or *iso*-nicotinamide (**17**) as ligands [33]. In **16**, one cation is linked to the next via two $R_2^2(8)$ hydrogen-bonded dimers [34] to give an infinite tape. In **17**, such an arrangement is not possible as the amide groups are now in the *para* position on the ring. Nevertheless, an infinite tape of similar topology is still formed, with the amide groups on neighbouring molecules now bridged by chloride anions and by a chain of hydrogen-bonded water molecules, as shown. Analogous work by Aakeröy and coworkers has also led to low-dimensional solids linked *via* coordinative bonds to silver cations and *via* hydrogen bonds [35].

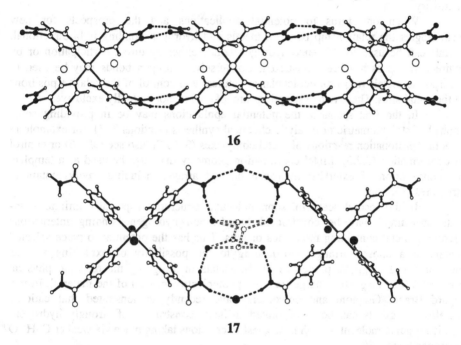

16

17

Figure 11. Crystal structures of [Pt(nicotinamide)₄]Cl₂ **16** and [Pt(*iso*-nicotinamide)₄]Cl₂•7H₂O **17** showing part of the hydrogen bonded chains formed by linking amide groups of neighbouring cations.

Zawarotko and coworkers have employed a similar strategy in the design of diamantoid networks based upon O–H⋯N hydrogen bonding between linear diamines and metallocubane-type compounds ([M(CO)₃(μ₃-OH)]₄; M = Mn, Re) that have tetrahedrally disposed hydroxyl ligands [36]. In a slightly different approach, Mingos and coworkers have constructed new ligands that have at their periphery arrangements of hydrogen bond donors and acceptors analogous to those involved in the base-pairing of DNA. These ligands have been used to prepare metal complexes that can undergo self-assembly to yield a variety of arrangements including tapes and sheets in the solid-state [3g].

Each of these examples also illustrates the general idea of transferring organic supramolecular synthons to organometallic systems. Further exploration of this idea has also been conducted by Braga, Grepioni, Desiraju and coworkers [37] using the CSD.

208

5. Conclusions and Prospects for New Technologies.

In this article the role of metals in hydrogen bonding has been discussed. While this presentation is not intended to be comprehensive, a variety of examples from our own work and that of other groups has been used to give an overview of the possible roles that metals may play. Interested readers are also recommended to read the well-referenced review by Raymo and Stoddart on strategies and applications in second-sphere coordination at transition metal centres [3h], as well as an excellent chapter by Dance [38], which provides a broader discussion of inorganic supramolecular chemistry.

When one turns to potential applications and the prospects for new technologies the relationship between metals and hydrogen bonding is bi-directional. Metals can be used to influence hydrogen bonds either by direct participation or by indirect means as has been illustrated. Conversely, hydrogen bonds may be used to incorporate metals into supramolecular structures with control of the relative positions of the metal sites. These two descriptions are, of course, not mutually exclusive.

In the first scenario, the potential applications may be in providing some control of stoichiometric or catalytic chemical synthesis (sections 2, 3), for example in directing protonation reactions of metal complexes (3.2.1,2, also see ref. 26) or control of conformation (3.2.3). Metal coordination geometry may also be used as a template for construction of extended hydrogen bonded arrays, including cavity-containing structures [33, 36b].

In the second scenario, supramolecular structures, especially infinite solid-state structures, may be constructed based upon hydrogen bonding interactions involving metal-containing molecules or ions. This has the potential to place selected metals in a chosen arrangement leading to the possibility of exploiting various electronic and magnetic properties of the metals in designing macroscopic physical properties (*e.g.* magnetic, charge-transfer, photoelectronic, *etc.*) of the material. In this regard Braga, Grepioni and coworkers have recently demonstrated that cationic metallocene guests can be incorporated in hosts constructed of strongly hydrogen-bonded organic molecules, with host-guest interactions taking place via weaker C–H···O hydrogen bonds [39].

6. Acknowledgements

The contribution of my current and former coworkers at University of Missouri-St. Louis, especially Juan Mareque, Eric Bruton, and Dong Zhao, and that of collaborators, especially Paul Sherwood (CLRC Daresbury Laboratory) and Morris Bullock (Brookhaven National Laboratory) is gratefully acknowleged. Funding from the ACS Petroleum Research Fund, NATO, and various grants awarded by the University of Missouri are also acknowledged.

7. References

1. For example, see (a) Pimentel, G. C. and McClellan, A. L. (1960) *The Hydrogen Bond*, W. H. Freeman, San Fransisco. (b) Schuster, P., Zundel, G., and Sandorfy, C. (eds.), (1976) *The Hydrogen Bond: Recent Developments in Theory and Experiment*, North-Holland, Amsterdam. (c) Jeffrey, G. A. and Saenger, W. (1991) *Hydrogen Bonding in Biological Structures*, Springer-Verlag, Berlin. (d) Jeffrey, G. A. (1997) *An Introduction to Hydrogen Bonding*, Oxford University Press, Oxford.

2. (a) Desiraju, G. R. (1989) *Crystal Engineering: The Design of Organic Solids*, Elsevier, Amsterdam. (b) Sharma C. V. K. and Desiraju, G. R. (1996) in Desiraju, G. R. (ed.), *Perspectives in Supramolecular Chemistry, Vol. II. The Crystal as a Supramolecular Entity*, Wiley, New York. (c) Aakeröy, C. B. (1997) *Acta Crystallogr.* **B53**, 569. (d) Subramanian, S. and Zawarotko, M. J. (1994) *Coord. Chem. Rev.* **137**, 357.

3. A recent symposium has been devoted to this topic: "Hydrogen Bonding in Inorganic and Organometallic Chemistry," ACS National Meeting, Boston, MA, August 1998 (L. Brammer and R. H. Crabtree, organizers). For pertinent reviews, see (a) Brammer, L., Zhao, D., Ladipo, F. T., and Braddock-Wilking, J. (1995) *Acta Crystallogr.* **B51**, 632. (b) Shubina, Ye. S., and Epstein, L. M. (1992) *J. Mol. Struct.* **265**, 367. (c) Epstein, L. M. and Shubina, E. S. (1992) *Metallorganich. Khim* **5**, 61 (English translation, *idem* (1992) *Organomet. Chem. USSR* **5**, 1). (d) Canty, A. J. and van Koten, G. (1995) *Acc. Chem. Res.* **28**, 406. (e) Crabtree, R. H., Siegbahn, P. E. M., Eisenstein, O., Rheingold, A. L., and Koetzle, T. F. (1996) *Acc. Chem. Res.* **29**, 348. (f) Braga, D., and Grepioni, F. (1997) *Acc. Chem. Res.* **30**, 81. (g) Burrows, A. D., Chan, C.-W., Chowdry, M. M., McGrady, J. E., and Mingos, D. M. P. (1995) *Chem. Soc. Rev.* **24**, 329. (h) Raymo, F. M. and Stoddart, J. F. (1996) *Chem. Ber.* **129**, 981.

4. (a) Adams, M. A., Folting, K., Huffman J. C., and Caulton, K. G. (1979) *Inorg. Chem.* **18**, 3020. (b) Epstein, L. Shubina, E. S., Krylov, A. N., Kreindlin, A. Z., and Ribinskaya, M. I. (1993) *J. Organomet. Chem.* **447**, 227. (c) Fairhurst, S. A., Henderson, R. A., Hughes, D. L., Ibrahim, S. K. and Pickett, C. J. (1995) *J. Chem. Soc., Chem. Commun.* 1569. (d) Peris, E. and Crabtree, R. H. (1995) *J. Chem. Soc., Chem. Commun.* 2179. (e) Braga, D. Grepioni, F., Tedesco, E., Biradha, K., and Desiraju, G. R. (1996) *Organometallics* **15**, 2692.

5. (a) Chatt, J., Duncanson, L. A. and Venanzi, L. M. (1958) *J. Chem. Soc.* 3203; Duncanson L. A. and Venanzi, L. M. (1960) *J. Chem. Soc.* 3841. (b) Baker, A. W. and Bublitz, D. E. (1965) *Spectrochim. Acta* **22**, 1787. (c) Roe, D. M., Bailey, P. M., Moseley, K., and Maitlis, P. M. (1972) *J. Chem. Soc., Chem. Commun.*, 1273. (d) Drago, R. S., Nozari, M. S., Klinger, R. J., and Chamberlain, C. S. (1979) *Inorg. Chem.* **18**, 1254.

6. (a) Brammer, L., Charnock, J. M., Goggin, P. L., Goodfellow, R. J., Orpen, A. G., and Koetzle, T. F. (1991) *J. Chem. Soc., Dalton Trans.* 1789, and refs. therein. (b) Brammer, L., McCann, M. C., Bullock, R. M., McMullan, R. K., and Sherwood, P. (1992) *Organometallics* **11**, 2339. (c) Brammer, L. and Zhao, D. (1994) *Organometallics* **13**, 1545. (d) Zhao, D, Ladipo, F. T, Braddock-Wilking, J., Brammer, L., and Sherwood, P. *Organometallics* **15**, 1441.

7. (a) Calderazzo, F., Fachinetti, G., Marchetti, F., and Zanazzi, P. F. (1981) *J. Chem. Soc., Chem. Commun.*, 181. (b) Cecconi, F., Ghilardi, C.A., Innocenti, P., Mealli, C., Midollini, S., and Orlandini, A. (1984) *Inorg. Chem.* **23**, 922. (c) Wehman-Ooyevaar, I. C. M., Grove, D. M., Kooijman, H., van der Sluis, P., Spek, A. L., and van Koten, G. (1992) *J. Am. Chem. Soc.* **114**, 9916. (d) Wehman-Ooyevaar, I. C. M., Grove, D. M., de Vaal, P., Dedieu, A., and van Koten, G. (1992) *Inorg. Chem.* **31**, 5484. (e) Kazarian, S. G., Hamley, P. A., and Poliakoff, M. (1993) *J. Am. Chem. Soc.* **115**, 9069. (f) Albinati, A., Lianza, F., Müller, B., and Pregosin, P. S. (1993) *Inorg. Chim. Acta* **208**, 119. (g) Albinati, A., Lianza, F., Pregosin, P. S., and Müller, B. (1994) *Inorg. Chem.* **33**, 2522. (h) Shubina, Ye. S., and Epstein, L. M. (1992) *J. Mol. Struct.* **265**, 367. (i) Shubina, E. S., Krylov, A. N., Kreindlin, A. Z., Rybinskaya, M. I., and Epstein, L. M. (1993) *J. Mol. Struct.* **301**, 1. (j) Shubina, E. S., Krylov, A. N., Kreindlin, A. Z., Rybinskaya, M. I., and Epstein, L. M. (1994) *J. Mol. Struct.* **465**, 259. (k) Braga, D., Grepioni, F., Tedesco, E., Biradha K., and Desiraju, G. R. (1997) *Organometallics* **16**, 1846. (l) Gao, Y., Eisenstein O., and Crabtree, R. H. (1997) *Inorg. Chim. Acta* **254**, 105.

8. Mareque Rivas, J. M. and Brammer, L. (1999) *Coord. Chem. Rev.* **183**, in press.

9. Dunitz, J. D. and Bürgi, H.-B. (1983) *Acc. Chem. Res.* **16**, 153.

10. Mareque Rivas, J. M. and Brammer, L. (1998) *Inorg. Chem.* **37**, 5512.

11. Brammer, L. and Sherwood, P. (1999) in preparation.

210

12. Feil, D. (1991) in G. A. Jeffrey and J. F. Piniela (eds.), *The Application of Charge Density Research To Chemistry and Drug Design*, Plenum Press, New York, *NATO ASI Series B: Physics* **250**, 103.

13. Yoshida, T., Tani, K., Yamagata, T., Tatsuno, Y. and Saito, T. (1990) *J. Chem. Soc., Chem Commun.*, 292.

14. Allen, F. H. and Kennard, O. (1993) *Chem. Design Automation News* **8**, 1 & 31.

15. Braga, D., Grepioni, F., Tedesco, E., Wadepohl, H., and Gebert, S. (1997) *J. Chem. Soc., Dalton Trans.*, 1727.

16. (a) For a study of organofluorine as a hydrogen bond acceptor, see Dunitz, J D. and Taylor, R. (1997) *Chem. Eur. J.* **3**, 89. (b) Aullón, G., Bellamy, D., Brammer, L., Bruton, E. A., and Orpen, A. G. (1998) *Chem Commun.*, 653.

17. Yap, G. P. A., Rheingold, A. L., Das, P., and Crabtree, R. H. (1995), *Inorg. Chem.* **34**, 3474.

18. Brammer, L. and Bruton, E. A., (1999) in preparation.

19. Mareque Rivas, J. M. and Brammer, L. (1998) *Inorg. Chem.* **37**, 4756.

20. For a review, see, Crabtree, R. H., Siegbahn, P. E. M., Eisenstein, O. Rheingold, A. L., and Koetzle, T. F. (1996) *Acc. Chem. Res.* **29**, 348.

21. The situation is perhaps more extreme for H (*cf.* halogens) as, while hydridic M–H can be moderately strong acceptor (hydrogen bonds of *ca.* 5 kcal/mol), its carbon-bound counterpart, C–H, obviously has the reverse bond polarity of that needed to be a hydrogen bond acceptor.

22. Peris, E., Lee, J. C., Rambo, J. R., Eisenstein, O., and Crabtree, R. H. (1995) *J. Am. Chem. Soc.* **117**, 3485.

23. For example, see Kubas G. J. (1988) *Acc. Chem. Res.* **21**, 120.

24. Lee, J. C., Rheingold, A. L., Müller, B., Pregosin, P. S., and Crabtree, R. H. (1994) *J. Chem. Soc., Chem Commun.*, 1021.

25. Albinati, A., Bakhmutov, V. I., Caulton, K. G., Clot, E., Eckert, J. Eisenstein, O., Gusev, D. G., Grushin, V. V., Hauger, B. E., Klooster, W. T., Koetzle, T. F., McMullan, R. K., O'Loughlin,, T. J., Pélissier, M., Ricci, J. S., Sigalas, M. P., and Vymenits, A. B. (1993) *J. Am. Chem. Soc.* **115**, 7300.

26. Kuhlman, R. (1997) *Coord. Chem. Rev.* **167**, 205.

27. Braga, D., Biradha, K., Grepioni, F., Pedireddi, V. R., and Desiraju, G. R. (1995) *J. Am. Chem. Soc.* **117**, 3156.

28. Mareque Rivas, J. M., Ph.D. Thesis, University of Missouri-St. Louis, 1999.

29. For example, see (a) Nishio, M. and Hirota, M. (1989) *Tetrahedron* **45**, 7201. (b) Nishio, M., Umezawa, Y., Hirota, M. and Takeuchi, Y. (1995) *Tetrahedron* **51**, 8665. (c) McMullan, R. K., Kvick, Å. and Popelier, P. (1992) *Acta Crystallogr.* B48, 726. (d) Mootz, D. and Deeg, A. (1992) *J. Am. Chem. Soc.*114, 5887. (e) Deeg, A. and Mootz, D. (1993) *Z. Naturforsch* 48b, 571. (f) Steiner, T. Starikov, E. B., Amado, A. M. and Teixeira-Dias, J. J. C. (1995) *J. Chem. Soc., Perkin Trans.* 2, 1323. (g) Weiss, H.-C. Bläser, D. Boese, R. Doughan, B. and Haley, M. M. (1997) *Chem. Commun.*, 2403. (h) Weiss, H.-C. Boese, R. Smith, H. L. and Haley, M. M. (1997) *Chem. Commun.*, 1703. (i) Allen, F. H., Howard, J. A. K., Hoy, V. J., Desiraju, G. R., Reddy, D. S., and Wilson, C. C. (1996) *J. Am. Chem. Soc.* **118**, 4081.

30. Lu, K.-L., Su, C.-J., Lin, Y.-W., Gau, H.-M., and Wen, Y.-S. (1992) *Organometallics* **11**, 3832.

31. For example, see (a) Zawarotko, M. J. (1994) *Chem. Soc. Rev.* **23**, 283. (b) Keller, S. W. (1997) *Angew. Chem., Int. Ed. Engl.* **36**, 247 and refs. therein.

32. (a) Stang, P.J. and Olenyuk (1997) *Acc. Chem. Res.* **30**, 502. (b) Stang, P. J. (1998) *Chem. Eur. J.* **4**, 19. (c) Fujita, M. and Ogura, K. (1996) *Coord. Chem. Rev.* **148**, 249.

33. Mareque Rivas, J. M. and Brammer, L. (1998) *New. J. Chem*, 1315.

34. For an explanation of the graph set notation and its application to describing hydrogen bonding patterns, see, (a) Etter, M. C. (1990) *Acc. Chem. Res.* **23**, 120; (b) Bernstein, J., Davis, R. E., Shimoni, L., and Chang, N.-L. (1995) *Angew. Chem., Int. Ed. Engl.* **34**, 1555.

35. Aakeröy, C. B. and Beatty, A. M. (1998) *Chem. Commun.*, 1067; (b) Aakeröy, C. B. Beatty A. M. and Helfrich, B. A. (1998) *J. Chem. Soc., Dalton Trans.*, 1943.

36. (a) Copp, S. B., Subramanian, S, and Zawarotko, M. J. (1992) *J. Am. Chem. Soc.* **114**, 8719. (b) Copp, S. B., Subramanian, S, and Zawarotko, M. J. (1993) *J. Chem. Soc., Chem. Commun.*, 1078.

37. (a) Braga, D., Grepioni, F., Sabatino, P., and Desiraju, G. R. (1994) *Organometallics* **13**, 3532. (b) Biradha, K., Desiraju, G. R., Braga, D., and Grepioni, F. (1996) *Organometallics* **15**, 1284.

38. Dance, I. (1996) in Desiraju, G. R. (ed.), *Perspectives in Supramolecular Chemistry, Vol. II. The Crystal as a Supramolecular Entity*, Chapter 5, Wiley, New York.

39. (a) Braga, D., Costa, A. L., Grepioni, F., Scaccianoce, L., and Tagliavini, E. (1996) *Organometallics* **15**, 1084. (b) Braga, D. Angeloni, A., Grepioni, F., and Tagliavini, E. (1997) *Organometallics* **16**, 5478.

SUPRAMOLECULAR ORGANIZATION IN ORGANOMETALLIC CRYSTALS

FABRIZIA GREPIONI and DARIO BRAGA
*Dipartimento di Chimica 'G. Ciamician', Università di Bologna,
Via Selmi 2, 40126 Bologna, Italy.*

1. Introduction

Structural variability and structural flexibility are important "molecular" characteristics of organometallic compounds which are reflected in the supramolecular organization at crystal structure level.

Organometallic compounds are those in which the carbon atoms of organic groups are bound to metal atoms *via* two-electron σ-bonds or *via* π-interactions of unsaturated groups with metal orbitals of appropriate orientation and symmetry. The organic residue bound to the metal centers is often a molecule itself, that is to say a stable and isolable chemical entity. Molecules and ions of great complexity can be obtained by increasing the number of metal atoms or by varying the number and type of ligands, the coordination geometry, and by taking advantage of the variable oxidation states of the metal atoms.

Most organometallic molecules are structurally non-rigid because of two distinctive features of the bonding between the metal center(s) and the ligands, *viz.* the availability of *almost* isoenergetic, though geometrically different, bonding modes for the same ligand and the delocalized nature of the bonding interactions between unsaturated π-systems (aromatic rings, alkenes, alkynes, *etc.*) and the metals.

Many organometallic molecules exist in several isomeric forms which can interconvert *via* low-energy processes (*viz.* reorientation, scrambling, fluxionality) both in the gas phase and in the condensed state. Organometallic structural variability and flexibility ought to be taken into account when approaching organometallic solids.

The combination of stereogeometrical constraints, imposed by metal coordination, with the intermolecular bonding capacity of the ligands and of the metal atoms results in profound differences between the characteristics of organometallic and organic crystals, even though the periphery of the constituent molecules are, in both cases, essentially organic in nature. The interactions between organometallic molecules in the crystal will be responsible for *collective* properties such as magnetism, conductivity and non-linear

J.A.K. Howard et al. (eds.),
Implications of Molecular and Materials Structure for New Technologies, 211–222.
© 1999 *Kluwer Academic Publishers. Printed in the Netherlands.*

optical response as well as for the behavior of the crystals with changing temperature and pressure.

This contribution is devoted to an analysis of the role played by metal atoms in intermolecular bonding and its relevance to the emerging new field of organometallic crystal engineering. As organometallic chemistry is a bridge between organic and inorganic chemistry, organometallic crystal engineering promises to bridge the traditional fields of organic crystal engineering and inorganic solid-state chemistry.

Many factors that influence and guide organic crystal packing are also important in the crystal engineering of organometallic compounds. This is because the metal centers in these compounds are mostly situated in the molecular cores and are well shielded from neighboring molecules. Since these peripheries are crucial in determining crystal structures, the packing problem reduces, in most instances, to the organic case. However, the interplay between ligand supramolecular bonding capacity and ligand-to-metal(s) coordination is responsible for distinct features of organometallic crystal architectures which will be discussed, for practical reasons, in the following order:

(i)   the effect of spatial geometry (topological effect),

(ii) the effect of coordination on the acid/base behavior of the ligands,

(iii) the direct participation of the metal atoms in intermolecular bonding, and

(iv) the effect of ionic charge

2. Topological Effect

Ligand geometry can be combined with coordination geometry around the metal center(s) *to preorganize in space* non-covalent linkages. For example, Mingos *et al.* have constructed solid-state sheets and tapes *via* cocrystallization of bifunctional transition metal complexes and organic bases by using complementary hydrogen bonds.

Walther *et al.*, in collaboration with us, have investigated the architecture of crystalline alkynol and alkynediol transition metal complexes of Pt and Ni in relation to that of crystalline organic alcohols. As an example the molecular row connected *via* (-O-H)$_4$ hydrogen bond rings on either side of the metal centers in the case of the crystalline complexes *bis*(t-butyl-methyl-alkynol)M(0) (M = Ni, Pt) is shown in the following figure:

SCHAKAL

One-dimensional aggregation in the solid state has been obtained by van Koten *et al.* for square-planar Pt-complexes linked by intermolecular hydrogen bonds of the -O-H···Cl or -C≡C-H···Cl types.

Zaworotko *et al.* have used cubane-like molecules such as $[M(CO)_3(OH)_4]$ (M = Mn, Re), which possess groups capable of hydrogen bond formation directed towards the vertices of a tetrahedron, to synthesize superdiamantoid networks with large empty space available for enclathration.

3. Effect of Coordination on the Acid/Base Behavior of Ligands

The acid/base behavior of ligands, hence their participation in hydrogen bonding interactions, can be *tuned* by changing the ligand-to-metal coordination mode, which, in turn, depends on the electronic nature of the metal.

For example, the hydrogen atoms belonging to metal coordinated methylidyne (μ_3-CH) and methylene (μ_2-CH$_2$) ligands have been shown to participate in intermolecular C-H---O hydrogen bonding networks. The acceptor atom belongs, in most cases, to a CO ligand. The average length of the H---O separations suggests that the μ_2-CH$_2$ ligand is less acidic that the μ_3-CH one. This is consistent with experimental and theoretical evidence indicating sp^2(=CH$_2$) and sp(≡CH) hybridization for the C-atom in the two types of coordination geometry. Hence, the hydrogen bonding capacity of the methylidyne and methylene ligands recalls that of *organic* ≡C-H and =CH$_2$ hydrogen bond donors which have been shown to be amongst the most acidic C-H systems.

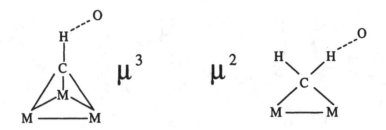

The hydrogen bond accepting capacity of the CO ligand has also been extensively studied. This ligand can adopt terminal, doubly (μ_2-CO) and triply (μ_3-CO) bridging coordination modes in transition metal complexes and clusters. The propensity of the CO ligand to adopt bridging bonding modes decreases on descending a group and increases on moving from left to right in the Periodic Table, according to the increasing need for electron delocalization and also to the contraction of the d-orbitals on increasing atomic number. The oxygen atom participates in acid-base interactions with Lewis acids (such as electropositive metal atoms in isocarbonyls and carbonylate anions) or Brønsted acids (such as water, -O-H and C-H groups). The order of basicity is terminal < μ_2-bridge < μ_3-bridge in keeping with the shift of stretching frequencies towards lower wavenumbers. The average length of the Tr-C-O--A interaction (where A is a Lewis or Brønsted acid) follows the reverse trend, *viz.* bridging ligands form shorter intermolecular bonds, and/or tighter ion-pairs with metal cations, than terminal ligands. In crystalline salts, Tr-C-O---H-X interactions, usually between carbonyl anions and organic-type cations, follow the same trend as above, though reinforced by the difference in charge (see below). The hydrogen bonding interactions of the terminal CO---H, μ_2-CO---H, and μ_3-CO---H are represented in the following scheme:

C-H---O bonds in organometallic compounds are also directional, with the C-O---H angle being, on average, around 140°, irrespective of the mode of bonding, *i.e.* both terminal and bridging ligands are approached in ketonic-like directions. This observation indicates that, at least in the solid state and in the presence of C-H donors, there is O-atom lone pair density in ketonic directions even when the CO-ligand is in the terminal bonding mode. An intriguing alternative explanation for the angularity of the C-H---O interaction in the case of terminal ligand may be the need to optimize the interaction between the X-H donor and both C-O triple bond system and the lone pair on the oxygen. This model was suggested by Daresbourg to explain the angularity of the interaction of CO with electrophiles. The sp-hybrid should favor a straight interaction, while the ketonic-type sp^2 structure is consistent with an angular interaction. This latter model is also in agreement with the lowering of the $v(CO)$ stretching frequency associated with the M-CO---Edge interaction. The two models are depicted in the following scheme:

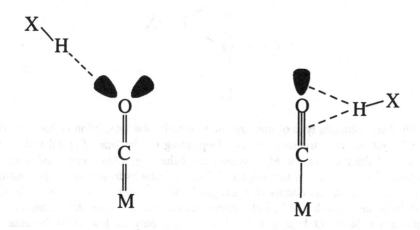

4. Effect of Direct Participation of Metal Atoms in Intermolecular Bonds

The direct involvement of metal atoms in intermolecular interactions is, obviously, one of the major differences between organic and organometallic crystal engineering. Once again the reference to the Periodic Table is immediate. Electron-deficient metal atoms on the left-hand side are able to accept electron density both intra- and intermolecularly from suitable Lewis bases (that could even be a C-H σ-bond in the case of agostic or pseudo-agostic interactions). Conversely, electron-rich metal atoms on the right hand side of the Periodic Table will be able to donate electron density and act, for example, as hydrogen bond acceptors towards organic or inorganic bases, as shown in the following scheme:

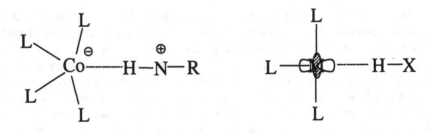

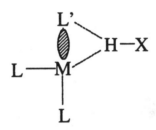

Another distinctive type of non-organic intermolecular interaction is that involving the Tr-H system (Tr = transition metal). Depending on the type of metal and on the nuclearity of the complex the M-H system can behave as a hydrogen bond donor or acceptor. It has been shown that edge and face bridging hydrogen atoms in transition metal clusters can act as donors in hydrogen bonds to "soft" bases such as the CO-ligand. In the large number of hydride clusters characterized to date, the formation of an intermolecular M-H---O hydrogen bond is evident only in few cases because the hydrogen ligands are usually sterically screened from the neighboring molecules by the other bulkier ligands.

Several examples of metal atoms acting as hydrogen bond acceptors are known. After the first report by Calderazzo *et al.* of a short N-H---Co interaction in crystalline $[NMe_3H^+][Co(CO)_4^-]$, Brammer has investigated systematically three-center four-electron N-H$\cdots$Co hydrogen bonds formed by the carbonyl anion $Co(CO)_4^-$ with counterions of the NR_3H^+ type (R = Me, Et).

The "agostic" concept has been extended to intermolecular interactions. It has been demonstrated that intermolecular pseudo-agostic (IPA) M---(H-X) interactions in which X-H σ-bonds (X = C, N, O) can donate electron density to electron-deficient metal atoms are a fairly common phenomenon. IPA interactions are usually, though not always, associated with the presence of tight ion pairs formed by an electron-deficient metal cation and a counterion carrying methyl or phenyl groups. Since C-H groups are more abundant than O-H or N-H groups, IPA interactions are found more frequently when X = C. M---(H-X) interactions were first observed in the ion-pair adducts between anionic Lewis acids and electron-deficient and coordinately unsaturated zirconocene compounds, such as $[(C_5Me_5)_2ZrMe][(C_6F_5)_3BMe]$. The Zr---(H-C) interactions are characterized by the presence of short Zr---H *and* Zr---C contacts due to the close approach of the B-atom bound CH_3 group to the electron deficient Zr-atom, as shown in the following scheme:

On closing this section, it should be mentioned that in a small number of cases "dihydrogen bonds" can be formed between two hydrogen atoms. One of these is bound to a metal atom and the other bound to an electronegative main group atom (C, N, O, S), as shown in the following scheme:

$$X = O, N, C, S$$

This interaction manifests itself not only with an H---H distance below the usual sum of van der Waals distance (*i.e.* less than 2.34 Å), but also with specific spectroscopic properties. The Ir-H---H-X interaction in mononuclear Ir complexes has also been the subject of theoretical studies. The existence of attractive M-H---H-X interactions demonstrates the duality of the M-H system with regard to its participation in hydrogen bonding interactions. These interactions involve, in general, terminally bound H-atoms. This intriguing type of hydrogen bonding interaction is specific to organometallic systems and has no parallel in the neighboring organic chemistry field.

5. Effect of the Ionic Charge

As mentioned above, the role of ionic charge in organometallic and inorganic crystal engineering is important since many organometallic building blocks are ions, with variable metal oxidation states and/or non-neutral ligands.

The effect of charge in crystal engineering depends on the number of atoms forming the ions. While the aggregation of large molecular ions (*e.g.* many transition metal clusters) carrying a small ionic charge follows essentially the same criteria as those for co-crystals (though formed by species of different polarity), this is not the case for low nuclearity complexes where the charge is "shared" by a smaller number of atoms. In contrast to hydrogen bonding interactions, electrostatic interactions are non-directional and active at long range as they fall off very smoothly with distance.

Organometallic anions and cations are often crystallized with large organic-type counterions, such as PPh_4^+, PPN^+, BPh_4^- *etc.* The ionic charge is usually small (-1 or -2, rarely higher) and is distributed over a large number of atoms. Carbonyl cluster anions, for example, form crystals in which the counterion size and shape is the prinicpal factor determining the anion aggregation pattern in the crystal.

The general rule is relatively simple: small cations favor one-dimensional or two-dimensional aggregation of the anions. When both ions are of comparable size and are nearly spherical in shape, the crystal is constructed as a mixed system in which anions and cations are distributed as van der Waals particles.

This information can be utilized to engineer anisotropic arrangements of the particles in the solid state. It has been shown that the size of the cations becomes a critical factor as the dimension of the anion increases. In fact, high-nuclearity transition metal clusters can be more easily isolated as crystalline materials when the anion charge is high, *viz.* when a large number of cations are "brought" in the crystal by a single anion. In some cases solvent of crystallization is also needed to fill in interstices in the crystals and improve cohesion.

The effect of charge becomes crucial when hydrogen bond donor/acceptor groups are present since the hydrogen bond is basically electrostatic in nature, hence all factors influencing the charge distribution on the donor/acceptor system affect the strength of the hydrogen bond. This consideration has been exploited in the preparation of organic-organometallic crystalline aggregates by means of co-operative strong O-H---O and weak C-H---O hydrogen bonds.

The synthetic procedure is founded on the reaction between organic or inorganic molecules possessing acidic protons and organometallic molecules that can be easily oxidized to yield organometallic hydroxides, such as $[(\eta^6\text{-arene})_2Cr][OH]$ (arene = η^6-C_6H_6, η^6-C_6H_5Me) and $[(\eta^5\text{-}C_5H_5)_2Co][OH]$ prepared *in situ*. The products of these reactions are supramolecular salts formed *via* hydrogen bonding interactions.

The design criteria can be summarized as follows:

(i) strong donor/acceptor hydrogen bonding groups must be present on the organic but not on the organometallic fragments for the selective self-assembling of the organic fragments;

(ii) a large number of acidic C-H groups, such as those carried by arene and cyclopentadienyl ligands, is needed to make use of the 'free' hydrogen bonding acceptor sites on the organic framework;

(iii) opposite charges on the two types of fragment reinforce the C-H---O hydrogen bonding interactions which hold together the organic superstructure around the organometallic cations.

According to this crystal synthesis strategy several organic-organometallic aggregates have been obtained. The most relevant aspect is that one can utilize enantiomerically pure polycarboxylic acids to obtain chiral organic networks in which the organometallic cations can be accommodated. Examples of this type are provided by the crystalline materials obtained with L-benzoyl tartaric acid and L-tartaric acid. In this latter species, of formula $[(C_5H_5)_2Co]^+[L\text{-}HTA]^-$, the formation of a tridimensional anionic network has been observed, with channels which are filled by piles of cobalticinium cations, as shown in the following figure:

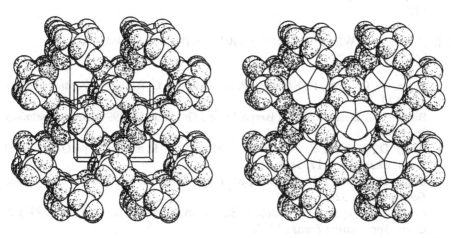

SCHAKAL

These results open up a convenient route to the engineering of chiral crystals containing organometallic ions or molecules carrying delocalized electron systems and/or aligned spin systems.

6. Acknowledgments

Financial support by MURST and by the University of Bologna (projects: *Intelligent molecules and molecular aggregates 1995-1997* and *Innovative Materials 1997-1999*), the ERASMUS program 'Crystallography', the CNR-JNICT bilateral project, the Deutscher Akademischer Austauschdienst, Bonn, and the Conferenza Nazionale dei Rettori, Roma (German-Italian scientific exchange grant: Vigoni Program) are acknowledged.

7. Relevant References

7.1. SUPRAMOLECULAR ORGANIZATION OF ORGANOMETALLIC MOLECULES AND IONS

1. Braga, D. and Grepioni, F. (1994) *Acc. Chem. Res.* **27**, 51.
2. Braga, D. and Grepioni, F. (1996) *Chem. Commun.*, 571.
3. Braga, D. and Grepioni, F. (1997) *Comments Inorg. Chem.* **19**, 185.
4. Braga, D., Grepioni, F., Milne, P. and Parisini, E. (1993) *J. Am. Chem. Soc.* **115**, 5115.

7.2. STRONG AND WEAK HYDROGEN BONDS IN ORGANOMETALLIC CRYSTALS

1. Braga, D., Grepioni, F., Sabatino, P. and Desiraju. G. R. (1994) *Organometallics* **13**, 3532.
2. Biradha, K., Desiraju, G. R., Braga, D. and Grepioni, F. (1996) *Organometallics* **15**, 1284.
3. Braga, D., Grepioni, F., Biradha, K., Pedireddi, V. R. and Desiraju, G. R. (1995) *J. Am. Chem. Soc.* **117**, 3156.
4. Braga, D., Byrne, J. J., Calhorda, M. J. and Grepioni, F. (1995) *J. Chem. Soc., Dalton Trans.*, 3287.
5. Braga, D., Grepioni, F., Tedesco, E., Gebert, S. and Wadepohl, H. (1997) *J. Chem. Soc., Dalton Trans.*, 1727.
6. Wadepohl, H., Braga, D. and Grepioni, F. (1995) *Organometallics* **14**, 24.
7. Braga, D., Grepioni, F., Tedesco, E., Biradha, K. and Desiraju, G. R. (1997) *Organometallics* **16**, 1846.
8. Braga, D., Grepioni, F., Tedesco, E., Biradha, K. and Desiraju, G. R. (1996) *Organometallics* **15**, 2692.

9. Braga, D., Grepioni, F., Biradha, K. and Desiraju. G. R. (1996) *J. Chem. Soc., Dalton Trans.*, 3925.

10. Braga, D. and Grepioni, F. (1997) *Acc. Chem. Res.* **30**, 81.

11 Braga, D., Grepioni, F. and Novoa J.J. (1998) *Chem. Commun.*, 1959.

7.3. ORGANOMETALLIC CRYSTAL ENGINEERING

1. Braga, D., Grepioni, F. and Desiraju, G.R. (1998) *Chem. Rev.*, 1375.

2. Braga, D. and Grepioni, F. (1999) *J. Chem. Soc., Dalton Trans.*, 1.

3. Braga, D. and Grepioni, F. (1997) *Coord. Chem. Rev.*, in press.

4. Braga, D., Costa, A. L., Grepioni, F., Scaccianoce, L. and Tagliavini, E. (1997) *Organometallics* **16**, 2070.

5. Braga, D., Grepioni, F., Byrne, J. J. and Wolf, A. (1995) *J. Chem. Soc., Chem. Commun.*, 1023.

6. Braga, D., Costa, A. L., Grepioni, F., Scaccianoce, L. and Tagliavini, E. (1996) *Organometallics* **15**, 1084.

7. Braga, D., Angeloni, A., Grepioni, F. and Tagliavini, E. (1997) *Chem. Commun.*, 1447.

8. Braga, D., Angeloni, A., Grepioni, F. and Tagliavini, E. (1997) *Organometallics* **16**, 5478.

7.4. RECENT REFERENCES ON NON-COVALENT INTERACTIONS IN CRYSTALS

1. Desiraju, G. R. (1995) *Angew. Chem., Int. Ed. Engl.* **34**, 2311.

2. Dunitz, J. and Taylor, R. (1997) *Chem. Eur. J.* **3**, 89.

3. Jeffrey, G. A. (1997) *An Introduction to Hydrogen Bonding,* Oxford University Press, New York.

4. Aakeröy, C. B. and Seddon, K. R. (1993) *Chem. Soc. Rev.*, 397.

5. Burrows, A. D., Chan, C.-W., Chowdry, M. M., McGrady, J. E. and Mingos, D.M. P. (1995) *Chem. Soc. Rev.*, 329.

6. Bernstein, J., Davis, R. E., Shimoni, L. and Chang, N.-L. (1995) *Angew. Chem., Int. Ed. Engl.* **34**, 1555.

7. Bernstein, J., Etter, M. C. and Leiserowitz. L. (1994) in Bürgi, H.-B. and Dunitz, J. D. (eds), *Structure Correlation*, VCH, Weinheim, p. 431.

8. Desiraju, G. R. (1996) *Acc. Chem. Res.* **29**, 441.

9. Steiner, T. (1996) *Cryst. Rev.* **6**, 1.

10. Steiner, T. (1996) *Chem. Commun.*, 727.

11. Brammer, L., Zhao, D., Ladipo, F. T. and Braddock-Wilking, J. (1995) *Acta Crystallogr.* **B51**, 632.

12. Pyykkö, P. (1997) *Chem. Rev.* **97**, 597.

7.5. SOME RECENT REFERENCES ON CRYSTAL ENGINEERING

1. Sharma, C. V. K. and Desiraju, G. R. (1996) in Desiraju, G. R. (ed.), *Perspectives in Supramolecular Chemistry. The Crystal as a Supramolecular Entity,*. Wiley, Chichester.
2. Desiraju, G. R. (1989) *Crystal Engineering: The Design of Organic Solids*, Elsevier, Amsterdam.
3. Aakeröy, C. B. (1997) *Acta Crystallogr.* **B53**, 569.
4. Etter M.C. (1990) *Acc. Chem. Res.* **23**, 120.
5. Subramanian, S. and Zaworotko, M. J. (1994) *Coord. Chem. Rev.* **137**, 357.
6. Zaworotko, M. J. (1997) *Nature* **386**, 220.

THEORETICAL APPROACHES TO THE STUDY OF NON-BONDED INTERACTIONS

S.L. PRICE
Centre for Theoretical and Computational Chemistry
University College London
20 Gordon Street
London WC1H 0AJ

There have been two main approaches to modelling the weak intermolecular forces between closed shell molecules for simulating the properties of the molecular solid, liquid and gas. Quite detailed model intermolecular potentials, specific to each molecule, are used for small polyatomics, such as HF, Cl_2 and water. These model potentials are often derived, at least in part, from *ab initio* calculations, and are tested for their ability to reproduce the spectra of van der Waals dimers or molecular beam experiments as well as condensed phase simulations. In contrast, the models for intermolecular forces used to simulate organic and biochemical interactions are mainly derived by assuming that each intermolecular atom-atom interaction is transferable between different molecules. Such models are usually derived empirically, by fitting to a range of experimental data such as molecular crystal structures. Such model potentials are now used to improve our understanding of biochemical interactions, including drug design. The realism and reliability of such simulations depends fundamentally on the accuracy of the model for the intermolecular forces, and thus we seek more accurate model potentials for organic molecules. Most of the recent progress towards this goal has come from extending the ideas and techniques, which are used to develop accurate potentials for small polyatomics, to larger molecules.

1. Physical Origins of the Contributions to the Intermolecular Energy

1.1. THE PAIRWISE ADDITIVE APPROXIMATION

In calculating the lattice energy of a molecular crystal, or the potential energy of a liquid, or indeed the energy of any ensemble of N molecules relative to their energy when completely separated, it is usual to assume that it is equal to the sum of the interactions between every pair of molecules in the ensemble. However, formally this energy should be expressed as an expansion

$$U = \sum_{i<j}^{N} U_{ij} + \sum_{i<j<k}^{N} U_{ijk} + \sum_{i<j<k<l}^{N} U_{ijkl} + \cdots \tag{1}$$

J.A.K. Howard et al. (eds.),
Implications of Molecular and Materials Structure for New Technologies, 223–234.
© 1999 *Kluwer Academic Publishers. Printed in the Netherlands.*

where U_{ij} is the energy of the pair of molecules *i* and *j*, and the three body term U_{ijk} is the difference between the energy of the three molecules and the sum of the pair potentials $U_{ij}+U_{ik}+U_{jk}$, etc. Thus, when a true pair potential is used to calculate the properties of a liquid or solid, there is an error due to the omission of the non-additive contributions. Conversely, if the pairwise additive approximation is made in deriving the pair potential U_{ij} , then it will have partially absorbed some form of average over the many body forces present, which will produce an error in the calculated gas phase properties.

1.2. CONTRIBUTIONS TO THE INTERMOLECULAR PAIR POTENTIAL

Most methods of quantifying intermolecular forces identify the different physical phenomena that are responsible for the attraction and repulsion between molecules, determine which are most important for the molecules in question, and quantify them separately. A summary of the different contributions [1] is given below.

Table 1. Contributions to the energy of interaction between non-spherical molecules a distance R apart. The last column indicates which terms are normally explicitly included in models for the forces between organic molecules.

Contribution	Additive?	Sign	Comment	In model U_{ij}?
Long-Range	$U \sim R^{-n}$			
Electrostatic	Yes	±	Strong orientation dependence	Atomic multipoles
Induction	No	−		
Dispersion	approx.	−	Always present	$- C/R^6$
Resonance	No	±	Degenerate states only	
Magnetic	Yes	±	Very small	
Short-range	$U \sim e^{-\alpha R}$			
Exchange	No	−		
Repulsion	No	+	Dominates	$Ae^{-\alpha R}$
Charge Transfer	No	−	Donor-acceptor interaction	
Penetration	Yes	−	Can be repulsive at very short range	
Damping	approx.	+	Modification of dispersion and induction	

The long-range terms are amenable to theoretical treatment, through regarding the interaction between the molecules as a perturbation. This provides analytical expressions which can be quantified in terms of the properties of the individual molecular charge distributions. However, this approach breaks down when there is overlap between the two molecular charge distributions, so there is no rigorous analytical theory for the interaction energy at short range, where the repulsion dominates the quantum-mechanical exchange effects and the modification of the long-range terms.

2 Methods of Deriving Model Intermolecular Potentials

2.1. EMPIRICAL MODEL POTENTIALS FOR ORGANIC MOLECULES

Early work on intermolecular forces concentrated on spherical molecules, such as argon, where the pair potential is only a function of the separation, R, of the two molecules, and there are no electrostatic or induction forces. This work was intimately associated with the development of the quantitative theories relating the experimental properties of molecules in the solid, liquid and gaseous states, and their van der Waals complexes, to the intermolecular potential [2]. The search for a quantitative model intermolecular pair potential for argon was based on empirically fitting the potential to a wide range of experimental data: *i.e.* a functional form was assumed for the potential, using perturbation theory for the form and estimated coefficients for the long-range contribution, and then the potential parameters were fitted to optimize the agreement between the calculated and experimental properties. A model intermolecular pair potential which can reproduce a wide range of experimental data of the gas, (including scattering data and dimer spectra), the liquid and solid (with the addition of the Axilrod-Teller term for the three-body dispersion contribution) was finally obtained [2] in the early 1970s. It has the functional form:

$$U(R)/\varepsilon = \exp\left[\alpha(1 - R/\rho)\sum_{i=0}^{5} A_i\left(\frac{R}{\rho} - 1\right)^i\right] + \sum_{j=0}^{2} C^*_{2j+6}\bigg/\left(\delta + \left(\frac{R}{\rho}\right)^{2j+6}\right) \quad (2)$$

In this multiparameter form, three terms are used in the dispersion series expansion. The modification of the dispersion, due to overlap, is absorbed into the approximately exponential short-range repulsion term, with the minor parameter $\delta=0.01$ just used to prevent an unphysical maximum in the potential at short range.

The usual model intermolecular potential for organic molecules is based on the assumption that the interaction between the molecules is the sum of the interactions between their constituent atoms. For example, the interaction energy U between molecules M and N can be given by

$$U = \sum_{i\in M, k\in N} A_{\iota\kappa} \exp(-B_{\iota\kappa} R_{ik}) - \frac{C_{\iota\kappa}}{R_{ik}^6} + \frac{q_i q_k}{R_{ik}} \quad (3)$$

where atom i in M is of type ι and a distance R_{ik} from atom k in N of type κ. This model for the atom-atom repulsion and dispersion between the atoms is much simpler than that required for argon. The additional electrostatic interaction which arises for polyatomic molecules is modelled by an atomic charge model. The entire model assumes that the molecule is a superposition of spherical atoms, ignoring the redistribution of the valence electron density on bonding.

Many sets of parameters for this type of model potential have been derived by empirical fitting to different choices of experimental data. The range of values for hydrocarbons [3] eloquently demonstrates the problems in the empirical parameterisation of such a potential. When a simplified functional form is fitted to experimental data, with its associated errors, using an approximate theory, the parameters will inevitably absorb many of the errors in an ill-defined manner. This may be very beneficial when using the model potential as a form of extrapolation to calculate the same properties for closely related molecules. However, if too much confidence is placed in the correspondence of the terms to a true intermolecular potential, for example, by mixing parameters from different potential schemes, or using them to calculate properties which depend on different regions of the potential, the results can be very disappointing.

The model potentials most widely used for crystal structure modelling are therefore those that have been developed by fitting a wide range of crystal structures and heats of sublimation for common atomic types. Widely used examples are the Hagler-Lifson [4] potentials for amides and carboxylic acids and the Williams group [5] C,H,N,O,Cl,F potentials, where the repulsion-dispersion parameters are derived in conjunction with an externally derived electrostatic model. More recently, Gavezzotti and Filippini have parameterised an *exp-6* potential for C,H,N,O,S [6] and certain hydrogen-bonding protons [7], specifically for crystal structure modelling, using a large quantity of crystallographic data and heats of sublimation. These parameters appear very successful for estimating heats of sublimation, and for reproducing many crystal structures, though with some limitations. The model has effectively absorbed the electrostatic term into the *exp-6* parameters, which causes considerable deviations of the parameters from the usual combining rules.

Although these model potentials are adequate for many modelling studies, the limitations imposed by their empirical nature, and the assumptions of transferability and limited functional form, mean that they can only give a first estimate of interaction energies, and may be found inadequate for some simulations. Hence, we need to develop more accurate models, learning the lessons from small molecule potentials.

2.2. AB INITIO POTENTIAL ENERGY SURFACES

2.2.1. *Supermolecule calculations*
The conceptually simplest method of evaluating the intermolecular interaction energy between a pair of molecules is to use an *ab initio* method of calculating the total energy of the molecules at a given relative orientation (the supermolecule) and then subtract the energy of the isolated molecules. In practice, this is very demanding of the *ab initio* method used [8], because the intermolecular energy is so tiny compared with the total electronic energy of the molecule. Since the dispersion energy arises from the instantaneous correlation of the electron fluctuations, it is not included at all in a self-consistent-field (SCF) calculation. Large basis sets and a correlated method, such as MP2, are required to calculate a significant fraction of the dispersion energy. This problem is compounded by the basis set superposition error (BSSE), where the intermolecular interaction is overestimated when the basis set of the other molecule is used to improve the description of the intramolecular electronic interactions of each molecule, relative to the isolated molecule calculation. This error can be corrected by a counterpoise procedure, but this adds to the expense of the calculation. Considerable progress is being made with such calculations for the intermolecular interactions of small polyatomics, such as the water dimer. However, supermolecule calculations are usually only done to find the energy of a few selected geometries, as the calculation at each point is too expensive to permit the calculation at sufficient geometries to define the range of the potential energy surface that would be sampled in a condensed phase simulation.

2.2.2. *Intermolecular Perturbation Theory methods*

Since a major problem with supermolecule methods is in providing an equivalent description of the intramolecular effects in the supermolecule and the isolated molecule, there is a great attraction in evaluating the components of the interaction energy directly from the wavefunctions of the isolated molecules as a perturbation. This approach provides the analytical form for the long-range potential, in terms of the properties of the isolated molecules. However, when the molecular charge distributions overlap at short range, the electrons can no longer be assigned to a specific molecule, and the wavefunction for the supermolecule has to be antisymmetric with respect to interchange of any two electrons. This produces the exchange energy, a significant attractive contribution, as well as the repulsion arising from the overlap of the electrons from the two molecules. The requirement for antisymmetry produces many complications in developing a short-range perturbation theory, including non-orthogonality of the antisymmetrised zeroth-order wavefunctions and a lack of definition of the zeroth-order Hamiltonian. As a result, many different methods [1] can be used, with different detailed partitionings of the energy. For example, although all methods yield a quantity that becomes the induction energy at long-range, the modifications of the induction energy at short-range differ between methods, and short range effects such as charge transfer can be viewed as a separate effect or considered part of the short range induction energy. Nevertheless, despite these caveats, the division of the interaction energy into components is an attractive aspect of these calculations as it gives some insights into the nature of the intermolecular interactions. The division is also helpful in fitting model intermolecular potentials, which is easier done for each contribution separately.

The method that has been most widely applied to organic molecules is the non-orthogonal intermolecular perturbation theory (IMPT) of Hayes and Stone [9], which evaluates the intermolecular interactions to second order of perturbation theory between the SCF charge densities. It enables a separation to be made between the charge-transfer [10] and the induction energy, which gives the charge-transfer free of the BSSE which often led to an overestimate of the importance of this term when only small basis sets were used. More accurate calculations, corresponding to the interactions between the correlated charge densities of the molecules, are possible for small systems, using symmetry adapted perturbation theories [11].

One of the most important observations to come from intermolecular perturbation theory calculations on small polyatomics, was that the orientation dependence of the intermolecular potential parallels that of the electrostatic interaction quite closely around the minima. Since the electrostatic term is the only major contribution to the intermolecular forces between organic molecules which can be either attractive or repulsive, we would expect that considering the electrostatic term would show which sterically accessible regions of the potential energy surface are most favourable. The IMPT results show that accurate models for the electrostatic forces alone can be very useful in locating the minimum energy structures. This was first demonstrated by Buckingham and Fowler [12] , who showed that optimizing the electrostatic interaction between two small polyatomic molecules, within accessible orientations defined by a hard-sphere potential, was remarkably successful at predicting the geometries of the van der Waals complexes, (e.g. H-F..N=C-H is linear, but H-F..H-F is bent by about 70°). Thus the directionality of the electrostatic interaction was able to account for the observed "rules" [13] that the hydrogen bond forms to the direction of the conventional lone pair density on the acceptor, or in the absence of lone pair density to the π bonding density. IMPT calculations on several small hydrogen bonded systems [14] accounted for this empirical observation, by showing that the total energy curve was usually approximately parallel to the electrostatic energy for angular displacements of the hydrogen bond, because the orientational variation of the other terms, although significant, tended to cancel.

This observation has also been found to be true for hydrogen bonds to organic molecules, such as aromatic heterocycles [15], as exemplified below. The electrostatic energy mirrors the total energy as O-H is rotated around the nitrogen, at least until there is significant repulsion from the α hydrogen atoms.

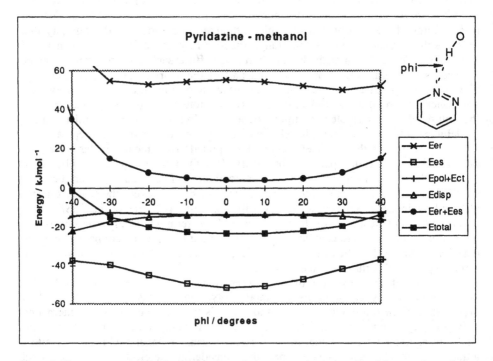

Figure 1. The components of the interaction energy between pyrazine and methanol for a coplanar, linear N..H-O hydrogen bond of length d(N..H)=1.85Å contrasting the exchange-repulsion Eer, electrostatic Ees, dispersion Edisp, induction Epol and charge transfer Ect contributions to the total energy Etotal estimated by IMPT calculations using a 6-31G** basis set.

IMPT calculations have often been used to rationalize the favoured relative orientations of functional groups found in molecular and protein crystal structures. They reveal the variation in the strengths of hydrogen bonds to a given atom (O or N) with the bonding environment of that atom [15], thus explaining, for instance, the relative scarcity of hydrogen bonds to the oxygen of furan [16]. The estimates of relative interaction energies are also useful for drug design, for assessing the effect of functional group replacement on protein binding energies, which is why IMPT estimates of the interaction energy for model systems accompany the ISOSTAR library [17] of the experimental distribution of functional groups in the Cambridge Structural Database.

2.3. SYSTEMATIC POTENTIALS

The above discussion shows that the number of points on a potential energy surface that can be calculated *ab initio* will always be rather limited, and that therefore deriving model potentials by fitting to a set of *ab initio* points suffers from the same disadvantage as empirical fitting, *i.e.* that you have to assume the functional form of the model potential, usually an isotropic atom-atom model type. If the functional form proves incapable of representing the data properly, then you can increase the flexibility by the addition of more parameters, but this in turn requires more fitting data and increases the problems of parameter correlation and ill-defined fits. Thus, it will always be advisable to derive the functional form, and as many of the parameters as possible, from the quantitative theory of intermolecular forces. This is the rationale behind the systematic potential approach [18], where each contribution is represented and quantified separately, and then added to give the total potential. This approach is making considerable progress for small polyatomics, and is behind the advances in model potentials for organic molecules. To understand the possibilities, and current limitations, of the approach requires the separate consideration of each contribution.

2.3.1. *The electrostatic contribution- distributed multipoles*
The electrostatic interaction, which is defined as the classical Coulombic interaction between the undistorted charge distributions of the isolated molecules, is the easiest to derive from the wavefunctions. When there is no overlap of the charge distribution, all that is required is a representation of the molecular charge density. This is most conveniently done for organic molecules by representing the charge density by sets of point multipoles centered at each atomic nucleus. The convergence of the calculated electrostatic energy as a function of the number of terms in the multipole expansion of the electrostatic energy, or the number and distribution of the sites, can be checked. There are a variety of methods of splitting up the molecule charge density for representation at the difference sites, which should all give the same results provided sufficient terms in the multipole expansion are used. The most widely used is the Distributed Multipole Analysis of Stone [1], with atomic sites, as this optimizes the convergence and range of validity of the expansion for molecules in van der Waals contact.

In less mathematical terms, the atomic dipoles and quadrupoles in the multipole series represent the electrostatic forces arising from the lone pair and π electron density, and other non-spherical features formed by the interaction of the s and p valence electrons, and the higher atomic multipoles refine this description to converge to an exact description of the molecular wavefunction. Calculation of the electrostatic interaction energy from these sets of atomic multipoles involves summing over known [1] anisotropic atom-atom potential terms in R_{ik}^{-n}, (n is usually 5, to include the atomic quadrupole-quadrupole terms, which are often important for $\pi-\pi$ interactions). The practical importance of these anisotropic electrostatic terms has been demonstrated extensively, with the most well-known being the need to use distributed multipole models to account for the structure of van der Waals complexes of small polyatomics [12].

The atomic point charge model, most commonly used in simulating organic systems, is thus a very severely truncated multipole expansion. The now usual practice of deriving these atomic charges by fitting to the electrostatic potential , as calculated by integration over the wavefunction, at a grid of points around the molecule, allows the charges to partially absorb the effects properly represented by the anisotropic atomic multipoles, and will give the best possible atomic point charge model. However, a survey [19] of different methods of deriving atomic charges, concluded that no atomic charge model was capable of giving a completely satisfactory representation of the electrostatic potential around a range of molecules. However, the use of a distributed multipole model requires special computer programs, as the anisotropic atomic multipoles give rise to non-central forces and torques between the interaction centres. Hence, there is considerable merit behind the pragmatic approach of using point charges away from the nuclear sites, to represent the effects of the atomic dipoles and quadrupoles on atoms where their effect is significant. This is gradually being adopted in molecular mechanics force-fields [20, 21].

2.3.2. The induction and dispersion contributions - distributed polarisabilities
The induction energy at long range can be related [1] to the permanent multipoles and polarisabilities of the individual molecules by the long range perturbation theory at second order. Similarly the coefficients in the perturbation theory expansion of the dispersion energy can be related to the integrals over the individual molecular polarisabilities at imaginary frequency. These properties can be derived from experimental data for atoms [2] and small molecules in favourable cases, but *ab initio* calculations are often the best source. Since polarisabilities measure the ease of distortion of the molecular charge density, their calculation requires a basis set with high angular momentum basis functions, and the inclusion of electron correlation.

For organic molecules, there is the additional problem that the polarization and dispersion must be distributed, usually to atomic centres, for the perturbation theory expressions to be valid and useful. An accurate treatment, must thus allow for the movement of charge between the atoms under the influence of the electric field due to another molecule, and involves a double sum over all intermolecular pairs of sites in the molecule. Thus although there are schemes for distributing polarisabilities and dispersion coefficients [1], they have yet to be extended to organic molecules.

2.3.3. The short-range terms
As already noted, there is no rigorous analytical theory for deriving the short range energy terms (except penetration energy) from the charge densities of the individual molecules. Nevertheless, there is considerable evidence that for some atoms [22], most notably chlorine [23], the anisotropy of the atom-atom repulsion is sufficient to affect the crystal packing of organic molecules. Hence, we need a method of determining the functional form of the atomic repulsion anisotropy, which will depend on the

orientation of the non-spherical features in the atomic charge density, such as lone pairs, relative to the intramolecular bonds, as well as its parameters. One approach is to use an *ab initio* potential energy surface for the molecule interacting with a test particle (a spherical atom such as helium or ^4S nitrogen), and fit the anisotropic form to these points, and then assume combining rules to estimate the repulsion between atoms in the two molecules. Tests of this approach for N_2, Cl_2, C_2H_2 and H_2S [24] show that it is promising. A second method assumes that the repulsion is proportional to the overlap of the undistorted molecular charge distributions. Analytical expressions can be obtained for the overlap between Gaussian functions centered on two atoms, and thus it is possible to analyze molecular charge densities (calculated using Gaussian basis sets) to give the overlap in an analytical atom-atom form for each atom-atom separation. This has been shown to give a reasonable model for the repulsion anisotropy for F_2, Cl_2 and N_2 [25], and the method is currently being extended to organic molecules.

The modification of the long range terms due to overlap, though dwarfed by the exchange-repulsion energy, is still significant in the van der Waals region of the potential sampled in simulations. The penetration energy, the difference between the electrostatic energy (as evaluated by an IMPT calculation) and that evaluated from a converged point multipole expansion (such as a DMA of the corresponding wavefunction), can be estimated using Gaussian Multipoles [26], which are an atomic multipolar representation of the molecular charge density which retains its spatial extent. However, the damping of the dispersion energy has only been investigated for very small systems, mainly atoms, and so there is little guidance on how the empirically-based isotropic forms could be modified. This is a pity, because when a model potential is systematically constructed from good models for all the component sections, the experimental predictions can be disappointingly sensitive to the choice of damping function.

2.3.4. *Examples of systematic potentials*

A major example of this systematic construction of a model potential, aimed at achieving the best accuracy in each component, is the water potential of Millot and Stone [27]. This used distributed multipoles, up to quadrupole on oxygen, charge on hydrogen, and central polarisabilities up to quadrupole-quadrupole, and anisotropic dispersion coefficients up to C_{10}. The repulsion was described by an anisotropic atom-atom exponential function, fitted to the exchange-repulsion plus penetration energy calculated by IMPT. This potential gave a good account of the equilibrium geometry, second virial coefficient, and the spectra of water clusters. However, it is worth noting that it has subsequently been improved by including a charge transfer term, and updating the dispersion coefficients. It is also far too complicated to be used in any Molecular Dynamics simulation, although this is a polyatomic molecule with only one non-hydrogenic atom, thus allowing a one site description of the induction and dispersion.

Another systematic potential derived for chlorine [28] required two atomic sites for the distributed multipoles and dispersion coefficients. The anisotropic atom-atom repulsion potential was derived from the overlap model, with the proportionality constant and one major anisotropic coefficient being adjusted by empirical fitting to the crystal structure. This empirical adjustment appeared to effectively absorb the missing contributions, including the many body effects, as the potential was able to reproduce a wide range of properties of the solid and liquid using Monte-Carlo simulations.

2.4. MIXED METHODS

The above account helps explain why we are still working towards accurate potentials for small polyatomics, such as water. Even when quite sophisticated and accurate potentials are available, they will often be too complex for use in many simulations, without being approximately fitted by a simpler functional form. Some known loss of accuracy in the potential will be acceptable for simulating experimental properties with a significant experimental error in their measurement, using an approximate theory.

The use of theoretical methods for those contributions that can be accurately estimated, and the fitting of the other main contributions explicitly (and thereby absorbing the missing contributions and non-additivity errors to some extent) seems to be the most promising approach for organic molecules. This is the method that we have been using for modelling molecular crystal structures. The electrostatic term is represented by a DMA of an *ab initio* charge distribution of the molecule, and the repulsion-dispersion potential is empirical. This potential scheme gives a reasonable reproduction [29], for static minimization, of the room temperature crystal structures of over forty simple rigid organic molecules (mainly aromatic and heterocyclic rings, with amine, nitro or amide substituents). The electrostatic forces represented by the anisotropic multipole moments play an important role in many crystal structures; several molecular crystal structures minimized to qualitatively different structures when the atomic dipoles and higher multipoles were removed. Thus, the extra computational expense, and program development required, in using a realistic electrostatic model seems well justified in molecular crystal structure modelling.

3. Future Model Potentials

Empirical isotropic atom-atom potentials have been widely used for modelling the crystal structures of organic molecules, and in the simulation of biological interactions. However, for molecules and simulations where these are either just not adequate, or the scientific argument requires the best possible estimate of the intermolecular interaction energies, there are now alternatives, at least for the electrostatic term. Using distributed multipoles has led to great improvements in our ability to predict the structures of van der Waals complexes and molecular crystal structures. However, to further improve our model intermolecular potentials, we need to put both the repulsion and dispersion models on a better theoretical footing, both in allowing an anisotropic form and to obtain the parameters. This will not be easy, but the work on small polyatomics suggests that it will be possible.

4. References

1. A. J. Stone, (1996) *The Theory of Intermolecular Forces*, vol. 32, 1st ed., Oxford: Clarendon Press, Oxford.

2. G. C. Maitland, M. Rigby, E. B. Smith and W. A. Wakeham, (1981) *Intermolecular Forces. Their Origin and Determination.*, vol. 3. Oxford: Clarendon Press, Oxford.

3. A. J. Pertsin and A. I. Kitaigorodsky, (1987) *The Atom-Atom Potential Method. Applications to Organic Molecular Solids*, vol. 43. Berlin: Springer-Verlag, Berlin.

4. A. T. Hagler, E. Huler and S. Lifson, (1974) Energy functions for peptides and proteins I Derivation of a consistent force-field including the hydrogen bond from amide crystals, *J. Amer. Chem. Soc.*, **96**, 5319-5327.

5. D. E. Williams and S. R. Cox, (1984) Nonbonded Potentials For Azahydrocarbons - the Importance of the Coulombic Interaction, *Acta Crystallographica Section B-Structural Science*, **40**, 404-417.

6. G. Filippini and A. Gavezzotti, (1993) Empirical Intermolecular Potentials For Organic-Crystals - the 6-Exp Approximation Revisited, *Acta Crystallographica Section B-Structural Science*, **49**, 868-880.

7. A. Gavezzotti and G. Filippini, (1994) Geometry of the Intermolecular X-H...Y (X, Y=N, O) Hydrogen-Bond and the Calibration of Empirical Hydrogen-Bond Potentials, *Journal of Physical Chemistry*, **98**, 4831-4837.

8. J. G. C. M. van Duijneveldt-van de Rijdt and F. B. van Duijneveldt, (1997) Ab Initio methods applied to hydrogen-bonded systems, in D. Hadzi, (ed.)*Theoretical Treatments of Hydrogen Bonding*. Chichester: John Wiley and Sons, Chichester, pp.13-47.

9. I. C. Hayes and A. J. Stone, (1984) An Intermolecular Perturbation-Theory For the Region of Moderate Overlap, *Molecular Physics*, **53**, 83-105.

10. A. J. Stone, (1993) Computation of Charge-Transfer Energies By Perturbation-Theory, *Chemical Physics Letters*, **211**, 101-109.

11. B. Jeziorski, R. Moszynski and K. Szalewicz, (1994) Perturbation-Theory Approach to Intermolecular Potential-Energy Surfaces of van der Waals Complexes, *Chemical Reviews*, **94**, 1887-1930.

12. A. D. Buckingham, P. W. Fowler and A. J. Stone, (1986) Electrostatic Predictions of Shapes and Properties of van der Waals Molecules, *International Reviews in Physical Chemistry*, **5**, 107-114.

13. A. C. Legon and D. J. Millen, (1987) Directional Character, Strength and Nature of the Hydrogen-Bond in Gas-Phase Dimers, *Accounts of Chemical Research*, **20**, 39-46.

14. G. J. B. Hurst, P. W. Fowler, A. J. Stone and A. D. Buckingham, (1986) Intermolecular Forces in van der Waals Dimers, *International Journal of Quantum Chemistry*, **29**, 1223-1239.

15. I. Nobeli, S. L. Price, J. P. M. Lommerse and R. Taylor, (1997) Hydrogen bonding properties of oxygen and nitrogen acceptors in aromatic heterocycles, *Journal of Computational Chemistry*, **18**, 2060-2074.

16. I. Nobeli, S. L. Yeoh, S. L. Price and R. Taylor, (1997) On the hydrogen bonding abilities of phenols and anisoles, *Chemical Physics Letters*, **280**, 196-202.

17. I. J. Bruno, J. C. Cole, J. P. Lommerse, R. S. Rowland, R. Taylor and M. L. Verdonk, (1997) IsoStar: A library of information about nonbonded interactions, *J. Comp. Aided Molecular Design*, **11**, 525-537.

18. S. L. Price, (1996) Anisotropic Atom-Atom Potentials, *Philosophical Magazine B-Physics of Condensed Matter Statistical Mechanics Electronic Optical and Magnetic Properties*, **73**, 95-106.

19. K. B. Wiberg and P. R. Rablen, (1993) Comparison of Atomic Charges Derived Via Different Procedures, *Journal of Computational Chemistry*, **14**, 1504-1518.

20. J. G. Vinter, (1996) Extended Electron Distributions Applied to the Molecular Mechanics of Some Intermolecular Interactions. 2. Organic-Complexes, *Journal of Computer-Aided Molecular Design*, **10**, 417-426.

21. R. W. Dixon and P. A. Kollman, (1997) Advancing beyond the atom-centered model in additive and nonadditive molecular mechanics, *Journal of Computational Chemistry*, **18**, 1632-1646.

22. A. J. Stone and S. L. Price, (1988) Some New Ideas in the Theory of Intermolecular Forces - Anisotropic Atom Atom Potentials, *Journal of Physical Chemistry*, **92**, 3325-3335.

23. S. L. Price, A. J. Stone, J. Lucas, R. S. Rowland and A. E. Thornley, (1994) The Nature of -Cl-...Cl- Intermolecular Interactions, *Journal of the American Chemical Society*, **116**, 4910-4918.

24. A. J. Stone and C. S. Tong, (1994) Anisotropy of Atom-Atom Repulsions, *Journal of Computational Chemistry*, **15**, 1377-1392.

25. R. J. Wheatley and S. L. Price, (1990) An Overlap Model For Estimating the Anisotropy of Repulsion, *Molecular Physics*, **69**, 507-533.

26. R. J. Wheatley and J. B. O. Mitchell, (1994) Gaussian Multipoles in Practice - Electrostatic Energies For Intermolecular Potentials, *Journal of Computational Chemistry*, **15**, 1187-1198.

27. C. Millot and A. J. Stone, (1992) Towards an Accurate Intermolecular Potential For Water, *Molecular Physics*, **77**, 439-462.

28. R. J. Wheatley and S. L. Price, (1990) A Systematic Intermolecular Potential Method Applied to Chlorine, *Molecular Physics*, **71**, 1381-1404.

29. D. S. Coombes, S. L. Price, D. J. Willock and M. Leslie, (1996) Role of Electrostatic Interactions in Determining the Crystal- Structures of Polar Organic-Molecules - a Distributed Multipole Study, *Journal of Physical Chemistry*, **100**, 7352-7360.

INTERMOLECULAR INTERACTIONS IN MOLECULAR CRYSTALS STUDIED BY AB INITIO METHODS:
From isolated interactions to patterns and crystals

J. J. NOVOA

Dept. de Química Física, Facultat de Química,
Universitat de Barcelona, Av. Diagonal 647,
08028-Barcelona (Spain)

1. Introduction

The rationalization of the properties of molecular crystals is a necessary step towards the design of crystals which present enhanced technological interest, such as conductivity [1] or magnetism [2], to cite some. Some molecular crystals are known to present interesting technological properties and these properties depend on the crystal structure. Thus, being able to design and grow crystals which present specific relative orientations among the constituent molecules is becoming increasingly interesting for the design of new materials and this goal requires a rationalization of the crystal packing of the known structures and of the ways one can force that packing to change [3,4]. One can only arrive to that degree of control if, firstly, the properties of the intermolecular interactions are fully understood and, secondly, one has a model to rationalize the way in which intermolecular interactions work collectively in the crystal. In this chapter we will describe how *ab initio* computations can help in both fields.

The forces responsible for the existence of molecular crystals are those generated by the intermolecular interactions [3-6]. When the intermolecular interaction energy is stronger than the thermal energy at a given temperature the molecules aggregate to form a liquid or a solid [5,6]. At sufficiently low temperature and under special conditions one can obtain a crystal, characterized by the presence of long range order in the condensed phase. The geometrical disposition of the molecules within the crystal, the crystal packing, characterizes the structure of the crystal. Some molecules can adopt many different stable crystal packing arrangements, each one being a polymorph [7]. Each of these polymorphs is one minimum in the crystal packing potential (E_p), defined as the sum of all different intermolecular interaction energy components for the molecules constituting the crystal. Under the usual pairwise approximation [3], the crystal packing potential E_p is computed as the sum of all different possible pairs of intermolecular interaction potentials E_{ij} between the molecules of the crystal (*i.e.*, $E_p = \Sigma_{ij}' E_{ij}$, where the prime indicates different pairs). This is an approximation and in some cases one has to include third order terms [5,6] to properly reproduce a crystal structure. From these considerations, it is obvious that the factors determining the packing of a crystal are the intermolecular interaction potentials E_{ij}. These E_{ij} potentials are just the intermolecular potentials for *each pair of atoms* located in different molecules. Their strength is the typical of intermolecular interactions, that is, much smaller than chemical bonds [5,6]. Their dependence on the

J.A.K. Howard et al. (eds.),
Implications of Molecular and Materials Structure for New Technologies, 235–250.
© 1999 Kluwer Academic Publishers. Printed in the Netherlands.

r_{ij} distance is given by a Morse-like potential, with a minimum around the sum of the van der Waals radii of the atoms i and j. In most cases, the E_{ij} potentials are approximated as radial atom-atom potentials which only depend on the distance between the atoms, although there are cases in which the angular dependence is also included [6].

The crystal packing, being one of the minima of the E_p function, is a compromise between all possible E_{ij} terms. The minimum energy geometrical arrangements for a pair of molecules or a molecular aggregate will be dominated by the stronger E_{ij} terms. Consequently, not all the intermolecular distances are at their minimum energy values: the E_{ij} curves are shallow and the r_{ij} distance can be increased if this allows the molecules to make more energetically favorable contacts for other stable E_{ij} pairs. At the same time, there is no reason for all the short contacts to be attractive: there may be short intermolecular distances associated with repulsive interactions if they are compensated by the existence of other stable attractive interactions. All of this complicates the analysis of the crystal packing in terms of the intermolecular interactions, as there is no simple connection between the r_{ij} distance and the E_{ij} energy: *one cannot safely deduce from the interatomic distance, whether short or not, the energetics of a given intermolecular contact.* Therefore, for a proper understanding of crystal packing one has to compute the form of the E_{ij} potential or, at least, evaluate E_{ij} at the distance of interest. These may easily be done by solving the Schrödinger equation for an appropriate model system. The methods which solve the non-relativistic Schrödinger equation without imposing any simplification on the Hamiltonian operator, besides the Born-Oppenheimer separation of the electronic and nuclear motion, are called *ab initio* methods [8].

In principle, using *ab initio* methods one could optimize the geometry of a crystal and obtain all possible polymorphs. This has been done for some simple molecules [9]. For larger ones, one can use empirical atom-atom potentials to obtain packing geometries in good agreement with the experimental ones [10]. However, such an optimization does not rationalize why the crystal packs in a particular form, in the same way that by optimizing a molecule one does not obtain a rationalization of its structure, which must be obtained by looking at the wavefunction with the help of some model which justifies how it changes (as these are based on the variation of the orbital overlap). *Here we will show that by looking at the possible intermolecular bonds each molecule can form with its neighbors and at the energetics of these bonds its possible to understand the way crystals pack.* Different polymorphs implicate intermolecular bond breaking and creation, similar to those seen in chemical reactions. The main difference between these two processes lies in the strength of the bonds and the size of the barriers separating the minima. We only need data from the intermolecular interaction curves (energy, minimum energy geometry, directionality) for a proper rationalization. *Ab initio* methods have been shown to provide very good estimates of these intermolecular interaction curves. In the rest of this chapter we will describe how this rationalization process can be carried out.

2. Principles of Intermolecular Association

At sufficiently low temperatures all the molecules demonstrate the overall attractive nature of their intermolecular interactions in the formation of energetically stable aggregates [5]. A clear manifestation of this is the existence of liquid and solid phases of all chemical compounds at low enough temperature. What is of interest to us in this section is the preferred geometrical arrangements formed during the association and how to rationalize them, as this will give us a way of understanding the crystal packing. In this section it will be shown that one can rationalize the molecular association by looking at the intermolecular bonds made by the functional groups of the molecules. This idea may be illustrated with an example.

One of the simplest intermolecular aggregates is the water dimer. This system has been widely studied at the experimental [11] and theoretical level [12] with the same conclusion: the dimer is stable by about 5 kcal/mol, and the geometry of its only minimum is the hydrogen-bonded one depicted in Figure 1.

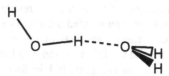

Figure 1. Optimum geometry of the water dimer

This optimum geometrical arrangement of $(H_2O)_2$ is also reproduced by the pairwise atom-atom potential discussed above, using expressions for the E_{ij} terms from empirical atom-atom potentials. Similar results are obtained by using the E_{ij} expression from an electrostatic model based on distributed multipoles [6], or by correcting this electrostatic model by the addition of van der Waals terms to avoid the collapse of the dimer which takes place when only electrostatic components are included in E_{ij}.

One can also rationalize the water dimer structure by looking at the intermolecular bonds present. The only intermolecular bond present is the H···O one shown with a broken line. How can one be certain of this? One can establish the presence of bonds in any system (molecules, molecular aggregates) by analyzing the topology of the electron density at the optimum geometry of the aggregate. This is done by searching for the points in which the gradient of the density is zero. These points are called critical points in the methodology developed by Bader [13], known as the Atoms-in-Molecules (AIM) method because it allows, amongst other things, atomic regions within a molecule to be defined. In the AIM method, *a bond is characterized by the existence of a bond critical point* along the line of maximum density linking the nuclei. Bond critical points are those critical points having two negative and one positive curvatures (eigenvalues in the Hessian matrix of the density) [13]. Notice, however, that the existence of a bond critical point is a *necessary but not sufficient* condition for the presence of a bond. In order to be a bond, the bond critical point has to be associated with an energetically stable interaction. Analyzing the critical points in the water dimer, one finds two intramolecular O-H bonds within each water molecule, and one intermolecular H···O bond. All of them correspond to energetically stable interactions, as manifested by the geometry optimization. These two types of bond critical points can be distinguished by the values of the density at the critical point (about 0.3 e in the intramolecular bonds, that is, one order of magnitude larger than in the intermolecular case) and by the values of their Laplacian, defined as the sum of the trace of the Hessian matrix, being negative for the intramolecular bonds and positive for the intermolecular bonds. However, ionic bond critical points behave like intermolecular interactions in the AIM method.

The intermolecular H···O bond of the water dimer has all the characteristics which many authors associate with an A-H···B hydrogen bond. This bond links the O-H group of one water with the lone pair region sitting on the oxygen atom of the other water. The stability of the water dimer can be associated with the presence of a stable intermolecular bond. In doing so, we have substituted atom-atom potentials by intermolecular bond potentials. Now we need to find the method of understanding and predicting why the dimer in Figure 1 is the preferred geometrical arrangement for that system.

The optimum geometrical arrangements of the water dimer can be predicted by overlapping the molecular electrostatic potential (MEP) maps [14] of two isolated water fragments. This overlap of the MEP maps is the qualitative equivalent of carrying out a computation of the electrostatic component of the interaction energy, which, as was shown by Buckingham and others [15], gives good results when used to predict the minimum energy geometry of hydrogen bonded dimers. The MEP map overlap analysis is much more powerful than a simple analysis of the point charges on the shortest contact atoms. For instance, for the CrO_4CH_3 anion, the localized point charges on the methyl hydrogens are positive, but the potential in this region is negative [16]. A simple point charge analysis of the shortest contacts gives the wrong idea: one could expect to obtain stable dimers by approaching the CH_3 and the CrO_3 ends of two $CrO_4CH_3^-$ molecules, as these groups present localized charges of the opposite sign. When the interactions with the other atoms of the molecules are taken into account, the interaction energy is found to be repulsive, in agreement with the *ab initio* result [16].

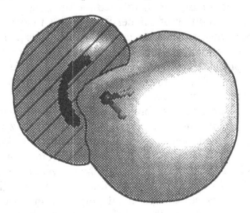

Figure 2. MEP map of the water molecule (isosurfaces plotted, left deep dark: -57 kcal/mol; left shadowed dark: -20 kcal/mol; right light region: +10 kcal/mol)

The MEP map for the water molecule, computed at the HF/6-31G(d,p) level [17] is shown in Figure 2 in the form of equipotential surfaces. It contains a strong banana-shaped negative region on the oxygen atom, indicating an accumulation of electronic charge there. The MEP minimum, of -61 kcal/mol, lies within that region. Conversely, there is a positive region surrounding all the atoms. This positive region becomes larger on the hydrogens, indicating a depletion of electronic charge. The +1 charge is repelled in the positive regions (positive MEP values), while it is attracted in the negative regions (negative MEP values).

The overlap of positive and negative regions corresponds to the overlap of positively charged zones of electronic density with negatively charged ones, that is, arrangements which give rise to a stable electrostatic interaction. This electrostatic interaction becomes more stable as more of the negative region is overlapped by the positive region. In the water dimer, the overlap of positive and negative regions is only possible by overlapping the hydrogens of one water fragment with the oxygen of the other fragment. One possible form of doing so is at the geometry shown in Figure 1. Another one, according to the overlap analysis, is that shown in the center of Figure 3, in which two simultaneous H···O contacts are made. The MEP map overlap analysis does not give any indication as to which conformation is more stable. One has to perform *ab initio* calculations to find that information. In fact, the central configuration in Figure 3 is a transition state connecting two single-contact conformers like those in Figure 1. Notice also that the MEP overlap analysis is also useful in allocating the hydrogen bond acceptor atoms (those with high negative potentials) and the hydrogen bond donor atoms (the hydrogens around which the MEP becomes positive away from the nuclei).

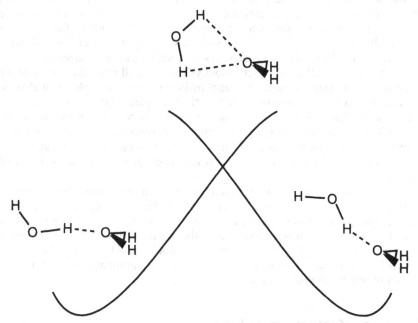

Figure 3. Energetics for the intermolecular transformation connecting the two minimum energy conformations of the water dimer.

The process presented in Figure 3 is one of the simplest examples of an intermolecular bond-breaking-bond-making transformation (one O-H···O intermolecular bond is broken and another O-H···O bond created when going from left to right or *vice-versa* in Figure 3). We cannot call this transformation a reaction, because no changes in the structure of the molecules occur. However, from the quantum chemical point of view, it has many similarities with a chemical reaction, the main difference being the intermolecular nature of the bonds broken and created and their strength. Consequently, one can use many of the tools developed to understand intramolecular transformations to analyze intermolecular transformations like those in Figure 3. Of particular interest is the use of *diabatic potential curves* (in Figure 3 shown for the O-H···O bonds), each diabatic being the potential energy curve for the creation or breaking of an isolated bond along the selected coordinate.

Diabatics help to rationalize the relative stability of conformers. One can go from one conformer to the other in Figure 3 by bending their O-H⋯O vectors away from colinearity. This is the reaction coordinate. As the conformer on the left hand side bends its O-H⋯O angle, the energy rises (a result from accurate *ab initio* computations which is also supported by statistical analysis of crystals). The same is true for the conformer on the right hand side. The point at which the two diabatics cross is the transition state. The well known collinear preference of the O-H⋯O hydrogen bond is the reason for the two-contact conformation to be a transition state.

Diabatics also help one to understand why some stable conformers are not minimum energy conformations: if one of the diabatics becomes more stable than the other, the crossing point between them becomes closer to the minimum of the less stable diabatic and, consequently, the transition state becomes smaller in energy. At one point, the transition state disappears and the higher energy conformer is no longer a minimum. This approach has been very helpful in understanding the structure of molecular clusters, such as the conformers of the $OH^-(H_2O)_n$ cluster [18]. We will see later that it can also be used to understand crystal packing and polymorphism.

The previous MEP map overlap analysis works well when the dominant forces in the structure are electrostatic, as in neutral hydrogen-bonded complexes. It also works well for understanding the structure of π-π stacked complexes [19]. Some recent studies [16] have also shown that the MEP map overlap is successful in rationalizing the relative orientation of A(-)⋯B(+) dimers, although here the main driving force driving the structure of these compounds is the charge-charge Coulombic component. However, it can fail when the dispersion component is important, as this component is not taken into account.

By performing the MEP map analysis one localizes bonds and associates the stability of the dimers to these bonds. We no longer look at atom pairs but at pairs of functional groups. Both approaches are energetically equivalent, as the interaction energy of an intermolecular bond results from the combined effect of all the atoms close to the two atoms making the shortest contacts in that bond. Thus, it will be interesting to review how, using *ab initio* methods, one can obtain information on the nature and energetics of intermolecular interactions.

3. The Nature of Intermolecular Bonds

At the outset, one should be aware that there is no general agreement in the literature on the definition and classification of intermolecular bonds [5, 6, 20]. Sometimes, all intermolecular bonds are called van der Waals bonds and hydrogen bonds are just a special class within them [20]. Furthermore, even the term *bond* is not always defined for polyatomic molecules in many books. Here we will shed some light into the intermolecular bond concept, and then we will analyze the most important classes of intermolecular bonds, classifying them into *ionic*, *hydrogen bond*, and *van der Waals* according to the leading energetic term in the intermolecular interaction energy and the bond topology.

The basic definition of the concept of a bond is due to Pauling [21], who established that "there is a chemical bond between two atoms or groups of atoms in case that the forces acting between them are such as to lead to the formation of an aggregate with sufficient stability to make it convenient for the chemist to consider it as an independent molecular species". Notice that this definition identifies a bond for the diatomic case, but does not allow the number of bonds in polyatomic systems to be identified. However, the usual practice applied to polyatomic molecules is to define bonds as *the set of dominant attractive interactions within the molecule*. Accordingly, in the water molecule we know that the dominant attractive interactions are the O-H ones, although the H-H interactions are also attractive, but much weaker. Therefore, we just define two O-H bonds in this molecule. Similarly, in the benzene molecule we define C-C bonds just between the closest carbon atoms, although there are many other attractive C-C interactions across the ring. In using this definition of a bond, one is introducing a useful concept which is a simplification of a more complex reality, as the stability of a molecule does not *only* come from the bonded atoms.

One can extend the previous ideas to the intermolecular bonds and define them as the set of dominant attractive intermolecular interactions present between two or more interacting molecules. This is also a simplification of a complex reality but, as in the intramolecular case, it is a very useful one. Intermolecular bonds can be characterized by the presence of their bond critical points, a necessary though not sufficient condition for the existence of a bond. Intermolecular bonds also have an interaction energy curve of a characteristic Morse-type shape, with a minimum at an energy below that of the isolated molecular fragments. When these two conditions are met, we can talk about the presence of an intermolecular bond. Notice that short contacts are not necessarily stable and therefore cannot always can be associated with the presence of a bond. Therefore, one should be careful in the use of the terms *short contacts*, *interactions* and *bonds*, since they are not equivalent. The term contact is based on a geometrical criterion, and is normally used to indicate some distance shorter than a given value, *e.g.* the sum of the van der Waals radii. The term interaction refers to those contacts whose interaction energy is different from zero. Only these interactions which are attractive are good candidates for being a bond. The set of dominant attractive interactions are the ones one normally identifies as intermolecular bonds.

The general expression for the interaction energy E_{int} of a dimer can be written without loss of generality as [5,6]

$$E_{int} = E(repulsion) + E(electrostatic) + E(induction) + E(dispersion) \qquad (1)$$

where E(repulsion) is the repulsion wall at very short distances caused by the Pauli exclusion principle (normally represented by an exponential term), E(electrostatic) comes from the interaction between the permanent multipoles, including the monopole term of the Coulombic interaction, E(induction) is due to the presence of induced multipoles, and E(dispersion) is the non-classical term caused by the instantaneous interactions between the electrons, recovered within the dispersion component in *ab initio* computations. If the dimer fragments have permanent charges on them, the leading term is the electrostatic one, in particular, the charge-charge Coulombic one, which has a distance dependence of the form $1/r_{ij}$. In these cases, we talk about *ionic interactions* and the bonds they make are called *ionic bonds*. Note that both of the fragments must be charged: bonds like those in the F-H-F⁻ system are not ionic, since they result from the interaction of FH and F⁻ fragments. *Hydrogen bonds*, in agreement with previous literature [15,22,23] are those bonds having a topology of the type A-H···B, in which an A-H intramolecular bond points towards a high electronic density region sitting in the B atom (normally, non-bonding or π electrons). The interaction energy of the bond is dominated by the permanent dipole-dipole interaction, although the induction term can be also important [12c]. It becomes stronger with increasing electronegativity of the A and B atoms. All the other intermolecular interactions are identified as *van der Waals interactions*. Of those, some are driven by the dispersion terms, as in the $O_2 \cdots O_2$ case, but for others, including the $CO \cdots CO$ dimer, the dipole term plays a very important role.

A simple rule of thumb to classify the intermolecular interactions is obtained by the following rules:

(1) Look for charges on the fragments. If the two fragments are charged, the interaction is ionic. Otherwise it is either a hydrogen bond or a van der Waals interaction.

(2) For non-ionic interactions, if the bond possesses an A-H$\cdots$B topology and, consequently, has a bond critical point linking the H and O atoms, then it is a hydrogen bond. Otherwise, it is a van der Waals bond. They also have a A:$\cdots$:B topology in which two high density regions, located on the A and B atoms, are at short distance of each other (thus the importance of the dispersion term). We have indicated by the symbol ":" the high density region, but in many cases this symbol is dropped out. Hydrogen bonds can be classified into weak (*e.g.* C-H$\cdots$O), moderate (*e.g.* O-H$\cdots$O) and strong (those in which one fragment has a net charge in it, *e.g.* O-H$\cdots$O$^-$, or the hydrogen bond gains extra stabilization by resonance) [24].

(3) To be an intermolecular bond, it must be a stable interaction and must present a bond critical point connecting the intermolecular fragments. This discards short repulsive contacts as bonds, a situation found in some cases.

Notice the importance in these rules of the critical point analysis: we classify as van der Waals bonds very bent A-H$\cdots$B topologies in which the critical point lies between the A and B atoms, that is, has an H-A$\cdots$B topology, while for other A-H$\cdots$B angles (collinear, for instance) the topology is classified as a hydrogen bond [25]. The presence of the critical point provides an exact way of establishing the existence and nature of each intermolecular bond. In many cases, the critical point analysis gives the same answers than the usual sum of van der Waals radii criterion. However, it is more powerful and precise. For instance, it solves the problems of defining the limiting H-A$\cdots$B angle at which a hydrogen bond becomes a van der Waals interaction and determining whether outliers, those contacts which are larger than a cut-off distance, are hydrogen bonds [26].

At this point, it is worth providing one more example which stresses the importance of the energetic stability of the interaction in identifying the presence of a hydrogen bond. This example is present in many ionic crystals involving salts of partially deprotonated polyprotic acids, such as the potassium hydrogen oxalate salt, KHC_2O_4. In this salt, chains of $HC_2O_4^-$ anions form short OH$\cdots$O contacts at H$\cdots$O distances which are shorter than the sum of the respective van der Waals radii. Consequently, they could be taken as being hydrogen bonds, according to the usual distance criterion. A search on the Cambridge Structural Database [27] revealed that this is a general trend for this type of salt. In fact, as shown in Figure 4, the O$\cdots$O distance in these salts is even shorter than the O$\cdots$O distance in their parent neutral acids. In the following, we will identify the first type of contacts as OH$^-\cdots$O$^-$, to distinguish them from these found in the neutral crystals, referred as OH$\cdots$O contacts. According to the commonly accepted energy-distance relationship, the OH$^-\cdots$O$^-$ contacts within the anion chains should be energetically more stable than the OH$\cdots$O contacts.

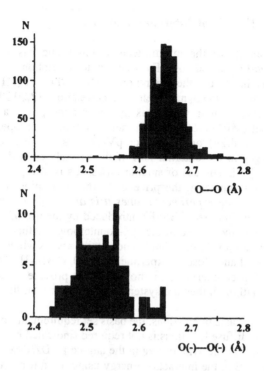

Figure 4. Histograms of the O···O separations for neutral O-H···O and O-H⁻···O⁻ contacts. The search has been performed using the CSD on systems containing carboxylic/carboxylate groups, based on a cut-off distance of 2.8 Å (from ref. 28).

A MEP map overlap analysis indicates that this is not the case, as the potential around the $HC_2O_4^-$ anions is always negative [28]. Accurate *ab initio* computations (carried out at the HF/6-31+G(2d,2p) and B3LYP/6-31+G(2d,2p) levels on the $HC_2O_4^-$ dimers at their crystal geometry) confirms that two anions are repelled in this geometry [28]. Full optimization of the geometry of these two anions shows no minimum at any distance or orientation [28]. All of this indicates that, as in the sodium chloride crystal, the stability of the KHC_2O_4 crystal is due to the anion···cation interactions which compensate the anion···anion and cation···cation repulsions. If this is so, we cannot say that the OH⁻···O⁻ is a bond, as it does not fulfill the stability criterion, even if it is shorter than the sum of the respective van der Waals radii. Furthermore, there is a clear violation of the energy-distance relationship, as the neutral OH···O contacts, which are longer in average, are more stable (their interaction energy is attractive) than the shorter OH⁻···O⁻ contacts. This suggests that the interpretation many structures containing OH⁻ ···O⁻ contacts, previously identified as being hydrogen bonded, should be revised.

Finally, it is worth considering that hydrogen bonds are, in most cases, stronger than van der Waals bonds, and their directionality and strength makes them very useful for the design of molecular crystals with specific properties. In fact, they are considered as the most helpful tool in that process. However, the energies of very weak hydrogen bonds are similar to that of strong van der Waals bonds [12b,29,30]. Therefore, in some cases, we cannot rule out the importance of van der Waals bonds.

4. Computing the Energy of Intermolecular Bonds

To compute the energy of the intermolecular bonds one has to use an adequate combination of *ab initio* methodology. For an accurate description of moderate and weak hydrogen bonds one has to use the second order Moller-Plesset (MP2) method, which includes a large amount of the dynamic electron correlation [12,20,29]. However, when more accuracy is desired, as in very weak hydrogen bonds, one has to use the fourth order Moller-Plesset (MP4) method. Both methods require a properly built basis set, such as the 6-31+G(2d,2p) or the aug-cc-pVDZ sets, to describe with sufficient precision the changes in the external part of the orbitals upon the formation of the molecular complex [31]. The use of smaller bases sets is also possible for very large complexes as a compromise, at the price of losing some precision. In addition, *to obtain results close to the experimental ones it is absolutely essential to correct for the basis set superposition error* (BSSE) introduced by the use of truncated basis sets [31]. This can be done by the use of the full-counterpoise method [32]. By doing this correction, if the basis set is large enough, one obtains results which are independent of the basis set employed and close to experiment [12,20,30,31,33]. If the basis set is not large enough, the full-counterpoise method does not provide results as close to the experimental ones, although they are systematically closer than the uncorrected-BSSE values.

For van der Waals interactions, the basis set requirements are similar, although now the inclusion of diffuse functions is not required (indicated by the symbol + in the 6-31+G(2d,2p) basis and the symbol *aug* in the aug-cc-pVDZ case). Here one also has to correct for the BSSE in the interaction energy using the full-counterpoise method to obtain results close to the experimental ones. Concerning the method, it is recommended to use MP4 to reproduce the interaction energy of systems which have no near degeneracy in their fragments' highest occupied orbitals (the usual case). If a near degeneracy is presumed (as in the Be···Be dimer), the one has to use multi-reference methods for an accurate description [20,30b,31].

Using the aforementioned methodology, one can mimic the qualitative details of the interaction energy potential surface [33], and even reproduce the experimental results with errors less than 10% in the energy and the geometries [12,20,29-31]. Therefore, one can compute with adequate precision the intermolecular energy for each bond of interest. One just needs a good model dimer, in which the intermolecular bond of interest and its environment are properly reproduced (for instance, the inductive effect generated by nearby groups). Thus, the methane-water complex is the simplest model dimer one can devise to study the strength of the weakest C-H···O interactions.

5. Molecular Association in Complex Systems

As the molecule becomes more complex and presents, for instance, many possible functional groups which compete to produce hydrogen bonds, the use of the MEP map overlap analysis becomes more useful as it allows one to obtain reasonable minimum energy conformers and an estimate of their interaction energy. More precisely, the direct observation of the MEP maps indicates the regions of localization of electronic charge, thus allowing us to locate potential acceptor atoms for the hydrogen bond. These maps also tell us which hydrogen atoms have more positive charge, that is, the better hydrogen bond donor groups. Then, the MEP overlap gives us reasonable starting geometries in which the hydrogen bond donor and acceptor groups can be combined in stable molecular conformations. An AIM analysis fully characterizes the intermolecular bonds present. Finally, using adequate model complexes, one can accurately compute the strength of their bonds. Adding the strengths for all the bonds one obtains a good estimate of the energy of each conformer. In fact that addition would be exact if the bonds were independent, but there is always a cooperative effect, not included when the bond strengths are added. Comparing the bond strengths one can identify the dominant bonds and rationalize the crystal packing.

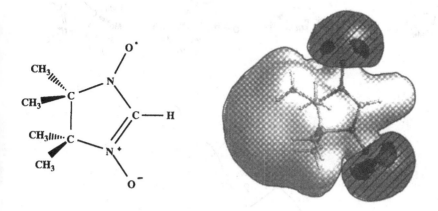

Figure 5. (a) Structure of the HNN molecule. (b) MEP potential (surfaces: -45 kcal/mol (deep dark), -20 kcal/mol (shadowed dark), +10 kcal/mol (light)

In a typical case, one finds more minimum energy conformations than for the water dimer. Each one has at least one diabatic which can cross with that of the nearest minimum, in the coordinate defining their transformation. This is illustrated for the 2-hydro nitronyl nitroxide (HNN, for short) molecule, whose structure and MEP is shown in Figure 5. We have two acceptor groups in the NO oxygens and donor groups in each CH hydrogen. The MEP overlap analysis of the HNN dimer indicates many possible minimum energy conformations, some in which the ONCNO groups of each fragment are collinear are indicated in Figure 6. These three conformations are related to each other by the angle between the center of mass of the second HNN fragment relative to $C(sp^2)$ and center of mass of the first HNN. We have also shown in Figure 6 the diabatics for each conformation. *Ab initio* computations show that the $C(sp^2)$-H···O interaction is more stable than the $C(sp^3)$-H···O one [34]. Thus, these conformers have interaction energies $E(1)<E(2)<<E(3)$, as reflected in the diabatics. Conformer **3** is not likely to be a minimum energy conformation due to its much smaller stability.

Figure 6 also helps to illustrate a fact that can be important in understanding polymorphic transformations in crystals: as there are no discontinuities in the potential energy surface of any system, one cannot go from **1** to **3** without passing through **2**, except by a much higher energy path which is not as probable statistically. As we will see later, in crystals each of the minimum energy conformers is associated with a primary packing pattern or synthon [35] and, consequently, not all polymorphs can be obtained by direct transformation from a given one, except by going through high energy pathways.

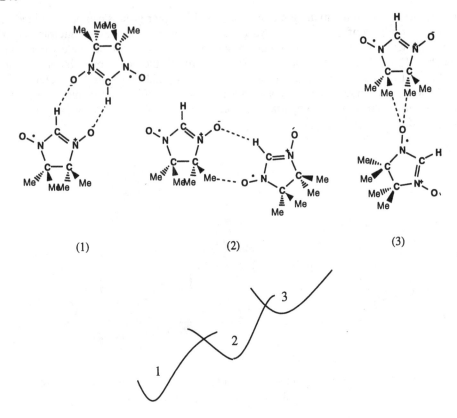

Figure 6. Above: Minimum energy conformations if the HNN dimer.
Below: Diabatics for the transformation (see text).

6. Molecular Association in Crystals

Molecular crystals are similar to molecular aggregates, the main difference being the long range order present in their association. From this perspective, a crystal is just an ordered supramolecular aggregate. Here, we will show that the structures of the molecular crystals can be rationalized using the tools described above for aggregates.

The molecular crystal is formed by the propagation, along the three crystallographic axes, of smaller aggregate units (not necessarily dimers). This propagation is done using the symmetry elements of the crystal in a periodic manner. Therefore, we can rationalize the crystal structure by looking at its molecular building units, that is, we could follow and approach similar to the synthon one introduced by Desiraju. Alternatively, we could look at the crystal from the opposite perspective and realize that molecular crystal are the results of the formation of characteristic tri-dimensional networks of intermolecular contacts, which allow to classify them in topological terms as $R_2^2(8)$, for instance. This perspective, first introduced by Etter [36], has been employed successfully by Bernstein and others as a form of distinguishing polymorphs [37].

Motifs or synthons are just the result of the association of intermolecular interactions in specific geometrical arrangements, following the principles described above for molecular aggregates. The properties of motifs and synthons can thus be understood by applying the principles of molecular association described above. A full rationalization of the crystal packing is achieved when one justifies the geometry of the building units and has a knowledge of their relative stability and their dynamics. This allows one to answer to questions like why some motifs show up in some crystals but do not in others, or why structurally different molecules pack using the same motifs, among others.

The presence of crystalline order in the molecular crystal imposes restrictions on the structure of the building units: each unit must be able to propagate its intermolecular bonds to the nearby units to allow the crystal to form. This is a serious restriction, not fulfilled by many of the minimum energy conformers which can be built in gas phase or in solution. An example of this is shown in schematic form in Figure 7 for a two dimensional ordered plane composed of a molecule having three donor groups and one acceptor group located in the same spatial position as those in the HNN molecule studied above. For the same molecule, pattern **a** allows an ordered two dimensional plane to be obtained, while **b** does not. The only difference between them is the relative orientation of the two nearby columns. Note here that it is not only the nature of the molecule which is important, but also the spatial arrangement of thegroups capable of forming intermolecular bonds. One will obtain an identical pattern from any molecule presenting the same number of similar acceptor and donor functional groups arranged in space, even though its chemical structure (represented in Figure 7 by the circle) is different, that is, *similar patterns are produced from molecules which have similar MEP map overlaps*, a principle that explains the similarity in the motifs obtained from structurally different molecules.

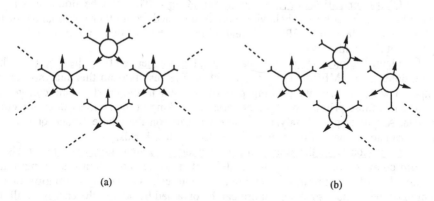

(a)   (b)

Figure 7. Packing patterns for a molecule like HNN, with three hydrogen-bonded donor groups (arrows) and two acceptor groups (inverted arrows).

At this point, we are able to indicate how this methodology, used above to rationalize the structure of molecular aggregates, can be applied to rationalize molecular crystals. The way in which crystals pack follows these principles:

(1) Molecules aggregate using their functional groups to form intermolecular bonds, thus increasing their stability;

(2) We can predict these bonds by performing a MEP overlap analysis to characterize their functional groups;

(3) There are many possible minimum energy aggregates (not necessarily dimeric), which form the crystal building units, each can be called a *primary packing pattern*, as these are the blocks that comprise the crystal packing, or also synthons;

(4) The molecular aggregate forming the crystal building units possess some intermolecular bonds which can be identified by a critical point analysis;

(5) The relative energy of each primary packing pattern is important to know its relative statistical probability of occurrence;

(6) The most stable crystals are grown by using the most stable primary packing patterns. The primary packing patterns (for instance, a dimer) associate to form a secondary packing pattern (for instance, a plane of dimers), which can associate further to form higher structures (stacks of planes). This allows us to describe the crystal in terms of primary, secondary, *etc.* structure.

These principles allow us to rationalize the existence of any crystal from its constituent parts, the molecules. The crystal structure can subsequently be characterized by a topological analysis [36,37]. At this point, one has to be aware that experimental crystals, due to the way the are grown, are not always in their most stable thermodynamic state. This means that the observed crystal structure does not always exhibit the most stable primary packing patterns.

The practical application of the above rules is performed using the procedure described below, which can be called *functional group-crystal packing analysis*, because it rationalizes the packing by looking at the bonds made by the functional groups present in the molecules. It consists in the following steps:

1)   Characterization of the functional groups. This can be done by simple inspection, extrapolating the behavior already obtained from other compounds, or by studying the molecular MEP map, searching for the regions of concentration or depletion of electronic density.

2)   Identification of the primary packing patterns generated from these groups. This can be done with a MEP map overlap analysis. The results from this analysis can be compared with the primary packing patterns present in the crystal. One can get an indication of these primary packing patterns by looking at the shortest intermolecular contacts. A critical point analysis of these patterns (on the whole crystal or pairs of dimers) and an estimate of their energy will identify their bonds.

3)   Computation of the strength of the primary packing patterns of interest. Using *ab initio* computations on adequate model systems, one can obtain the strength and directionality of the intermolecular bonds of interest. A first approximation of the strength of the primary packing pattern can be obtained by adding the energy of all the intermolecular bonds present. When possible, one can carry out the *ab initio* computation on pairs of dimers or the whole aggregate.

4)   Rationalization of the structure. Using the information from the previous steps, one can define the primary, secondary and other higher structures present in the crystal.

We have successfully applied these rules to various molecular crystals [16,38]. As an example, we will show how they rationalize the crystal structure of HNN [31]. This crystal can be described as stacks of planes, each plane resulting from the association of dimers in a T-like motif. Using the crystal packing functional group analysis one knows that the molecule can form aggregates by making $C(sp^2)$-H···O-N and $C(sp^3)$-H···O-N intermolecular hydrogen bonds. The crystal has a primary structure of $R_2^2(8)$ HNN dimers, each involving the formation of two strong $C(sp^2)$-H···O-N contacts (each one of -3.71 kcal/mol). This dimer is much more stable than any other possible primary packing pattern. These dimers then aggregate among themselves making weaker $C(sp^3)$-H···O-N contacts (-0.40 kcal/mol of interaction energy each). As a result, they form planes, as this maximizes the number of $C(sp^3)$-H···O-N contacts. These planes are the secondary structure of the HNN crystal. No short NO···ON distances are found in this crystal because *ab initio* computations show that the NO groups repel each other when coplanar [34]. Once the planes are formed, there are still some unused potential intermolecular contact groups. For instance, some methyl groups can make contacts against the NO groups already involved in the $C(sp^2)$-H···O-N contacts. In this way, the planes can be linked to each other. The result is the formation of ordered stacks of planes, piled up in such a way that they maximize the inter-plane interaction energy. This is the tertiary structure of the crystal. No higher-order structures are required to rationalize the HNN crystal.

7. Acknowledgements

I would like to thanks the work of Dr. F. Mota, Dr. M. C. Rovira, Dr. M. Planas and M. Deumal, who did many of the computations and helped to shape the ideas described here. I also wish to express my gratitude to D. Braga and F. Grepioni for their permission to use Figure 4. Finally, many thanks to DGICYT (project PB95-0848-C02-02) for their financial support and to CESCA and CEPBA for the allocation of computer time in their facilities.

8. References

1. Ferraro, J. R. and Williams, J. M. (1987) *Introduction to synthetic electrical conductors*, Academic Press, Orlando.
2. Kahn, O. (1993) *Molecular magnetism*, VCH, Weinheim.
3. Kitaigoroddsky, A. I. (1973) *Molecular crystals and molecules*, Academic Press, New York.
4. Desiraju, G. R. (1989) *Crystal engineering. The design of organic solids*, Elsevier, Amsterdam.
5. Maitland, G. C., Rigby, M., Smith, E. B. and Wakeham, W. (1981) *Intermolecular forces. Their origin and determination*, Clarendon Press, Oxford.
6. Stone, A. J. (1996) *The theory of intermolecular forces*, Clarendon Press, Oxford.
7. See, for instance: (a) Dunitz, J. D. and Berbstein, J. (1995) *Acc. Chem. Res.* **28**, 193; (b) Dunitz, J. D. (1995) *Acta Cryst.* B**51**, 619.
8. Szabo, A. and Ostlund, N. S. (1982) *Modern quantum chemistry*, Macmillan, New York (there is a recent Dover edition).
9. See, for instance: Enoit, M., Bernasconi, M., Focher, P. and Parrinello, M. (1996), *Phys. Rev. Lett.* **76**, 2934.
10. (a) Karfunkel, H. R. and Gdanitz, R. J. (1992) *J. Comput. Chem.* **13**, 1171; (b) Perlstein, J. (1994) *J. Am. Chem. Soc.* **116**, 455; (c) Gavezzotti, A. (1994) *Acc. Chem. Res.* **27**, 309.
11. (a) Dyke, T. R., Mack, K. M. and Muenter, J. S. (1979) *J. Chem. Phys.* **66**, 498; (b) Reimers, J., Watts, R. and Klein, M. (1982) *Chem. Phys.* **64**, 95.

12. (a) Feller, D. (1992) *J. Chem. Phys.* **96**, 6104; (b) Novoa, J. J., Planas, M. and Rovira, M. C. (1996) *Chem. Phys. Lett.* 251, 33; (c) van Duijneveldt-van de Rijdt, J. G. C. M. and van Duijneveldt, F. (1997) *Ab initio* methods applied to hydrogen-bonded systems, in D. Hadzi (ed.), *Theoretical treatments of hydrogen bonding*, John Wiley, Chichester.

13. Bader, R. F. W. (1990) *Atoms in Molecules. A Quantum Theory*, Clarendon Press, Oxford.

14. (a) Scrocco, E. and Tomasi, J. (1978) *Adv. Quantum Chem.* **11**, 115; (b) Politzer, P. and Murray, J. S. (1991) Molecular electrostatic potentials and chemical reactivity, in Lipkowitz, K. B. and Boyd, D. B. (eds.) *Reviews in Computational Chemistry II*, VCH, New York., ch. 7.

15. Buckingham, A. D. (1997) The hydrogen bond: An electrostatic interaction?, in D. Hadzi (ed.), *Theoretical treatments of hydrogen bonding*, pp.1-12. and references therein.

16. Braga, D., Grepioni, F., Tagliavini, E., Novoa, J. J. and Mota, F. *New. J. Chem.* (1998) 755.

17. It means a Hartree-Fock computation using the 6-31G(d,p) basis set.

18. Novoa, J. J., Mota, F., Perez del Valle, C. and Planas, M. (1997) *J. Phys. Chem. A* **101**, 7842.

19. Fowler, P. W. and Buckingham, A. D. (1991) *Chem. Phys. Lett.* **176**, 11.

20. Hobza, P. and Zahradnik, R. (1988) *Intermolecular complexes*, Elsevier, Amsterdam.

21. L. Pauling, (1960)*The nature of the chemical bond*, 3rd ed., Cornell University Press, Ithaca.

22. Pimentel, G. C. and McClellan, A. L. (1960) *The hydrogen bond*, Freeman, San Francisco.

23. Kollman, P. A. and Allen, L. C. (1972) *Chem. Rev.* **72**, 283.

24. Jeffrey, G. A. and Saenger, W. (1991) *Hydrogen bonding in biological structures*, Springer-Verlag, Berlin.

25. Novoa, J. J., Mota, F. and Lafuente, P. (1998) *Chem. Phys. Lett.* **290**, 519.

26. Bernstein, J. and Novoa, J. J. (submitted).

27. Allen, F. H., Kennard, O. (1993) *Chem. Des. Aut. News* **8**, 31.

28. Braga, D., Grepioni, F. and Novoa, J. J. (1998) *Chem. Commun.* 1959-1960.

29. (a) Novoa, J. J., Tarron, B., Whangbo, M.-H. and Williams, J. M. (1991) *J. Chem. Phys.* **95**, 5179; (b) Novoa, J. J. and Mota, F. (1997) *Chem. Phys. Lett.* **266**, 23; (c) Rovira, C. and Novoa, J. J. (1997) *Chem. Phys. Lett.* **279**, 140.

30. (a) Novoa, J. J. and Whangbo, M.-H. (1990) Nature of chalcogen...chalcogen contact interactions in organic donor-molecule salts, in *Organic superconductivity*, Kresin, V. Z. and Little, W. A. (eds.), Plenum, New York; (b) Novoa, J. J., Whangbo, M.-H. and Williams, J. M. (1991) *J. Chem. Phys.* **94**, 4835.

31. van Duijneveldt, F. B., van Duijneveldt-van de Rijdt, J. G. C. M. and van Lenthe, J. H. (1994) *Chem. Rev.* **94**, 1873.

32. Boys, S. F. and Bernardi, F. (1970) *Mol. Phys.* **19**, 553.

33. Novoa, J. J. and Planas, M. (1998) Chem. Phys. Lett. **285**, 186.

34. Deumal, M., Cirujeda, J., Veciana, J., Kinoshita, M., Hosokoshi, Y. and Novoa. J. J. (1997) *Chem. Phys. Lett.*, **165**, 190.

35. Desiraju, G. R. (1995) *Angew. Chem. Int. Ed. Engl.* **34**, 2311.

36. Etter, M. C. (1990) *Acc. Chem. Res.* **23**, 120.

37. Bernstein, J., Davis, R. E., Shimoni, L. and Chang, N.-L. (1995) *Angew. Chem. Int. Ed. Engl.* **34**, 1555.

38. (a) Novoa, J. J., Rovira, M. C., Rovira, C., Veciana, J. and Tarrés, J. (1995) *Adv. Mater.* **7**, 233; (b) Veciana, J., Cirujeda, J., Rovira, C., Molins, E. and Novoa, J. J. (1996) *J. Phys. I France* **6**, 1967.

SYSTEMATIC STUDY OF CRYSTAL PACKING

CAROLYN PRATT BROCK
Department of Chemistry
University of Kentucky
Lexington, KY 40506-0055
USA

1. Abstract

Systematic studies of crystal packing have included analyses of classes of molecules (*e.g.*, aromatic hydrocarbons, carboxylic acids) and of space-group frequencies. Combinations of the two approaches are especially informative. Once the basic rules are known, classes of exceptions can be identified and investigated.

2. Introduction

The successful prediction of crystal structures is a long-standing, albeit elusive, goal; classic references include books by Pauling [1] and Kitaigorodskii [2]. More recent important references include Desiraju's book [3] and Gavezzotti's essay [4]. The structures of simple salts and metals can be predicted with some confidence, but the structures of molecules usually cannot. Given the enormous amount of crystallographic data now available for molecular crystals [5,6] it seems as if it ought to be possible to deduce rules for the packing of at least some classes of molecules. The problem is to discover those rules.

Two complementary approaches have been tried. Some investigators have used databases (especially the Cambridge Structural Database, hereafter, the CSD [5,6]) to look for patterns common to the crystal structures of sets of related molecules. Other groups have compiled space-group frequencies and looked for patterns related to the crystallographic symmetry elements. In a few studies these approaches have been combined.

This review is, of necessity, incomplete. Topics covered elsewhere in this volume have been omitted. For a more complete list of the many publications that have used the CSD to study crystal packing see [3,4] and the bibliography available within the CSD itself.

J.A.K. Howard et al. (eds.),
Implications of Molecular and Materials Structure for New Technologies, 251–262.
© *1999 Kluwer Academic Publishers. Printed in the Netherlands.*

3. Studies of the Packing of Classes of Molecules

The following is limited to studies in which the fragment dominates the crystal packing or in which the shape of the molecule is of paramount importance.

3.1. AROMATIC HYDROCARBONS

Robertson used a private database, which was generated mostly by his own group, to discover that planar aromatic hydrocarbons usually crystallize in a herringbone pattern of molecular stacks [7]. The interplanar angle between molecules in adjacent stacks varies with the area and the eccentricity of an ellipsoid that encloses the 2D molecular surface. Gavezzotti and Desiraju used the CSD to extend and quantify this idea. They described four kinds of packing motifs [8] that could be predicted from molecular dimensions [9]. References to extensions of this work can be found in [4]. Notable is Gavezzotti's proposal of the descriptor $C_{self} = V_{molecule}/V_{bounding\ parallelepiped}$, where $V_{molecule} = V_{cell}/Z$ and $V_{bounding\ parallelepiped}$ is determined by the van der Waals surface of the molecule [10]. Low values of C_{self} are associated with poor crystal packing and the formation of solvates.

3.2. ORGANIC ACIDS AND AMIDES; HYDROGEN BONDING

Leiserowitz's examination of structures of carboxylic acids [11] is a classic study of crystal packing. Monocarboxylic acids RCOOH form hydrogen-bonded dimers unless (1) R is small, in which case chains are also found, or (2) R is enantiomerically pure, in which case chains along a 2_1 axis are common. Diacids HOOC-R-COOH form extended chains in which adjacent molecules are linked by -COOH dimers. Leiserowitz also investigated the packing of amides [12], which almost always form H-bonded dimers that associate to form ribbons, or sometimes sheets, that are held together by a second set of N-H···O=C bonds.

These two papers [11, 12] laid the groundwork for a large body of work describing H-bonding patterns, *e.g.* [13], and using those patterns to gain some control over crystal packing (see [14]).

3.3. "PORPHYRIN SPONGES"

Strouse noticed that many tetraarylporphyrins have very similar packing arrangements and that solvent is often included in the crystal. He went on to discover a series of 65 solvates having essentially the same crystal packing [15]. The key to the series is the inability of the porphyrin to fill space efficiently. The aryl groups at the periphery of the molecule must be twisted relative to the porphyrin core. Gavezzotti's descriptor C_{self} [10] is small.

3.4. ADDITIONAL EXAMPLES

The interactions between pairs of $PPh_4{}^+$ ions, which are common counterions that often dominate crystal packing, have been described by Dance [16]. The packing of nitrobenzene derivatives has been categorized by André, Foces-Foces, Cano, and Martinez-Ripoli [17, 18].

4. Studies of Space-Group Frequencies

Work on space-group frequencies to *ca.* 1993 has been reviewed by Brock and Dunitz [19]. Other recent papers on the subject include those by Belsky, Zorkaya and Zorky [20], Wilson [21], Baur and Kassner [22] (for inorganic compounds), and Wukovitz and Yeates [23] (for macromolecules). All of these groups compiled tables of space-group frequencies that confirmed results obtained previously. What was new, however, was the explicit consideration of the role of Z', *i.e.*, of the number of formula units in the unique part of the unit cell. Belsky *et al.* [20] went farther by investigating the actual symmetry imposed on the packing units (molecules or ions). Those authors discriminated between *e.g.*, Z' = 1 and Z' = 2(Ω) in group P$\overline{1}$ and between structures in C2/c, Z' = Ω with imposed symmetry $\overline{1}$ and imposed symmetry 2. Their work, which was done with a restricted database, was extended by Cole to the entire CSD [24].

4.1. MIRROR PLANES

An important result [19, 21] of these studies is the observation that space groups containing mirror planes occur only if all the mirror planes are occupied. Group Pnma, which is the seventh most common in the CSD, occurs almost exclusively with Z' = Ω [19,20,21,24]. If Z' = 1, then there are almost certainly two independent formula units in the unit cell, each of which is located on a mirror plane {*i.e.*, Z' = 1 = 2(Ω), which is Z = 8(m^2) in the notation of Belsky *et al.* [20]}. Cole noticed [24] that structures in space groups including mirror planes are especially likely to be disordered (*i.e.*, to be flagged as disordered in the CSD). If the disorder flag is set then the deviations from mirror symmetry within a single unit cell are large enough that twinning must be suspected. Other structures are apparently "ordered", but have displacement ellipsoids that are elongated perpendicular to the mirror plane.

4.2. ROTATION AXES

The 3-, 4-, and 6-fold axes of high-symmetry space groups do not usually occur unless molecules or ions of the appropriate symmetry are located on at least some of them. Tetragonal, trigonal, hexagonal, and cubic space groups do not usually occur unless the symmetries of the packing units are also high. Two-fold rotation axes are sometimes occupied and sometimes not [19, 21, 24].

4.3. INVERSION CENTERS

The new studies confirm that inversion centers are favorable for crystal packing. Only about 25% of the structures in the CSD occur in noncentrosymmetric space groups. About 80% of the noncentrosymmetric structures (or, 20% of the total structures) occur in space groups (the Sohnke groups) without improper symmetry operations (inversion centers, mirror planes, and improper rotation axes). Assume for the sake of argument that about half of the structures in the Sohnke groups correspond to crystals that were grown from enantiomerically pure material. Compounds not restricted to Sohnke groups then crystallize in centrosymmetric space groups about 83% of the time.

Another argument for centrosymmetry being favorable is the large number of unoccupied inversion centers in common space groups. The two most common groups are $P2_1/c$ and $P\bar{1}$. If $Z'=1$, which is normal, then there are four independent sets of inversion centers in the former group and eight in the latter. If $Z'=\Omega$ so that one set of inversion centers is occupied there are still three ($P2_1/c$) or seven ($P\bar{1}$) unoccupied sets.

4.4. Z'

Crystallization with $Z'>1$ is rare. Only about 6% of well-determined structures in a large sample of the CSD had $Z'>1$ and only about 0.4% had $Z'>2$ [19]. The percentages in other studies (e.g., [25]) are higher, but are no greater than 10% for $Z'>1$ and 1% for $Z'>2$. Most of the structures with $Z'>1$ are found in the very low-symmetry groups (especially P1). The independent molecules usually have very similar conformations [25,26].

Many of the $Z'>1$ structures are pseudosymmetric. Davis and Wheeler [27] estimate that about a quarter of the $Z'=2$ structures in seven common space groups have approximate inversion, screw, rotation, or translation operations that relate the independent molecules. Marsh, Schomaker, and Herbstein [28] observed that many of the structures in $Pca2_1$ and $Pna2_1$ have two independent molecules that are related by an approximate inversion center. (These two are the only groups that contain screw axes and glide planes but no other symmetry element). The pseudo-centers tend to occur in well defined positions (cf. [31]).

4.5. COMPOUND TYPE

Space-group frequencies differ with compound type because the dominant interactions between packing units vary. The packing of ionic compounds is determined primarily by electrostatic considerations. The packing of molecular compounds is determined by their van der Waals surfaces and by the interactions of functional groups. Protein crystals, however, contain such large amounts of solvent that there are few direct contacts between macromolecules. The importance of contacts between individual functional groups is therefore small.

The more ionic the compound the greater the probability of crystallization in a high-symmetry space group. The immediate environment of a simple ion is usually highly symmetric (octahedral, tetrahedral, trigonal, *etc.*). This symmetry is expected to propagate through the crystal unless there is a conflict with the packing requirements of the counterion. Crystallization in high-symmetry groups is also more likely for salts (and solvates) than for single molecules because the presence of two different covalently bonded units relieves the "like-like" interactions that would be generated by the symmetry operations if only one type of packing unit were present [19].

Space group frequencies for the Protein Data Bank [29] are also different than for the CSD and Inorganic Database [30] because improper symmetry operations are not normally allowed.

4.6. POSSIBLE BIAS IN THE DATABASES

It is worth noting that the database of all structures published or deposited is not the same as the database of all structures attempted. High-symmetry and $Z' > 1$ structures are almost certainly underrepresented in the CSD. Structures with large values of Z' can be difficult to solve and refine because of pseudosymmetry and correlations. High-symmetry structures are quite often twinned because the most common high-symmetry space groups are in the lower-symmetry Laue group(s) of the crystal system (*e.g.*, 4/m rather than 4/mmm, $\bar{3}$ and 3m rather than $\bar{3}$ m) [19].

5. Distribution of Molecules and Ions Within the Unit Cell

Motherwell looked in detail at the seven most common space groups and discovered that there are clear patterns in the distributions of the molecular centers [31]. The centers of mass tend to be located midway between inversion centers and screw axes.

6. Exceptions to the Rules

Exceptions to the general rules of crystal packing can be very informative. This line of inquiry has been stressed by Zorky and Belsky (20,32).

6.1. EXCEPTIONS TO THE STRUCTURE-CORRELATION PRINCIPLE: BIPHENYLS WITHOUT *ORTHO* SUBSTITUENTS

Molecules usually crystallize in a low-energy conformation. This observation forms the basis for the structure-correlation method [33,34]. There are, however, exceptions, of which biphenyls with H atoms in all four *ortho* positions are a classic example. Brock and Minton showed [35] that the distribution of torsion angles around the central C-C bond does not match the Boltzmann distribution expected for the intramolecular potential-energy surface that was derived from electron-diffraction data. In the crystal there is a second cluster of structures that are planar or nearly so. It must be that the (nearly) planar molecules fill space more efficiently than the twisted molecules, probably because the value of C_{self} [10] increases as the molecule becomes more planar.

Another indication that biphenyls have a packing problem is the relatively large number of such structures with $Z' > 1$.

6.2. ANOMALOUS SPACE-GROUP FREQUENCIES: STRUCTURES OF MONOALCOHOLS, C_nH_mOH

A chance encounter with a sterol structure in $P2_1$ with $Z' = 3$ led Brock and Duncan to the observation [36] that monoalcohols very often crystallize in either high-symmetry (trigonal or tetragonal) space groups or in low-symmetry groups with $Z' > 1$. This anomalous pattern is a direct consequence of the hydrogen-bonding requirements of the hydroxyl groups, which require that each -OH group be within *ca.* 2.8 Å each of -OH groups on two adjacent molecules. The three interacting molecules cannot be related by glide or 2_1 screw operations or by pure translations if the molecule is large in the directions perpendicular to the C-O bond.

Some of the trigonal and tetragonal structures of monoalcohols have the -OH groups arranged in rings around sites of $\bar{3}$ or $\bar{4}$ point symmetry. These structures are not really exceptions to the rule that high-symmetry space groups occur only when the high-symmetry sites are occupied because the $\bar{3}$ and $\bar{4}$ sites are "occupied" by molecular aggregates.

The structures of phenols have been examined by Perrin *et al.* [37] (see also [3]) and by Prout *et al.* [38]. The rules given above are not followed so strongly by phenols as by alcohols in general because simple phenols are so "thin" that three H-bonded molecules can be related by translation, a 2_1 axis, or a glide plane. Perrin *et al.* [37] noted, however, that close to half of the simple phenols examined crystallize in noncentrosymmetric space groups.

6.3. STRUCTURES WITH $Z' > 1$

Structures having more than one molecule in the asymmetric unit are sufficiently rare that either a packing problem or a strong intermolecular interaction should be suspected. Steroids are known for crystallizing with large values of Z' (*e.g.*, [39]), probably because the periodicities needed for the relatively rigid ring system and the more flexible side chain are in conflict. The 1,2,3,5-dithia- and 1,2,3,5-diselenadiazolyl radicals (RCN_2E_2, E= S or Se) often crystallize with $Z' > 1$, perhaps because of conflicts between the optimum distance for radical-radical interactions and the steric requirements of the R groups [40,41].

6.4. RETENTION OF $\bar{4}$ SYMMETRY: STRUCTURES CONTAINING MAR_4 MOLECULES AND IONS

Rigorous retention of the highest possible molecular symmetry in the crystalline state is the exception rather than the rule. It has long been known, however, that molecules MAr_4 often retain their $\bar{4}$ (or S4) symmetry (see, *e.g.*, [2]), as do simple salts containing MAr_4 ions [42]. Lloyd and Brock have shown recently [43] that the persistence of the symmetry is a consequence of the low energy of the $\overline{4}$ conformation, the "nesting" of the units around *empty* $\overline{4}$ sites to form columns, and the interleaving of aryl rings in adjacent columns to form a herringbone arrangement.

Salts of MAr_4 ions with counterions that can occupy $\overline{4}$ sites nearly always crystallize in group $I\overline{4}$ [43]. This rule is so strong that it is often followed even when the counterion must be disordered.

6.5. SPACE GROUPS DOMINATED BY A SINGLE TYPE OF STRUCTURE

Most of the space groups represented in the CSD have entries for a variety of types of compounds. Some groups, however, have entries dominated by a specific type of material [44].

Of the entries in the CSD for group P4/n, about 40% have the formula $[MPh_4^+][M'XY_4^-]$ or $[MPh_4^+][M'X_2Y_4^-]$ (cation site symmetry $\overline{4}$; anion symmetry 4). Another 25% of the entries have peripheral phenyl groups that surround an empty $\overline{4}$ site. The probability that salts $[MPh_4^+][MY_4^-]$ will crystallize in the related group $I\overline{4}$ was discussed above.

About half of the entries in space group $P6_3/m$ have the formula $[M(H_2O)_9^{3+}][C_2H_5OSOS_3^-]_3$, where M is a rare-earth element.

7. Other Examples of Systematic Studies of Crystal Packing

7.1. ARRANGEMENTS OF DIPOLES

Whitesell *et al.* [45] investigated the idea that molecules with large dipole moments should crystallize in centrosymmetric space groups even more preferentially than molecules with small dipole moments. They examined random samples of molecular structures in space groups P1, P$\overline{1}$, and $P2_1$; the dipole moment for each of the molecules in the samples was calculated numerically. They found that the average dipole moment did not vary with space group and that in the $P2_1$ structures the angle between the dipole moment and the screw axis was independent of the moment's magnitude. In any event, local interactions between bond dipoles are much more important to the crystal packing than are the interactions between overall molecular dipoles. Furthermore, inferences based on the dipole-dipole approximation, which requires a separation large relative to the size of the dipole, are inappropriate since intermolecular separations are nearly always shorter than the molecular dimensions.

7.2. POLYMORPHS

The comparison of polymorphs has long been a fruitful line of inquiry. For a recent example with references see [46]. All studies have concluded that the energy differences between polymorphs that crystallize under similar conditions is small.

7.3. WALLACH'S RULE

Brock, Schweizer, and Dunitz [47] used the CSD to locate pairs of matched structures for which one member was in a Sohnke group and one was not. The original purpose of the project was to test Wallach's Rule, which claims that racemic compounds (R) are denser than their enantiomerically pure counterparts (A). For each pair of structures the ratio V_A/V_R was calculated. If Wallach's Rule is valid the average value of the ratio should be greater than 1.

The problem turned out to be more subtle than anticipated. If the molecule is achiral or racemizes rapidly then the two structures are polymorphs. For this set of 64 temperature-matched pairs the mean value of V_A/V_R was found to be 1.002(3), which indicates that, on average, structures that have similar stabilities also have similar densities. Wallach, however, was concerned only with enantiomers that do not interconvert during crystallization. For this second set of 65 pairs the mean value of V_A/V_R was found to be 1.009(3), which indicates that, on average, the racemic structures are denser.

No racemic compound of a resolvable material can be observed unless the racemic compound is more stable than the conglomerate (or, eutectic) of the enantiomers. Since S for mixing macroscopic crystals is too small to measure, the racemic compounds in the list have, on average, lower energies than their enantiomerically pure counterparts. If higher density implies lower energy, as the comparison of polymorphs [46,47] indicates, then Wallach's Rule must be true. To the extent that it is based on comparisons of density, however, the Rule does not have any predictive power. No material for which it is false can appear on the list of pairs. Similar arguments apply to all comparisons of data (*e.g.*, of H_f values) for racemic and resolved material.

If Wallach's Rule were only true because the list of matched pairs of structures is biased, then there should be numerous enantiomers that fail to form racemic compounds, but that is not the case. Wallach's Rule does have predictive power because inversion centers are very favorable for crystal packing (see above). Molecules that must crystallize in Sohnke groups have a packing problem, which explains the higher rate of Z' > 1 structures in those groups.

There has been some thought [48,49] that molecules having potential two-fold rotational symmetry are more likely to undergo conglomerate crystallization (or, spontaneous resolution) than molecules in general. Cole's statistics [24] on space-group frequencies suggest otherwise. The most common group (*ca.* 67% of the total) for molecules retaining a two-fold axis is the centrosymmetric group C2/c.

The fraction in Sohnke groups of molecules retaining C2 symmetry is no greater than of molecules in general.

7.4. ISOSTRUCTURAL PAIRS AND SERIES

Molecules in which a simple change does not alter the van der Waals surface or functionality are expected to crystallize in the same space group. The series Ph_3X-$X'Me_3$, X and Ph_3X-$X'Et_3$, X, X' Si, Ge forms an example [50]. Sometimes, however, less closely related molecules form isostructural pairs (see, *e.g.*, [50]). There is no simple way to search for such sets, which are very interesting.

7.5. MOLECULES WITH ODD AND EVEN NUMBERS OF C ATOMS

A provocative study [51]shows that molecules with an even number of C atoms appear in several large databases (including the CSD) much more frequently than do molecules with an odd number of C atoms. The basis for this disparity is not fully understood.

8. Suggestions for future work

Continued investigation of the exceptions to general rules for crystal packing is likely to be productive.

Substances that show evidence of poor crystal packing (low melting point, Z' > 1, low C_{self}) are more likely to form solid-state compounds (including solvates and inclusion complexes) than substances that pack well.

The low frequency of structures with Z' > 1 may be related to the rarity of solid-state compounds between molecules A and B that do not interact strongly. If A and B can react by complete or partial transfer of an electron or proton then compound formation is expected. On the other hand, compound formation between stereoisomers is rare unless the isomers are enantiomers.

Scientists working to synthesize nonlinear optical materials might consider trying to make molecules that crystallize preferentially in crystal systems with greater than monoclinic symmetry. Commonly occurring space groups in the triclinic and monoclinic systems are *much* more likely to be centrosymmetric than groups in the higher-symmetry systems.

9. References

1. Pauling, L. (1960) *The Nature of the Chemical Bond*, Cornell Univ. Press, Ithaca, NY.

2. Kitaigorodskii, A. I. (1961) *Organic Chemical Crystallography*, Consultant's Bureau, New York.

3. Desiraju, G. R. (1989) *Crystal Engineering: The Design of Organic Solids*, Elsevier, Amsterdam.

4. Gavezzotti, A. (1994) Are Crystal Structures Predictable?, *Acc. Chem. Res.* **27**, 309-314.

5. Allen, F. H., Kennard, O. and Taylor, R. (1983) Systematic Analysis of Structural Data as a Research Technique in Organic Chemistry, *Acc. Chem. Res.* **16**, 146-153.

6. Allen, F.H. and Kennard, O. (1993) 3D Search and Research Using the Cambridge Structural Database, *Chemical Design Automation News*, **8**, 1 & 31-37.

7. Robertson, J. M. (1951) The Measurement of Bond Lengths in Conjugated Molecules of Carbon Centres, *Proc. R. Soc. London Ser. A*, **207**, 101-110.

8. Gavezzotti, A. and Desiraju, G. R. (1988) A Systematic Analysis of Packing Energies and Other Packing Parameters for Fused-Ring Aromatic Hydrocarbons, *Acta Cryst.* **B44**, 427-434.

9. Desiraju, G. R. and Gavezzotti, A. (1989) Crystal Structures of Polynuclear Aromatic Hydrocarbons. Classification, Rationalization and Prediction from Molecular Structure, *Acta Cryst.* **B45**, 473-482.

10. Gavezzotti, A. (1990) Crystal Packing of Hydrocarbons. Effects of Molecular Size, Shape and Stoichiometry, *Acta Cryst.* **B46**, 275-283.

11. Leiserowitz, L. (1976) Molecular Packing Modes. Carboxylic Acids., *Acta Cryst.* **B32**, 775-802.

12. Leiserowitz, L. and Hagler, A. T. (1983) The Generation of Possible Crystal Structures of Primary Amides, *Proc. R. Soc. London Ser. A* **388**, 133-175.

13. Etter, M. C. (1990) Encoding and Decoding Hydrogen-Bond Patterns of Organic Compounds, *Acc. Chem. Res.* **23**, 120-126.

14. Aakeröy, C. B. (1997) Crystal Engineering: Strategies and Architectures, *Acta Cryst.* **B53**, 569-586.

15. Byrn, M. P., Curtis, C. J., Khan, S. I., Sawin, P. A., Tsurumi, R. and Strouse, C. E. (1990) Tetraarylporphyrin Sponges. Composition, Structural Systematics and Applications of a Large Class of Programmable Lattice Clathrates, *J. Am. Chem. Soc.* **112**, 1865-1874.

16. Dance, I. and Scudder, M. (1996) Supramolecular Motifs: Concerted Multiple Phenyl Embraces between Ph_4P^+ Cations are Attractive and Ubiquitous, *Chem. Eur. J.* **2**, 481-486.

17. André, I., Foces-Foces, C., Cano, F. H. and Martinez-Ripoli, M. (1997) Packing Modes in Nitrobenzene Derivatives. I. The Single Stacks, *Acta Cryst.* **B53**, 984-995.

18. André, I., Foces-Foces, C., Cano, F. H. and Martinez-Ripoli, M. (1997) Packing Modes in Nitrobenzene Derivatives. II. The 'Pseudo-Herringbone' Mode, *Acta Cryst.* **B53**, 996-1005.

19. Brock, C. P. and Dunitz, J. D. (1994) Towards a Grammar of Crystal Packing, *Chem. Mater.* **6**, 1118-1127.

20. Belsky, V. K., Zorkaya, O. N. and Zorky, P. M. (1995) Structural Classes and Space Groups of Organic Homomolecular Crystals: New Statistical Data, *Acta Cryst.* **A51**, 473-481.

21. Wilson, A. J. C. (1993) Space Groups Rare for Organic Structures. III. Symmorphism and Inherent Molecular Symmetry, *Acta Cryst.* **A49**, 795-806. See also earlier papers referenced in [19].

22. Baur, W. H. and Kassner, D. (1992) The Perils of Cc: Comparing the Frequencies of Falsely Assigned Space Groups with their General Population, *Acta Cryst.* **B48**, 356-369.

23. Wukovitz, S. W. and Yeates, T. O. (1995) Why Protein Crystals Favour Some Space Groups Over Others, *Nature Structural Biology* **2**, 1062-1067.

24. Cole, J. C. (1995). Ph.D. Thesis, University of Durham, UK.

25. Gautham, N. (1992) A Conformational Comparison of Crystallographically Independent Molecules in Organic Crystals with Achiral Space Groups, *Acta Cryst.* **B48**, 337-338.

26. Zorkii, P. M. (1993) A New View on the Structure of Organic Crystals, *Russ. J. Phys. Chem.* **68**, 870-876.

27. Wheeler, K. and Davis, R. E. (1993). Approximate Symmetry in Crystal Structures with Multiple Molecules per Asymmetric Unit. Abstracts of the Albuquerque, NM Meeting of the American Crystallographic Association, pp. 109 (paper PH02).

28. Marsh, R. E., Schomaker, V. and Herbstein, F. H. (1998) Centrosymmetric Arrays in Space Groups *Pca2$_1$* and *Pna2$_1$*, preprint.

29. Bernstein, F. C., Koetzle, T. F., Williams, G. J. B., Meyer E F, J., Brice, M. D., Rodgers, J. R., Kennard, O., Shimanouchi, T. and Tasumi, M. (1977) The Protein Data Bank: a Computer-Based Archival File for Macromolecular Structures, *J. Mol. Biol.* **112**, 535-542.

30. Bergerhoff, G., Hundt, R., Sievers, R. and Brown, I. D. (1983) The Inorganic Crystal Structure Data Base, *J. Chem. Inf. Comput. Sci.* **23**, 66-69.

31. Motherwell, W. D. S. (1997) Distribution of Molecular Centres in Crystallographic Unit Cells, *Acta Cryst.* **B53**, 726-736.

32. Zorky, P. M. (1996) Symmetry, Pseudosymmetry and Hypersymmetry of Organic Crystals, *J. Mol. Struct.* **374**, 9-28.

33. Buergi, H.-B. and Dunitz, J. D. (1983) From Crystal Statics to Chemical Dynamics, *Acc. Chem. Res.* **16**, 153-161.

34. Buergi, H.-B. and Dunitz, J. D. (1994) Structure Correlation; the Chemical Point of View, in H.-B. Buergi and J. D. Dunitz (eds.), *Structure Correlation*, VCH Verlagsgesellschaft GmbH, Weinheim, pp. 163-204.

35. Brock, C. P. and Minton, R. P. (1989) Systematic Effects of Crystal-Packing Forces: Biphenyl Fragments with H Atoms in All Four Ortho Positions, *J. Am. Chem. Soc.* **111**, 4586-4593.

36. Brock, C. P. and Duncan, L. L. (1994) Anomalous Space-Group Frequencies for Monoalcohols CmHnOH, *Chem. Mater.* **6**, 1307-1312.

37. Perrin, R., Lamartine, R., Perrin, M. and Thozet, A. (1987). Solid State Chemistry of Phenols and Possible Industrial Applications, in G. R. Desiraju (ed.), *Organic Solid State Chemistry*, Elsevier, Amsterdam pp. 271-329.

38. Prout, K., Fail, J., Jones, R. M., Warner, R. E. and Emmett, J. C. (1988) A Study of the Crystal and Molecular Structures of Phenols with Only Intermolecular Hydrogen Bonding, *J. Chem. Soc. Perkin Trans.* 2, 265-284.

39. Craven, B. M. (1986). Cholesterol Crystal Structures: Adducts and Esters, in D. M. Small (ed.), *The Physical Chemistry of Lipids*, Plenum Press, New York, pp. 149-182.

40. Cordes, A. W. (1993) Personal communication.

41. Cordes, A. W., Bryan, C. D., Davis, W. M., de Laat, R. H.., Glarum, S. H., Goddard, J. D., Haddon, R. C., Hicks, R. G., Kennepohl, D. K., Oakley, R. T., Scott, S. R. and Westwood, N. P. C. (1993) Prototypal 1,2,3,5-Dithia- and 1,2,3,5-Diselenadiazolyl [HCN$_2$E$_2$]. (E=S,Se): Molecular and Electronic Structures of the Radicals and Their Dimers, by Theory and Experiment, *J. Am. Chem. Soc.* **115**, 7232-7239.

42. Mueller, U. (1980) Strukturverwandtschaften unter den EPh$_4$+-Salzen, *Acta Cryst.* **B36**, 1075-1081.

43. Lloyd, M. A. and Brock, C. P. (1997) Retention of $\overline{4}$ Symmetry in Compounds Containing MAr$_4$ Molecules and Ions, *Acta Cryst.* **B53**, 780-786.

44. Brock, C. P. (1996) Investigations of the Systematics of Crystal Packing Using the Cambridge Structural Database, *J. Res. Natl. Inst. Stand. Technol.* **101**, 321-325.

45. Whitesell, J. K., Davis, R. E., Saunders, L. L., Wilson, R. J. and Feagins, J. P. (1991) Influence of Molecular Dipole Interactions on Solid-State Organization, *J. Am. Chem. Soc.* **113**, 3267-3270.

46. Gavezzotti, A. and Filippini, G. (1995) Polymorphic Forms of Organic Crystals at Room Conditions: Thermodynamic and Structural Implications, *J. Am. Chem. Soc.* **117**, 12299-12305.

47. Brock, C. P., Schweizer, W. B. and Dunitz, J. D. (1991) On the Validity of Wallach's Rule: On the Density and Stability of Racemic Crystals Compared with their Chiral Counterparts, *J. Am. Chem. Soc.* **113**, 9811-9820.

48. Collet, A., Brienne, M.-J. and Jacques, J. (1972) Dédoublements Spontanés et Conglomérats d'Énantiomères, *Bull. Soc. Chem. Fr.*, 127-142.

49. Collet, A. (1990). The Homochiral versus Heterochiral Dilemma, in M. Simonyi (ed.), *Problems and Wonders of Chiral Molecules*, Akademiai Kiado, Budapest, pp. 91-109.

50. Kalman, A., Parkanyi, L. and Argay, G. (1993) Classification of the Isostructurality of Organic Molecules in the Crystalline State, *Acta Cryst.* **B49**, 1039-1049.

51. Sarma, J. A. R. P., Nangia, A., Desiraju, G. R., Zass, E. and Dunitz, J. D. (1996) Even-Odd Carbon Atom Disparity, *Nature (London)* **384**, 320.

MOLECULAR SHAPE AS A DESIGN CRITERION

RAYMOND E. DAVIS[a], JAMES K. WHITESELL[a] and KRAIG A. WHEELER[b]

[a]*Department of Chemistry and Biochemistry, University of Texas at Austin, Austin TX 78712-1167, U.S.A. and* [b]*Department of Chemistry, Delaware State University, Dover, DE 19901, U.S.A.*

1. Introduction

Many applications of crystalline materials depend not just on the underlying molecular structures, but equally on the alignment of molecules in the crystals. An understanding of molecular arrangements is critical to the design of materials with specific bulk properties. Crystal structure reports have traditionally contained little significant discussion (frequently no mention at all) of molecular arrangements in the crystal, presumably because the motivation for many studies was the elucidation of some molecular feature. The Cambridge Structural Database [1] makes available crystal data for many thousands of organic and organometallic crystal structures, together with continually evolving retrieval and analysis software. The commitment at the Data Centre to the development of increasingly powerful software for 3-dimensional search, analysis, and display is changing the landscape of the study of crystal packing. This superb set of tools facilitates many important and innovative programs of study that are slowly but surely increasing our understanding of molecular solid state structure. This systematic study is, in turn, fueling significant progress in crystal engineering, one of the major themes of this School. Useful reviews in this area include those by Desiraju [2] and by Braga and Grepioni [3] (both of these surveying a broad range of intermolecular interactions) and those by Aakeröy [4, 5] and Zaworotko [6] (focusing more specifically on the exploitation of hydrogen-bonding as a design tool).

We have been especially interested in the study of crystal structures that exhibit approximate symmetry, either between molecules that are chemically the same but are in different crystallographic environments ($Z' > 1$), or between molecules that are isosteric, that is, similar enough (at least in their shapes) that they might substitute for one another in the crystal. In the course of these studies, we have investigated some cases where similarities in shapes of molecules or portions of the same molecule can be utilized to influence molecular packing -- pseudosymmetry as a design element.

2. Some Background

2.1 SYMMETRY AND APPROXIMATE SYMMETRY

Crystallographers have long been interested in the relative frequencies of different space groups. In 1994, Brock and Dunitz [7] presented not only a thorough history of such

J.A.K. Howard et al. (eds.),
Implications of Molecular and Materials Structure for New Technologies, 263–274.
© 1999 *Kluwer Academic Publishers. Printed in the Netherlands.*

studies [8], but a detailed and thoughtful analysis of their own statistical study that emphasizes the role of molecular symmetry in influencing crystal packing. Their observations that 'structures in group $P2_1/c$ account for 38% of the 31770 entries' and that 'structures in group $P\bar{1}$ account for 20% of the entries', though not surprising to the practicing crystallographer, emphasize the favorable nature of centrosymmetric packing. A recent study of packing energies has shown on a quantitative basis the relative importance of inversion among the symmetry operators in many organic molecular crystals [9].

Various workers have also surveyed structures with more than one molecule in the crystallographic asymmetric unit [7, 10] ($Z' > 1$). Analyses by Gautham showed that the crystallographically independent molecules in 399 structures distributed among 65 chiral space groups [11] and in 767 structures distributed among 165 achiral space groups [12] displayed a strong tendency for conformational similarity. Further, packing energy calculations on such structures indicate that two such molecules have very similar interaction energy with their respective crystalline environments [13]. Crystallographically independent molecules often pack across pseudo-inversion centers. Desiraju, Calabrese, and Harlow [14] showed that pseudo-inversion centers were common in $P\bar{1}$ with $Z' = 2$. We have surveyed 2503 structures with $Z' > 1$ in seven common space groups (centrosymmetric: $P\bar{1}$, $P2_1/c$, $Pbca$; noncentrosymmetric: $P1$, $P2_1$, Pc, $P2_12_12_1$) [15]. We found that 27% of these structures have approximate $\bar{1}$, 2, 2_1 or translation operations relating the independent molecules. About 5.5% (140) of these structures have approximate inversion symmetry operators at positions suggesting at least a strong approximation to a higher symmetry space group, if not a misidentified space group; nearly 35% (56) of the 162 structures with $Z' > 1$ in $P2_1$ mimic $P2_1/c$ packing.

2.2 ISOSTERIC/ISOSTRUCTURAL SUBSTITUTION

Molecules that are sufficiently similar in size and shape are often observed to crystallize together, forming solid solutions at one extreme or fully ordered structures at the other [16]. The development by Lahav and Leiserowitz of 'tailor-made auxiliaries' for the control of nucleation, growth, and dissolution of crystals and their application to assignment of absolute configurations, resolution of racemic conglomerates, and control of morphology testifies to the importance of even small-scale substitution of one molecule for another, either in the bulk or at the crystal surface [17]. Topochemical reactions between chemically distinct species have been designed by Desiraju and coworkers [e.g., 18-20]. The latter studies were based on nearly equal sizes and shapes (isosterism) either of substituents (Cl, 19 Å^3; Me, 24 Å^3 in references 18 and 19) or of molecules (2,4-dichlorocinnamic acid and 6-chloro-3,4-methylenedioxycinnamic acid in reference 20).

2.3 QUASIRACEMATES

In 1899, Centnerszwer [21] observed that (+)-chlorosuccinic acid and (-)-bromosuccinic acid form a 'molecular compound' that he called a 'partial racemate'. The same behavior by mixtures of (+)-tartrate/(−)-malate salts and of (+)-tartramide/(−)-malamide had been described by Pasteur in one of his earliest papers on optical activity [22]. In general, cocrystallization of two different optically active compounds, A and A', is not likely unless there is a strong, specific interaction (e.g. hydrogen bond formation) between them [23]. However, if A and A' are similar enough in size and shape (isosteric), then special situations can arise. The like-handed compounds, e.g., (R)-A and (R)-A', could form solid solutions throughout the concentration range. Crystallization of an equimolar mixture of opposite near-enantiomers, e.g., (R)-A and (S)-A', can occur in several ways, analogous to those for mixtures of true enantiomers [25]. (1) They could form solid solutions (analogous to *pseudo*racemates). (2) They could form separate crystals (conglomerate crystallization). This could be signaled by a eutectic in the melting point diagram or by an X-ray powder diffraction pattern that is a composite of those for the two components. (3) They could form quasiracemic crystals. Melting point behavior (a maximum in the melting point curve, either lower or higher than the melting points of one or both of the pure substances) or X-ray powder patterns (similarity to those of either or both of the corresponding true racemates) can characterize the quasiracemate formation. Another way of describing a quasiracemate is as a molecular crystal related to a true racemate by a not-too-extensive change in the structure of one of the two enantiomeric components, e.g., (R)-A and (S)-A'.

Quasiracemate formation is discussed extensively in the classic book by Jacques *et al.* [25, pp. 100-103]. This cocrystallization phenomenon was developed into the powerful quasiracemate method for establishing the absolute configurations of optically active substances. This method and its application have been described by Fredga [26, 27], one of its leading practitioners. In the process of establishing their utility in determination of absolute configuration, Fredga and his collaborators determined and published hundreds of melting point phase diagrams for quasiracemic systems. His papers thus provide an extensive collection of isosteric pairs of chemical groups. Nevertheless, full crystal structural data on quasiracemates are sparse in the literature and characterization of their structural relationships to racemates even more so [28-32].

3. Strategies for Crystal Engineering

Optimization of many bulk properties requires tailoring of intermolecular orientations in addition to molecular properties. For example, the optimal relative orientation of the polarization of collinearly propagating waves and of molecular charge transfer axes for second-order nonlinear optics have been described [33]. These properties disappear for rigorously centrosymmetric crystals. Methods for crystal engineering, especially of polar crystals [34] because of their unique physical properties such as nonlinear optical effects and piezoelectricity, have received considerable attention from solid-state chemists. In the absence of general methods for controlling multimolecular arrays, we can only hope to alter the odds significantly in favor of an array that includes some desirable packing feature. We undertook the development of approaches to provide a bias away from centrosymmetry. In much of this work, we have intentionally avoided the directional control afforded by hydrogen bonding, focusing instead on possible roles of molecular shape and approximate molecular symmetry as tools for crystal design.

The bias for centrosymmetry is estimated to exceed 10:1 for crystals of achiral, nonpolar organic molecules and is about 4:1 for all structures [Table 8 in reference 7]. We believe that molecular shape represents the dominant factor in determining the high preference for centrosymmetric alignments in most molecular crystals [35]. We suggest that the high propensity for centrosymmetry (whatever its physical origin) can be turned to advantage to provide an alignment of molecules that, by virtue of the presence of groups of nearly identical shapes, possesses approximate symmetry either between pairs of molecules of within a single molecule. This approximate symmetry can allow tailoring of desirable orientational features.

We have developed two approaches. The first is based on the design of pairs of polarizable molecules that might cocrystallize as quasiracemates. In the second approach, which we have elsewhere termed "structural mimicry", the near-symmetry is designed into a single polarizable chiral molecule in such a way that the shape 'drive' for centrosymmetry (whatever its origin!) can be satisfied, while still leaving the crystal not rigorously centrosymmetric in full detail. Accounts of some of this work have appeared previously [37-39]. Much of the systematic study of quasiracemate formation by Fredga and by others [26, 27], as well as many current programs in crystal engineering [4-6], exploit the strongly directional aspects of hydrogen bonding. We wished to explore, and perhaps to utilize, the role of molecular shape in crystal packing and crystal engineering; therefore in much of this work we have avoided compounds with the potential for hydrogen bonding in favor of molecules that would form crystals that are 'based mostly on van der Waals forces' [40].

4. Design of Quasiracemic Crystals

In a true racemate, any vectorial property (say dipole moment or hyperpolarizability) in one molecule is exactly balanced by an equal and opposite vectorial property in the centrosymmetric mate (Figure 2a). We wish to introduce a subtle change in the structure of one hand of the molecule by substituting an isosteric group while altering the vectorial property of that molecule; if molecular shape is a sufficiently strong factor in determining the crystal packing, then the resulting arrangement could be quasicentrosymmetric by shape but perhaps not so in other properties.

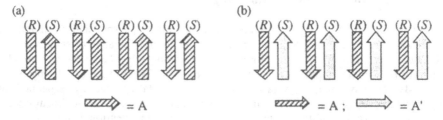

Figure 1.   Schematic representation of vectorial quantities in (a) centrosymmetric racemic and (b) pseudocentrosymmetric quasiracemic arrangements. In (a) the vectorial properties all cancel and the array is not polar; in (b) the vectorial properties do not cancel and the array is polar.

Our work has utilized two classes of hyperpolarizable compounds, benzoates and sulfoxides. In these, the carboxylate and sulfinyl groups are somewhat electron withdrawing. These groups are connected to electron-donating groups either by an aromatic core or by an extended conjugated stilbene framework. For the benzoates, various groups were introduced as chiral handles, whereas the sulfoxide group is intrinsically chiral.

4.1 BENZOATES

Crystallization experiments with enantiomeric and with racemic solutions of benzoates **1a-1c** were carried out, to determine the tendency of these compounds to form true racemate crystals. Melting points, crystal morphology, and X-ray powder patterns were used to characterize the results. Only for **1b** was formation of a true racemate suggested, though not conclusively. In the hope of forming quasiracemates of **1b** with one of the other compounds, e.g., (–)-**1a** and (+)-**1b**, equimolar solutions of opposite enantiomers of pairs of these benzoates were prepared and crystallized, but no evidence of quasiracemate formation was obtained. Numerous attempts to produce single crystals of X-ray quality of any of these benzoates either as enantiomers, racemates, or in combination as quasiracemates, for single-crystal structural studies were unsuccessful, yielding only poorly ordered rods or very thin needles or plates. Similarly equivocal results were obtained with other chiral benzoate esters, due mainly to the tendency of many of these to form oils or very poorly formed crystalline samples, either from solution or from the melt.

1a X = S ; 1b X = O ; 1c X = CH$_2$

4.2 SULFOXIDES

More success was achieved with a series of compounds based on diaryl sulfoxides, a chiral group for which one of us already had developed a considerable synthetic methodology [41]. Melting points, crystal morphologies, and X-ray powder diffraction patterns of various recrystallized samples of **2a** and/or **2b** indicated conglomerate crystallization, i.e., simultaneous but separate crystallization of the compounds or enantiomers in the solution, but neither racemate nor quasiracemate formation. When the isosteric modification X was buried more deeply within the molecule -- heptyl groups in place of methyl, **3a** and **3b** -- powder patterns indicated four different crystal structures for the two pure enantiomers and the two racemates. The crystalline sample from an equimolar mixture of (R)-**3a** and (S)-**3b** in CH$_2$Cl$_2$ gave and X-ray powder pattern that closely resembles that of racemic **3b**. This suggests a quasiracemate derived from rac-**3b** by replacing each (R)-**3b** molecule with a molecule of (R)-**3a**. Unfortunately, none of the enantiomeric, racemic, or quasiracemic crystalline phases obtained from this set of sulfoxides was of suitable quality for single-crystal analysis.

2a X = CH$_2$; 2b X = S   3a X = O; 3b X = S

We have had most success with isosteric substitutions of groups the ends of the molecule, when these groups match well enough both in volume and in conformational relationship to the rest of the molecule. The pair isopropenyl and dimethylamino, **4a** and **4b**, are especially well matched in several regards -- similar volumes (50.4 Å^3, 50.5 Å^3, respectively) and surface areas (64.1 Å^2, 66.8 Å^2, respectively), similar conformations relative to attached aromatic rings, and quite different electron-donating/withdrawing properties (advantageous for non-linear optics target systems).

4a   **4b**

Crystallization of an equimolar mixture of (*S*)-**5a** and (*R*)-**5b** gave crystals with higher melting points than either of the true racemates, which in turn each melted even higher than either enantiomer; this is strong evidence for racemate and quasiracemate formation. Both racemates and the quasiracemate formed crystals of excellent quality for single-crystal structure determination. Each racemate crystal is centrosymmetric, $P2_1/c$ with Z = 4; the quasiracemate crystal is noncentrosymmetric, $P2_1$ with two molecules each of (*S*)-**5a** and (*R*)-**5b** in a unit cell with similar dimensions to those of the two racemates. These two molecules are related by a pseudoinversion center, so the symmetry approximates $P2_1/c$. The structures are compared in Figure 2. Cocrystallization of (*S*)-**6a** and (*R*)-**6b** gave another quasiracemate with a packing arrangement quite similar to those shown in Figure 2.

5a R = ... ; 5b R = ...   6a R = ... ; 6b R = ...

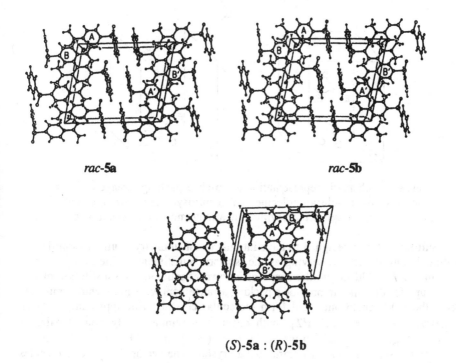

rac-**5a**　　　　　　　　　　*rac*-**5b**

(*S*)-**5a** : (*R*)-**5b**

Figure 2. Packing arrangements in crystals of racemic and quasiracemic sulfoxides.

5. Crystals of Molecules with Internal Quasisymmetry

The quasiracemate approach has the disadvantage of requiring synthesis and cocrystallization of two different chiral compounds, with little control over the relative orientations of the two isosteric groupings. We have also designed molecules with internal quasisymmetry by virtue of the presence of isosteric groups within the same chiral molecule. The rationale of this method is depicted in Figure 3. Each molecule is chiral by virtue of four different groups (C, D, E, E') about a tetrahedral center. In the centrosymmetric packing of such a unit with its enantiomer (Figure 3(a)) there is a pairwise antiparallel arrangement of *all* like vectors (X→C, X→D, X→E, X→E') in centrosymmetrically related molecules. A single enantiomer cannot crystallize in such a centrosymmetric arrangement; yet to the extent that groups E and E' are isosteric, a packing arrangement with the vector X→E of one molecule antiparallel to the X→E' vector of the other molecule might mimic centrosymmetry (Figure 3(b)). If the properties of E and E' differ in some way (e.g., polarizability), then this quasicentrosymmetric packing can result in net additivity, rather than cancellation, of molecular properties. We note that this arrangement would not require any disorder in the structure.

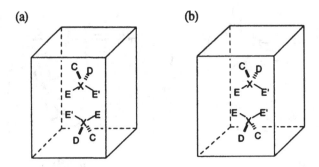

Figure 3. Schematic representation of possible packing arrangements for a quasisymmetric chiral molecule: (a) centrosymmetric arrangement of the racemate; (b) quasicentrosymmetric arrangement of the enantiomer.

Sulfoxide **7** was designed to have internal quasisymmetry, with the isopropenyl and dimethylamino groups serving as the isosteric substituents. The hope was that enantiomeric **7** would mimic the shape of its own enantiomer by substituting one of these groups for the other to provide a packing arrangement with quasicentrosymmetry. Indeed, the (S)-**7** enantiomer did form molecular crystals with approximate $P2_1/c$ symmetry, true space group $P2_1$, with quasicentrosymmetrically related pairs of molecules (Figure 4). The two molecular pairs (A,B) and (A',B') are related to one another by the screw axis in this chiral, polar crystal. The vectors N→S for molecules B and B' are (coincidentally) nearly perpendicular to the screw axis, and are thus virtually antiparallel; however, those for A and A' are inclined at 44.9° to the screw axis, so they combine to make a net polar direction for the crystal.

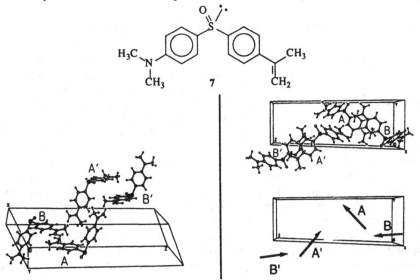

Figure 4. The four molecules in the unit cell of (S)-**7**. *Left:* Oblique view. *Right:* View along the x-axis, showing polarizability vectors N→S shown as arrows.

6. Continuing Studies on Quasiracemate Structures

In light of the scarcity of data for comparing racemate and quasiracemate crystals, we have a program of experimental determination and modeling studies of such structures. We have carried out structure determinations for several pairs of racemates related to quasiracemates reported by Fredga [26, 27] on the basis of melting point behavior. Crystal data for some of these appear in Table 1, where each pair of rows refers to the two racemates related to a single Fredga quasiracemate.

Table 1. Crystal data for some racemates related to Fredga quasiracemates.

	S.G.	a	b	c	α	β	γ	V	R
a1	$P\bar{1}$	7.264	7.268	11.061	99.04	104.56	101.31	541.0	0.035
a2	$P\bar{1}$	7.340	7.344	10.961	98.25	104.87	101.77	547.0	0.037
b1	$P\bar{1}$	4.472	10.536	11.426	80.60	82.07	82.45	522.8	0.050
b2	$P\bar{1}$	4.559	10.556	11.411	80.74	82.57	82.35	533.9	0.053
c1	$P2_1/n$	7.478	7.953	16.461	90	93.55	90	977.2	0.047
c2	$P2_1/n$	7.394	7.923	16.515	90	93.19	90	966.0	0.058
d1	$C2/c$	34.985	5.217	11.209	90	101.46	90	2004.8	0.052
d2	$Pna2_1$	11.513	5.009	33.314	90	90	90	1921.0	0.042
e1	$P2_1/n$	6.550	4.862	29.660	90	91.64	90	944.2	0.044
e2	$Pbca$	13.924	8.060	17.796	90	90	90	1997.2	0.055

We believe that comparisons of calculated crystal packing energies might help to clarify packing relationships and perhaps even assist in prediction of potential quasi-racemate candidates. Packing energies for known racemates can be compared to those for proposed quasiracemates simulated from these racemate arrangements. For example, packing energies of racemic 5a and 5b are -145 and -149 kcal/mol, respectively. The known quasiracemate described earlier (Section 4.2) can be simulated either by substituting one enantiomer of 5b into the known racemic structure 5a, or vice versa. Calculated packing energies for these two simulated structures, normalized per molecule, are -145 and -146 kcal/mol respectively.

7. Acknowledgments

Financial support from the Advanced Research Program of the Texas Higher Education Coordinating Board (Grant 277 to J.K.W. and R.E.D.), the Robert A. Welch Foundation (Grants F-233 to R.E.D. and F-626 to J.K.W.), the U. S. National Science Foundation (Grant DMR-9014026 to J.K.W.) and the Petroleum Research Fund, administered by the American Chemical Society (Grant ACS-PRF AC-20714 to J.K.W.) is gratefully acknowledged. Current work on quasiracemates is also supported by the U. S. Air Force Office of Scientific Research (Grant F49620-97-1-0263 to K.A.W and R.E.D.).

8. References

1. Allen, F. H. and Kennard, O. (1993) 3D Search and research using the Cambridge Structural Database, *Chemical Design Automation News*, **8**, 31-37.

2. Desiraju, G. R. (1995) Supramolecular synthons in crystal engineering -- a new organic synthesis, *Angew. Chem. Int. Ed. Engl.* **34**, 2311-2327; *Angew. Chem.* **107**, 2541-2558. See also reference 10.

3. Braga, D. and Grepioni, F. (1993) Intermolecular interactions and supramolecular organization in organometallic solids, *Chem. Commun.* 571-578.

4. Aakeröy, C. B. and Seddon, K. R. (1993) The hydrogen bond and crystal engineering, *Chem. Soc. Reviews*, **22**, 397-407.

5. Aakeröy, C. B. (1997) Crystal engineering: Strategies and architectures, *Acta Cryst.* **B53**, 569-586.

6. Subramanian, S. and Zaworotko, M. J. (1994) Exploitation of the hydrogen bond: Recent developments in the context of crystal engineering, *Coord. Chem. Reviews*, **137**, 357-401.

7. Brock, C. P. and Dunitz, J. D. (1994) Towards a grammar of crystal packing, *Chem. Mat.* **6**, 1118-1127.

8. See references 15-51 in Reference [7].

9. Filippini, G. and Gavezzotti, A. (1991) A quantitative analysis of the relative importance of symmetry operators in organic molecular crystals, *Acta Cryst.* **B48**, 230-234.

10. Wilson, A.J.C. (1993) Space groups rare for organic structures. III. Symmorphism and inherent molecular symmetry, *Acta Cryst.* **A49**, 795-806.

11. Sona, V. and Gautham, N. (1992) Conformational similarities between crystallographically independent molecules in organic crystals, *Acta Cryst.* **B48**, 111-113.

12. Gautham, N. (1992) A conformational comparison of crystallographically independent molecules in organic crystals with achiral space groups, *Acta Cryst.* **B48**, 337-338.

13. Karthe, P., Sadavasan, C. and Gautham, N. (1993) Packing interaction of crystallographically independent molecules in organic crystals, *Acta Cryst.* **B49**, 1069-1071.

14. Desiraju, G. R., Calabrese, J. C., and Harlow, R. L. (1991) Pseudoinversion centers in space group $P\bar{1}$ and a redetermination of the crystal structure of 3,4-dimethoxycinnamic acid. A study of non-crystallographic symmetry, *Acta Cryst.* **B49**, 77-86.

15. Wheeler, K. A. and Davis, R. E. unpublished results. Some of these results were reported at meetings of the American Crystallographic Association (Albuquerque, 1993, Abstract PH02; Atlanta, 1994, Abstract U02).

16. Desiraju, G. R. (1989) *Crystal Engineering. The Design of Organic Solids.* Elsevier Science Publishers B.V., Amsterdam.

17. Lahav, M. and Leiserowitz, L. (1996). Tailor-made auxiliaries for the control of nucleation, growth and dissolution of crystals, in Tsoucaris, G., Atwood, J. L., and Lipkowski, J. (eds.) *Crystallography of Supramolecular Compounds*, NATO ASI Series, Kluwer Academic Publishers, Dordrecht, pp. 431-507.

18. Jones, W., Theocharis, C. R., Thomas, J. M. and Desiraju, G. R. (1983) Structural mimicry and the photoreactivity of organic solids, *J. Chem. Soc., Chem. Commun.*, 1443-1444.

19. Theocharis, C. R., Desiraju, G. R. and Jones, W. (1984) The use of mixed crystals for engineering organic solid-state reactions: Application to benzylbenzylidenecyclopentanones, *J. Amer. Chem. Soc.*, **106**, 3606-3609.

20. Sarma. J. A. R. P. and Desiraju, G. R. (1986) Molecular discrimination in the formation of mixed crystals of some substituted chlorocinnamic acids, *J. Amer. Chem. Soc.*, **108**, 2791-2793.

21. Centnerszwer, M. (1899) On the melting points of mixtures of optical antipodes, *Z. Physk. Chem.*, **29**, 715.

22. Pasteur, L. (1853). *Ann. Chim. Phys.*, *Ser 3*, **38**, 437.

23. This point is discussed in reference 7, page 1127. A common type of exception, not explicitly mentioned there, would be the cocrystallization of solute and solvent to form solvated crystals, in which the solvent often plays only a space-filling role [24].

24. van der Sluis, P. and Kroon, J. (1989) Solvents and X-ray crystallography, *J. Cryst. Growth*, **97**, 645-656.

25. Jacques, J., Collet, A. and Wilen, S. H. (1981) *Enantiomers, Racemates, and Resolutions*, John Wiley & Sons, New York, pp. 100-103.

26. Fredga, A. (1960). Steric correlations by the quasi-racemate method, *Tetrahedron*, **8**, 126-144.

27. Fredga, A. (1973). Quasiracemic compounds and their use for studying the configuration of optically active compounds, *Bull. Soc. Chim. Fr.*, 174-182.

28. Karle, I. and Karle, J. (1966) The crystal structure of the quasi-racemate from (+)-*m*-methoxyphenoxypropionic acid and (-)-*m*-bromophenoxypropionic acid *J. Amer. Chem. Soc.*, **88**, 24-27.

29. Misra, R., Wong-Ng, W., Chang, P.-T., McLean, S. and Nyburg, S. C. (1980) Crystal structure of two quasi-racemates of (-)-podopetaline and (-)-ormoxanine isolated from *Podopetalum ormondii*; the absolute configuration of (-)-ormosanine, *J. Chem. Soc., Chem. Commun.*, 659-660.

30. Gillard, R. D., Payne, N. C. and Phillips, D. C. (1968) Optically active coordination compounds. Part XIII. An inorganic quasi-racemate, *J. Chem. Soc. A.*, 973-974.

31. Bilton, M. S. (1982) $(-)_{409}$-R,S-[(R-N(2-Aminopropyl)salicylaldiminato)-chromiumIII)] perchlorate, $[C_{20}H_{26}N_4O_2Cr]^+ClO_4^-$, *Cryst. Struct. Commun.*, **11**, 755-762.

32. Whuler, A., Brouty, C., Spinat, P. and Herpin, P. (1976) Structural study of the active racemate hydrate [(+)-Coen3(-)-Cren3]Cl6·6.1H2O, *Acta Cryst.*, **B32**, 194-198.

33. Chemla, D. S. and Zyss, J., (eds.) (1996) *Non-Linear Optical Properties of Organic Molecules and Crystals*, Vols 1 and 2, Academic Press, New York.

34. Curtin, D. Y. and Paul, I. C. (1981) Chemical consequences of the polar axis in organic solid-state chemistry, *Chem. Rev.*, **81**, 525-541.

35. The relationships between molecular shape and crystal packing were elegantly discussed in the classic book by Kitaigorodsky [36].

36. Kitaigorodsky, A. I. (1973) *Molecular Crystals and Molecules*, Academic Press, New York.

274

37. Whitesell, J. K., Davis, R. E., Wong, M.-S. and Chang, N.-L. (1993). Shape mimicry as a design tool in crystal engineering, *J. Phys. D: Appl. Phys.* **26**, B32-B34.
38. Whitesell, J. K., Davis, R. E., Wong, M.-S. and Chang, N.-L. (1994) Molecular Crystal Engineering by Shape Mimicry, *J. Amer. Chem. Soc.,* **116**, 523-527.
39. Davis, R. E., Whitesell, J. K., Wong, M.-S. and Chang, N.-L. (1996) Molecular shape as a design criterion in crystal engineering, Chapter 3 in G. R. Desiraju (ed.), *The Crystal as a Supramolecular Entity*, John Wiley & Sons, Chichester.
40. This terminology borrows the title of Desiraju's Chapter 4 in reference 16.
41. Whitesell, J. K. and Wong, M.-S. (1994) Asymmetric synthesis of chiral sulfinate esters and sulfoxides. Synthesis of Sulforaphane, *J. Org. Chem,* **59**, 597-601.

GRAPH SET ANALYSIS OF HYDROGEN BOND MOTIFS

J. BERNSTEIN
Department of Chemistry, Ben-Gurion University of the Negev
Beer Sheva, Israel 84105

R. E. DAVIS
Department of Chemistry and Biochemistry, University of Texas
Austin TX 78712-1167, USA

1. Introduction

Chemists are inveterate users of symbols. Elements are denoted by letters, compound formulae by combinations of letters and numbers, physical constants, wave functions, reaction mechanisms by additional combinations of letters and numbers. The language of chemistry also uses pictorial symbols to represent molecules. The familiar two-dimensional structural formula implies many things to the chemist, such as shape, symmetry, physical and chemical properties, even chemical reactivity.

In mathematical terms the two dimensional structural formula is a graph - a collection of vertices (the atoms) connected by a set of lines (the bonds). The meeting ground between mathematics and chemistry that involves graphs and graph sets has been explored and utilized by a number of people [1-4, for example]. Being essentially the strongest and most directional of intermolecular interactions, hydrogen bonds have been the subject of study since the 1920's [5,6]. In the language of graphs the patterns exhibited by hydrogen bonds in crystals may be considered in a way analogous to that of molecules, wherein molecules are treated as vertices and the hydrogen bonds that connect them as lines. Drawing upon these fundamental concepts and some initial ideas that had been proposed earlier regarding the patterns of hydrogen bonds, the late M.C. Etter developed the graph set ideas to characterize and analyze patterns of hydrogen bonds in crystals [7-9]. This chapter summarizes some of those ideas, how they have been refined and used over the last decade, and some potential developments and uses for the future.

2. Historical Development of Graph Sets Applied to Hydrogen Bonds

The notion of classifying and studying networks of hydrogen bonds appears to have been initiated for crystal structures by Wells [10], who recognized the importance of patterns

J.A.K. Howard et al. (eds.),
Implications of Molecular and Materials Structure for New Technologies, 275–290.
© 1999 *Kluwer Academic Publishers. Printed in the Netherlands.*

independent of the geometry, thermal or spectroscopic properties of the individual hydrogen bonds. The consequences of repetitive hydrogen-bond interactions through a crystal structure had not been addressed prior to Wells, and this topological point of view afforded the characterization of subsets of crystal structures or arrays, in particular the hydrogen-bonded subset.

Wells' model employs the fundamentals of graph sets, namely that the molecules can be considered as single points from which hydrogen bonds emanate as lines. He then proposed a classification scheme for describing the patterns that are developed. This idea was extended by Hamilton and Ibers [11], who characterized hydrogen bond networks with two indices (N,M), the number of hydrogen bonds per point (N) and the number of molecules to which a point is hydrogen bonded (M). A major advance came when Kuleshova and Zorky [12] recognized that these earlier classification schemes were in fact an application of graph theory, which is a mathematical formalism for analyzing graphs and networks [13]. They analyzed approximately 800 hydrogen-bonded organic crystal structures and identified finite sets (termed "islands"), and extended ones such as chains, layers and frameworks. These were designated by symbols, generically $G_m^n(k)$ where m was the number of other points to which a point was connected, n indicated how many points had to be broken to free the molecule from the network, and k indicated the size of a non-overlapping ring. They surveyed the crystallographic literature to determine the comparative frequency of these various patterns. Using these ideas to compare the hydrogen bonding within a set of polymorphic compounds they found that about half exhibited the same hydrogen bonding patterns in both polymorphs.

The symbol $G_d^a(n)$ with the same format was adopted by Etter [7,8,14], but the terms were used to describe characteristics of hydrogen bonds that were more compatible with the chemists' lexicon of hydrogen bonds. That nomenclature is described in the next section.

3. The Graph Set Nomenclature

3.1. THE DESIGNATOR

A remarkable feature of the graph set approach to analysis of hydrogen-bond patterns is the fact that most complicated networks can be reduced to combinations of four simple patterns, each specified by a designator: chains (C), rings (R), intramolecular hydrogen-bonded patterns (S), and other finite patterns (D). Specification of a pattern is augmented by a subscript d designating the number of hydrogen-bond donors (most commonly covalently bonded hydrogens, but certainly not limited to them), and a superscript indicating the number of hydrogen bond acceptors a. When no subscript and superscript are given, one donor and one acceptor are implied. In addition, the number of bonds n in the pattern is called the degree of the pattern and is specified in parentheses. The general graph set descriptor is then given as $G_d^a(n)$ where G is one of the four possible designators.

These four patterns and their descriptors are best illustrated by examples. A chain whose "link" is composed of four atoms as in **1** is specified as **C(4)**. Similarly, the intramolecular hydrogen bond in **2** would be specified as **S(6)**, for the six atoms comprising the intramolecular pattern. When the donor and acceptor are from two (or more) *discrete* entities (molecules or ions), as in **3**, the designation of the hydrogen bond is **D**. The entities may differ on chemical grounds (different ions or molecules) or on crystallographic grounds (chemically identical but not related by a crystallographic symmetry operation). In **3** there is only one donor and one acceptor, and the pattern involves only one hydrogen bond. The fourth possible pattern is the ring **4**. In the example shown, the two hydrogen bonds in the ring could be different but in this case they are related by a crystallographic inversion centre. The pattern contains a total of eight atoms, two of them donors and two acceptors, and hence is designated $\mathbf{R}_2^2(8)$.

1   C(4)   2   S(6)   3   D

4    $R_2^2(8)$ 5

3.2. MOTIFS AND LEVELS

A motif is a pattern containing only one type of hydrogen bond [14]. Specifying the motif for each different hydrogen bond in a network according to one of the four pattern descriptors above leads to a description of the network in the form of a list of the motifs. This is the *unitary*, or *first level*, graph set, noted as N_1.

The chemically interesting or topologically characteristic patterns of a system often appear when more than one type of hydrogen bond is included in the description, i.e. in higher level graph sets [14,17-19]. This will be true for **S**, **D**, and **C** patterns. Suppose a structure contains three distinct hydrogen bonds, designated *a*, *b*, and *c*. There will now be several possible *binary* (or *second level*) graph sets, each one describing a pattern formed by two of these H-bonds - that is $N_2(ab)$, $N_2(ac)$ and $N_2(bc)$. These concepts are illustrated in Figure 1. The *ternary* or *third level* graph sets (for which there might be more than one possible [20]) are those that involve three hydrogen bonds.

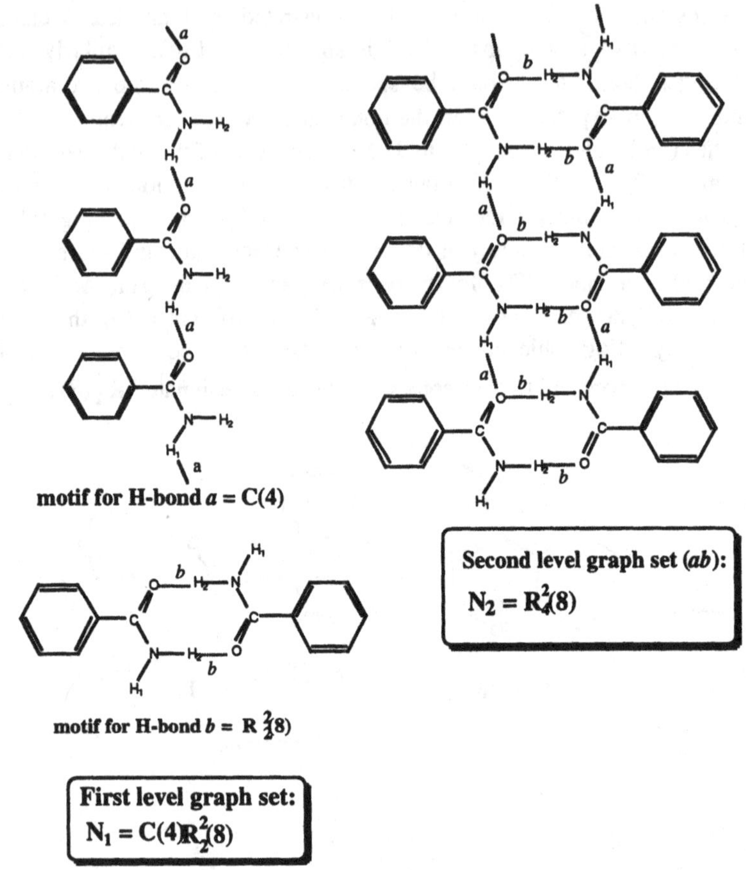

278

Figure 1. Examples of the use of graph set descriptors to define motifs and first and second level graph sets for schematic representations of the hydrogen bond patterns in the crystal structure of benzamide.

Different pathways might also be found that include the same set of H-bonds, but with different degrees (n). To describe this situation, we suggest the term *basic* to describe the graph set of the *lowest degree* and the term *complex* to describe ones of *higher degree*. Consideration of some of the choices for the binary graph set for α-glycine [21] illustrates this point (Figure 2). The shortest path involving H-bonds *a* and *b* gives the binary graph set $C_2^2(6)$; this is the *basic* binary graph set for *a* and *b*. A longer chain, $C_2^2(10)$, represents a *complex* binary graph set for the two H-bonds. Neither of these, however, describes the most obvious feature of this array, the ring structure, which is denoted by another complex binary graph set $R_4^4(16)$. In addition, in this structure it is possible to define an infinite number of increasingly larger ring systems. In such a case it remains for the chemist or crystallographer to choose a ring or those rings that characterize the particular structure in question [22].

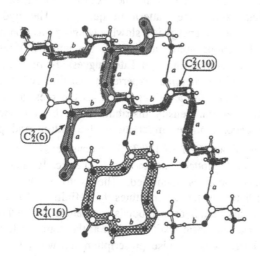

Figure 2. Graph set assignments for the binary level of α-glycine. As in Figure 1, different types of hydrogen bonds (solid lines) are distinguished by labeling with lower case italic letters; carbon and hydrogen atoms are shown as open circles, oxygen atoms as solid circles, and nitrogen atoms as shaded circles.

4. The Practicalities of Assigning Graph Sets

The assignment of the graph sets for any structure involves a few simple steps. One of the important aspects in the use of this method is in the standardization of the methodology, so that it can be used as a means of communicating the characteristics of structures. As the procedure for making the graph set assignment becomes more automated, inconsistencies will be reduced. Until that automation is achieved with its concomitant standardization, it is important to review the steps involved in making that assignment.

4.1. STEPS IN DETERMINING THE GRAPH SET ASSIGNMENT

The first step is to determine which hydrogen bonds are to be analyzed. While this may at first glance seem obvious, there are a number of sources of uncertainty, discrepancy and even dispute. First the positions of the hydrogen atoms must be known. In older structures many were not determined, even though the presence of hydrogen bonds is generally recognized. In more recently determined structures the hydrogen positions either have been found experimentally with e.s.d.'s considerably larger than those of the heavy atoms to which they are attached or their positions have been calculated and not refined. In either situation significant uncertainty is introduced. This can be reduced by correcting the hydrogen atom positions to correspond to accepted values of bond lengths [23].

In many cases the definition of a hydrogen bond is still a matter of some debate [24]. For instance, hydrogen bonds have historically been defined on the basis of geometrical criteria [25-27]. However, the use of a distance cutoff, for instance, may exclude some hydrogen bonds that exhibit the pattern in question. The hydrogen bond is an electrostatic interaction, so the energy is a slowly changing function of distance, with the consequence that interactions above and below the nominal threshold values for a hydrogen bond may have similar energies. Etter suggested a more intuitive definition of the hydrogen bond based on chemical experience and the recognition of patterns [8], whereas a quantum mechanical definition based on the presence or absence of bond critical points in the electron density distribution has been proposed [28,29]. Such considerations will become more important as the nature and patterns of weaker interactions (e.g. C-H$\cdots$O and C-H$\cdots$N [30,31]) are considered.

The structure is then plotted out and each chemically and crystallographically independent hydrogen bond is identified. The motif of each hydrogen bond is determined, and the listing of these constitutes the definition of the first level. Pairwise combinations of the individual hydrogen bonds are then determined to define the second level graph sets, first at the basic level and then at the complex level, and so forth for successive levels. There is no precise prescription for when this process ends; that depends on when the chemically useful information is obtained.

4.2. SOME HELPFUL METHODS AND TOOLS

The early studies and applications of graph sets to hydrogen-bonded structures were carried out in the manner described in the previous section. Even for structures containing three different hydrogen bonds, the manual counting can become cumbersome and is fraught with error. In addition, the *comparison* of the resulting graph sets requires considerable bookkeeping. Most of the standard programs for plotting crystal structures provide for simple and rapid generation of packing diagrams, usually including hydrogen bonds if desired, which may be viewed from any convenient direction. However, the identification of any particular hydrogen bond and the pattern it forms in that structure is usually not available as an option. This renders the graph set assignments considerably more complicated.

What is required is an algorithm for *building up* the crystal from a reference molecule (or molecules) by expanding the structure through individual hydrogen bonds, defined and chosen by the user if desired. The pattern of each hydrogen bond and hydrogen bond pattern (for first, second or higher levels) could then be color coded or otherwise highlighted and the graph set assignments made either automatically or by the user by counting as above.

The bookkeeping and the comparison of the graph sets for different crystal structures are considerably simplified by the preparation of a matrix-like table that summarizes the graph set assignments. The columns and rows are designated in identical order from top to bottom and left to right, respectively, each column and each row being labeled by one of the identified hydrogen bonds. The 'diagonal elements' of the resulting matrix will then contain first order graph set assignments, while the off-diagonal ones will contain the binary graph set assignments. Ordering the columns and rows the same way for, say, polymorphic structures or a family of compounds with similar hydrogen bonding capabilities greatly facilitates the comparison of these structures.

Fortunately, many of the desirable features described in this section have been or are being incorporated into the software of the Cambridge Structural Database (CSD). Some of these are included in the April, 1998 release, and those, with subsequent improvements, should greatly simplify and standardize the use of graph sets.

5. Current and Potential Uses of Graph Sets

5.1. CLASSIFICATION OF HYDROGEN BONDING PATTERNS IN CRYSTALS

5.1.1. *Characterization and Analysis of Hydrogen Bonds of a Single Crystal Structure*

To date by far the greatest use of the graph set notation has been for the characterization and analysis of patterns in individual crystal structures [32-37]. The shorthand notation is often used in conjunction with the traditional tabulation of the metrics of the individual bonds, but the convenience of summarizing the pattern and the information content of the descriptor has made the use of the notation increasingly popular. As a result, it is becoming part of the *lingua franca* of structural chemistry [38].

5.1.2. *Characterization and Analysis of Hydrogen Bonds for a Family of Structures*

As noted above the analysis and characterization of the hydrogen bonding patterns for a family of compounds is considerably facilitated by the use of graph sets. A number of recent examples indicate the variety of systems studied in this manner: the comparison of an anhydrous and monohydrate structure [39]; hydrogen bonding in coordination and organometallic chemistry [40,41]; surface chemistry [42]; and conformational studies [43].

5.2. COMPARISON OF POLYMORPHIC STRUCTURES

Among the key questions when considering polymorphic structures is understanding the structural similarities and differences between them. There are few tools available to do this, other than visually inspecting the two structures from the same perspective [44,45]. However, differences and similarities recognized from such a visual examination are often difficult to describe in a simple manner. However, when hydrogen bonding (and potentially any other pattern-definable intermolecular interaction) is involved in the packing it is possible to use the graph set notation to describe those similarities and differences. This application of graph sets was first demonstrated with iminodiacetic acid and glutamic acid [17,18], and has been employed with increasing frequency [17,46-50]. Perhaps not surprisingly, it is generally found that the hydrogen bond patterns do differ among polymorphs although the analysis may have to take into account higher complex or higher level graph sets in order to re-cognize the distinction [19,47].

5.3. DO CHEMICALLY IDENTICAL MOLECULES IN THE SAME STRUCTURE HAVE THE SAME HYDROGEN BONDING PATTERNS?

Crystal structures in which the asymmetric unit contains more than one molecule ($Z' > 1$ in CSD nomenclature) have always roused the curiosity of crystallographers. In the early days of crystallography the full structure analysis was usually not pursued because this situation essentially doubled the complexity of the problem. Things have changed, of course, and the fact that one structure determination yields two determinations of the molecular structure now makes these attractive examples for study. Along with the fundamental question of why a substance chooses to crystallize with $Z'>1$ there is the question of whether the molecules are the same or different on the molecular level and whether they are the same or different in terms of the crystal environment in that specific structure.

In particular for hydrogen-bonded systems there is the question of the similarities or differences in the independent molecules in the asymmetric unit. A preliminary survey of this question [46] was rather inconclusive, indicating that in many cases the hydrogen bonding is different, although there certainly are cases where it is identical. A particular curiosity of this situation is that in some structures, the hydrogen bonding networks for the independent molecules do not intersect at any level [51].

5.4. HYDROGEN BOND PATTERN CHARACTERIZATION AND FUNCTIONALITY

A few years experience of working with graph sets has led us to view hydrogen bonds increasingly in terms of patterns rather than in terms of the individual hydrogen bonds, much in the way that organic chemists tend to view a reaction in terms the role of functional groups and the general mechanism rather than the particulars of the reaction. This focus on the patterns has suggested a number of general areas of inquiry that might be pursued for any particular pattern. Two of them are posed here as examples of some of the ways that these patterns might be studied and characterized.

5.4.1. *Hydrogen-Bond Functionality: What is the Chemical Diversity of a Particular Hydrogen Bond Pattern?*

To chemists, one of the most familiar hydrogen bond patterns, perhaps *the* paradigmatic pattern, is the familiar $R_2^2(8)$ associated with carboxylic acids **4** and (earlier in this chapter) amides. It also appears in Watson-Crick base pairing [14]. It is possible to inquire as to the *chemical diversity* of this (or any other) prototypical pattern by posing a generic search question to the CSD. The results of such a search provide information on the *hydrogen bond functionality* of that pattern. In particular cases where one seeks information on the chemical diversity of the $R_2^2(8)$ pattern, the generic search would be based on the pattern **5**, in which D represents a donor, A an acceptor, X is any element, and the X-A and X-D bonds are of any order. Such a survey revealed some 3900 occurrences of this pattern in more than 3400 structures, including more than 80 different element combinations represented by D, X, and A. Some of these are *homo*-dimeric in nature while others are *hetero*-dimeric. The relative occurrence of such patterns is, of course, biased by compounds and crystal structures that usually were determined for other reasons. Nevertheless, a general picture of the diversity of this particular pattern within the CSD emerges from another matrix-type representation, shown in Figure 3 [52].

	H N C O	H O C O	H N C N	H N C S	H O P O	H N S O	H N P O
HNCO	1147						
HOCO	41	762					
HNCN	136	24	494				
HNCS	2	2	6	259			
HOPO	4	2			81		
HNSO			5			72	
HNPO	1					1	55

Figure 3. Frequency of occurrence of the generic $R_2^2(8)$ pattern **5** for a sampling of seven of the more than 80 different element combinations that exhibit the pattern. The numerical entry is the number of hits for the search described in the text. Diagonal elements can represent homo-dimers and off-diagonal elements represent hetero-dimers. Due to space limitations, only a portion of the larger than 80x80 diagonally symmetric matrix is shown.

5.4.2. What are the Metric Characteristics and Limitations of a Hydrogen Bond Pattern?

The ready recognition of a pattern and the ability to extract all the instances of its occurrence from the CSD open up the possibility to determine the metric limitations of that pattern. For instance, again using **5** as a point of reference, one can inquire about the 'conformation' of the pattern, the limits to the D$\cdots$A and/or H$\cdots$A distance, the frequency of crystallographically centrosymmetric or non-centrosymmetric patterns, *etc.* [53]

6. Some Prospects and Problems

The graph sets are a convenient way of visualizing, defining and transmit-ting information about hydrogen bond patterns. We have already shown that these concepts are becoming part of the language of chemistry and crystallography. As they do, we expect that they will be applied to an every greater variety of problems. There are both challenges and opportunities in the continued development and use of the method, some of which are noted here.

6.1. THE DEFINITION OF A HYDROGEN BOND

Hydrogen-bond patterns exist only where there are hydrogen bonds. In spite of the fact that hydrogen bonds have been part of the chemical lexicon for nearly 80 years, there is still considerable debate about when an interaction should or should not be called a hydrogen bond. In general, there is agreement regarding the "classic" strong hydrogen bonds (*i.e.* D-H$\cdots$A; D,A=O,N) within "accepted" geometric bounds [8,16,24,28]. The controversy increases beyond those geometric bounds and becomes especially sharp when weaker hydrogen bonds are considered [54-58]. Nevertheless, even if a standard definition is not possible, a consensus as to what constitutes a hydrogen bond and how it may be identified automatically is very desirable in advancing the application of graph sets and of our understanding of hydrogen-bond patterns.

6.2. MATHEMATICAL FOUNDATIONS OF THE GRAPH SET THEORY

Although she was aware of the fundamental mathematical principles of graph theory, Etter's original ideas were based very much on her long experience with hydrogen-bonded structures and her incredible scientific intuition. In spite of the fact that those basic ideas about four fundamental patterns and the parameters necessary to define them have stood the tests of time and close examination, it was clear that it was necessary to establish a firmer mathematical foundation to this approach. Considerable progress has been made in this area recently [59,60] and the dialog that has been established between mathematicians and chemical crystallographers will undoubtedly serve as a catalyst for solving a number of the outstanding problems and opening up new possibilities for the use of the graph sets.

6.3. AUTOMATION OF THE SEARCH FOR AND ASSIGNMENT OF GRAPH SETS -PROSPECTS

Although there has been considerable progress in the mechanics of assigning graph sets since the original Etter papers, as of this writing the process still has more of an element of tediousness than convenience. However, imminent software developments, some of which were described above, will make that procedure more efficient and precise. It will allow the rapid, essentially interactive, assignment of the graph sets for various compounds with a consistent ordering to enable comparisons of large groups of compounds. Such developments should lead to the use of the graph set notation among a much wider community of chemists.

6.4. SOME ADDITIONAL POTENTIAL USES

The simplified mechanical and bookkeeping aspects of the assignment and use of graph sets creates the potential for a wide variety of applications. Some of these are already starting to appear in the literature, and to illustrate the potential uses of the graph sets as well as to provide an entry into this literature a few examples of those recent applications are noted here.

The hydrogen bond is arguably the single most important structural tool in crystal engineering [61,62. Therefore, knowledge of the patterns of hydrogen bonds may be used to design desired structures. This is potentially one of the greatest uses for graph sets, especially in terms of the definition of patterns to obtain desired structural motifs and hydrogen bond functionality to modify the chemical and physical properties of that motif, while maintaining structural integrity. Some of those uses include: the predilection of some hydrogen bonds to lead to alignment of molecules [63-65], the correlation of hydrogen bonding properties to thermodynamic properties [66,67] and the general classification and utilization of specific patterns in generating one- two- and three-dimensional networks [68-70].

The generation of crystal structures based on hydrogen-bond patterns can also be used as an aid in the search for crystal structure prediction and solution (including possible polymorphs) [71,72] and in preparing trial structures for structure solution by Rietveld refinement of powder patterns [73]. It has also been used to as an aid in understanding the process of spontaneous resolution [74-76], in developing models for protein binding and active site [77-79], and in identifying possible lead compounds in drug design [80]. Finally, in a very recent study graph sets have been of considerable assistance in unraveling the structural similarities and differences in the four polymorphic forms of sulfathiazole and have provided a key to understanding the role of solvent and additives in the formation of those polymorphic forms [51].

286

7. Acknowledgments

Some of the work described here was supported by a grant from the U.S.-Israel Binational Science Foundation (Jerusalem) to both authors. JB wishes to thank the staff of the Cambridge Crystallographic Data Centre for their hospitality during a sabbatical leave.

8. References

1. Merrifield, R.E and Simmons, H.E. (1989) *Topological Methods in Chemistry,* John Wiley, New York.
2. Babaev, E.V. and Zefirov, N.S. (1992) Ring transformation graphs in hetereocyclic chemistry, *Bull Soc. Chim. Belges.* **101**, 67-84.
3. Ugi, I., Bauer, J., Bley, K., Dengler, A., Dietz, A., Fontain, E., Gruber, N., Herges, R., Knauer, M., Reitsam, K., and Stein, N. (1993) The computer-assisted solution of chemical problems - The historical development and the present state of the art of a new discipline of chemistry. *Angew. Chem. Int. Ed. Eng.* 32, 201-227.
4. Babaev, E.V., Lushnikov, D.E. and Zefirov, N.S. (1993) Novel graph-theoretical approach to ring-transformation reactions - hierarchical classification and computer design of heterocyclic rearrangements. (1993), *J. Am. Chem. Soc.* 115, 2416-2427.
5. Latimer, W.M and Rodebush, W.H. (1920) Polarity and ionization from the standpoint of the Lewis theory of valence, *J. Am. Chem. Soc.* **42**, 1419-1433.
6. Pimentel, G.C. and McClellan, A.L. (1960) *The Hydrogen Bond*, Freeman, San Francisco.
7. Etter, M.C. (1985) Aggregate structures of carboxylic acids and amides, *Isr. J. Chem.* **25**, 312-319.
8. Etter, M.C. (1990) Encoding and decoding hydrogen-bond patterns of organic compounds, *Acct. Chem. Res.* **23**, 120-126.
9. Etter, M.C. (1991) Hydrogen bonds as design elements in organic chemistry. *J. Phys. Chem.* **95**, 4601-4610.
10. Wells, A.F. (1962) *Structural Inorganic Chemistry*, Clarendon Press, Oxford, pp. 294-315.
11. Hamilton, W.C. and Ibers, J.A. (1968) *Hydrogen Bonding in Solids,* W.A. Benjamin, New York, pp. 19-21.
12. Kuleshova, L.N. and Zorky, P.M. (1980) Graphical enumeration of hydrogen-bonded structures, *Acta Cryst.* **B36**, 2113-2115.
13. Zorky, P.M. and Kuleshova, L.N. (1980). Comparison of the hydrogen bonds in polymorphic modifications of organic substances, *Zh. Strukt. Khim.* **22**, 153-156.
14. Etter, M.C., MacDonald, J.C. and Bernstein, J. (1990) Graph-set analysis of hydrogen-bond patterns in organic crystals, *Acta Cryst.* **B46**, 256-262.
15. In earlier descriptions [*e.g.* 8,14,16,17] **n** was given as the number of *atoms* (vertices in mathematical graph set language) in the path. Recent work on establishing the mathematical basis for the graph set approach indicated that defining **n** in terms of the *bonds* (edges in mathematical graph set language) would be more consistent with the mathematical conventions.

16. Bernstein, J., Etter, M.C. and Leiserowitz, L. (1994) The role of hydrogen bonding in molecular assemblies, Chapter 11 in Bürgi, H.-B. and Dunitz, J.D. (eds.) *Structure Correlation*, Volume 2. VCH, Weinheim, pp. 431-507.

17. Bernstein, J., Davis, R.E., Shimoni, L. and Chang, N.-L.(1995) Patterns in hydrogen bonding: Functionality and graph set analysis in crystals, *Angew. Chem. Int. Ed. Engl.* **34**, 1555-1573.

18. Bernstein, J. Etter, M.C. and MacDonald J.C. (1990) Decoding hydrogen-bond patterns. The case of iminodiacetic acid, *J. Chem. Soc. Perkin Trans. II*, 695-698.

19. Bernstein, J. (1991) Polymorphism of L-glutamic acid: Decoding the α-β phase relationship *via* graph-set analysis, *Acta Cryst.* B47, 1004-1010.

20. See Section 12 of reference 17.

21. Power, L.F., Turner, K.E. and Moore, F.H.(1976) The crystal and molecular structure of α-glycine by neutron diffraction - a comparison, *Acta Cryst.* B32, 11-16; Legros, P.-P. and Kvick, A. (1980) Deformation electron density of α-glycine at 120 K.*Acta Cryst.* B36, 3052-3059.

22. A summary table of thse definitions may be found in reference 17.

23. Allen, F.H., Kennard, O., Watson, D.G., Brammer, L., Orpen, A.G. and Taylor, R. (1987) Tables of bond lengths determined by x-ray and neutron diffraction. Part 1. Bond lengths in organic compounds, *J. Chem. Soc. Perkin. Trans. II*, S1-S19.

24. Jeffrey, G.A., and Saenger, W. (1991) *Hydrogen Bonding in Biological Structures*, Springer, New York.

25. Donohue, J. (1968) in Rich, A. and Davidson, N. (eds.) *Structural Chemistry and Molecular Biology*. W.H. Freeman, San Francisco, pp. 443-465.

26. See also pp. 29-30 of reference 24.

27. Stout, G.H. and Jensen, L.H. (1968) *X-ray Structure Determination. A Practical Guide*, McMillan, London.

28. Bader, R.F.W. (1990) *Atoms in Molecules: A Quantum Theory*, Clarendon Press, Oxford.

29. Novoa, J.J., Mota, F., Perez del Valle, C. and Planas, M. (1997) Structure of the first solvation shell of the hydroxide anion. A model study using OH-$(H_2O)_n$ (n = 4,5,6,7,11,17) clusters, *J. Phys. Chem.* **A101**, 7842-7853. See also the chapter in *this volume* by J.J. Novoa.

30. Steiner, T., Tamm, M., Grzegorzewski, A., Schulte, N., Veldman, N., Schreurs, A.M.M., Kanters, J.A., Kroon, J., Vandermassa, J. and Lutz, B. (1996) Weak hydrogen bonding. 5. Experimental evidence for the long-range nature of $C\equiv C$-H$\cdots\pi$ interactions - crystallographic and spectroscopic studies of three terminal alkynes. *J. Chem. Soc. Perkin II*, 2441-2446, and previous papers in the series.

31. See also the chapter in *this volume* authored by T. Steiner.

32. It is not possible to give a comprehensive list of references for this or other uses of the graph sets to characterize hydrogen bonds. We will provide a few literature examples. Others may be readily identified through the ISI Citation Index, especially citations of references 8,9,14 and 17.

33. Palmore, G.T.R., and McBride-Weiser, M.T. (1997) Pyridinium trifluoroacetate: Spoked columns of hydrogen-bonded cyclic dimers. *Acta Cryst.* C53, 1904-1907.

34. Huang, K.S., Stowell, J.G. and Byrn, S.R. (1997) 3-(4-nitroanilino)-2-cyclopenten-1-one. *Acta Cryst.* C53, 1717-1719.

35. Kubicki, M., Szkaradzinska, M.B. and Codding, P.W. (1997) 3,4,-dihydro-2,6,7-trimethyl-4-oxo-3-(2-pyridyl)quinazoline-8-carboxylic acid, *Acta Cryst.* C53, 1291-1293.

288

36. Ebisuzaki, Y., Boyle, P.D. and Smith, J.A. (1997) Methylxanthines. 1. Anhydrous theophylline. *Acta Cryst.* **C53**, 777-779.

37. Black, D.S., Craig, D.C. and McConnell, D.B. (1997) Self-assembly of an indolyl-glyoxylamide by unusual hydrogen bonding *Tetrahedron Letters* **38**, 4287-4290.

38. See, for instance, the caption to the cover illustration for the *International Union of Crystallography Newsletter*, Volume 5 Issue 3 (1997), p. 14, where the $R_2^2(8)$ notation is used without any further explanation or reference.

39. Lalancette, R.A., Brunskill, A.P.J. and Thompson, H.W. (1997) (+/-)-2,3-dihydro-3-oxo-1H-indenecarboxylic acid: Hydrogen-bonding patterns in a γ-keto acid and its monohydrate, *Acta Cryst.* **C53**, 1838-1842.

40. Falvello, L.R., Pascual, I., Tomas, M. and Urriolabeitia, E.P. (1997). The cyanurate ribbon in structural coordination chemistry: An aggregate structure that persists across different coordination environments and structural types, *J. Am. Chem. Soc.* **119**, 11894-11902.

41. Benyei, A.C., Glidewell, C., Lightfoot, P., Royles, B.J.L. and Smith, D.M. (1997) Functionalized acyl ferrocenes: Crystal and molecular structures of 4-aminobenzoylferrocene, 4-hydroxylbenzoyl ferrocene and 1,1'-bis (4-hydroxybenzoyl) ferrocene, *J. Organomet. Chem.* **539**, 177-186.

42. Rudert, R., Andre, C., Wagner, R. and Vollhardt, D. (1997). A consideration of the hydrogen bonding schemes of the surfactant N-tetradecyl-(2,4-dihydroxyl)-butanoic acid amide and some related amphiphilic compounds. *Z. für Krist.* **212**, 752-755.

43. Gawronski, J., Gawronska, K., Skowronek, P., Rychlewska, U., Warzajtis, B., Rychlewski, J., Hoffmann, M. and Szarecka, A. (1997) Factors affecting conformation of (R,R)-tartaric acid acid ester, amide and nitrile derivatives. X-ray diffraction, circular dichroism, nucler magnetic resonance and ab initio studies, *Tetrahedron* **53**, 6113-6144.

44. Bernstein, J. (1993) Crystal growth, polymorphism and stucture-property relationships in organic crystals, *J. Phys. D: Appl. Phys.* **26**, B66-B76. This prescription is followed for conformational polymorphs in reference 45.

45. Bernstein, J. (1987) Chapter 13 in Desiraju, G.R. (ed.) *Organic Solid State Chemistry* Volume 32 of *Studies in Organic Chemistry*, Elsevier, Amsterdam.

46. Shimoni, L. (1992). *M.Sc. Thesis* Ben-Gurion University of the Negev, Beer Sheva, Israel.

47. MacDonald, J.C. (1994) *Ph.D. Thesis*, University of Minnesota, Minneapolis, Minnesota, USA.

48. Griesser, U.J., Burger, A. and Mereiter, K. (1997) The polymorphic drug substances of the European pharmacopoeia. 9. Physicochemcial properties and crystal structure of acetazolamide crystal forms, *J. Pharm. Sci.* **86**, 352-358.

49. Nakano, K., Sada, K., and Miyata, M. (1996) Novel additive effect of inclusion crystals on polymorphs of cholic acid crystals having different hydrogen-bonded networks with the same organic guest, *Chem. Comm.* 989-990.

50. Ceolin, R., Agafonov, Louer, D., Dzyabchenko, V.A., Toscani, S., and Cense, J.M. (1996) Phenomenology of polymorphism. 3. p,t diagram and stability of piracetam polymorphs. *J. Sol. State Chem.* **122**, 186-194.

51. Blagden, N., Davey, R.J., Lieberman, H.F., William, L., Payne, R., Roberts, R., Rowe, R., and Docherty, R. (1998) Crystal chemistry and solvent effects in polymorphic systems. Sulfathiazole, *J. Chem. Soc. Faraday Trans.* **94**, 1035-1044.

52. Davis, R.E., Pelosof, L.C., Balasubramaniam, S., Bernstein, J. and Sharabi, C., unpublished.

53. Shimoni, L., Glusker, J.P. and Bock, C.W. (1996) Energies and geometries of isographic hydrogen-bonded networks. 1. The $\mathbf{R_2^2(8)}$ graph set, *J. Phys. Chem.* **100**, 2957-2967.

54. Donohue, J. (1968) in Rich, A. and Davidson, N. (eds.) *Structural Chemistry and Molecular Biology*, Freeman, San Francisco, pp. 443-465.

55. Desiraju, G.R. (1991) The C-H···O hydrogen bond in crystals: What is it?, *Acc. Chem. Res.* **24**, 290-296.

56. Steiner, T., Tamm, M., Grzegorzewski, A., Schulte, N., Veldman, N., Schreurs, A.M.M., Kanters, J.A., Kroon, J., Vandermassa, J. and Lutz, B. (1996) Weak hydrogen bonding. 5. Experimental evidence for the long-range nature of C-C-H... p- interactions - crystallographic and spectroscopic studies of three terminal alkynes, *J. Chem. Soc. Perkin. II*, 2441-2446, previous papers in the series, and the chapter in *this volume* by T. Steiner.

57. Cotton, F.A., Daniels, L.M., Jordan, G.T., and Murillo, C.A. (1997) The crystal packing of bis(2,2'-dipyridylamido) cobalt (II), Co(dpa)$_2$, is stabilized by C-H···N bonds: Are there any precedents?, *Chem. Comm.* 1673-1674.

58. Evans, T.A. and Seddon, K.R. (1997) Hydrogen bonding in DNA - a return to the status quo, *Chem. Comm.* 2023-2024.

59. Bernstein, J., Ganter, B., Grell, J., Hengst, U., Kuske, D. and Pöschel, R. (1997) Mathematical basis of graph set analysis, *Technische Universität Dresden Preprint Math-A1-17-1997*, to be submitted for publication.

60. Grell, J. and Bernstein, J. (1998) Graph set analysis of hydrogen bond patterns. Some mathematical concepts, *in preparation*.

61. Desiraju, G.R. (1989) *Crystal Engineering*, Elsevier, Lausanne.

62. See also chapter by Desiraju in *this volume*.

63. Shimoni, L. Carrell, H.L., Glusker, J.P., and Coombs, M.M. (1994) Intermolecular effects in crystalts of 11-(trifluoromethyl)-12,16-dihydrocyclopenta[a]phenanthrene-17-one, *J. Am. Chem. Soc.* **116**, 8162-8168.

64. Shimoni, L. Glusker, J.P. and Bock, C.W. (1995). Structures and dissociation energies of the complexes CH$_m$F$_n$···NH$_4^+$ (n+m=4, n=1-3) - an *ab initio* molecular orbital study, *J. Phys. Chem,* **99**, 1194-1198.

65. See also the chapter by R.E. Davis in *this volume*.

66. Larsen, S. and Marthi, K. (1997) Structures of optically active monofluoro-substituted mandelic acids: Relation to their racemic counterparts and thermochemical properties,*Acta Cryst.* **B53**, 280-292.

67. Belochradsky, M., Raymo, F.M. and Stoddard, J.F. (1997) Template-directed syntheses of catenanes. *Coll. Czech. Chem. Comm.* **62**, 527-557.

68. Coupar, P.I., Glidewell, C., and Ferguson, G. (1997) Crystal engineering using bisphenols and trisphenols. Complexes with hexamethylenetetramine (HMTA): Strings, multiple helices and chains-of-rings in the crystal structures of the adducts of HMTA with 4,4'-thiodiphenol(1/1), 4,4'-sulfonyldiphenol (1/1), 4,4'-isopropylidenediphenol (1/1), 1,1,1-tris(4-hydroxyphenyl)ethane(1/2) and 1,3,5-trihydroxybenzene (2/3), *Acta Cryst.* **B53**, 521-533.

69. Aakeroy, C. (1997) Crystal engineering: Strategies and architectures, *Acta Cryst.* **B53**, 569-586.

70. Ashton, P.R., Matthews, O.A., Menzer, S., Raymo, F.M., Spencer, N., Stoddard, J.F. and Williams, D.J. (1997). Molecular meccano, 27 - A template-directed synthesis of a molecular trefoil knot. *Liebigs Annal.-Rec.*, 2485-2494.

71. Payne, R.S., Roberts, R.J., Rowe, R.C. and Docherty, R. (1998) Generation of crystal structures of acetic acid and its halogenated structures, *J. Comput. Chem.* **19**, 1-20.

72. Tran Qui, D., Raymond, S., Pecaut, J., and Kvick, A. (1997) Crystal structure study of (E)-3-(4-nitrobenzyloxyimino)hexahydroazepin-2-one using synchrotron radiation, *J.Synch. Rad.* **4**, 78-82.

73. Kariuki, B.M., Zin, D.M.S., Tremayne, M., and Harris, K.D.M. (1996) Crystal structure solution from powder x-ray diffraction data - the development of Monte Carlo methods to solve the crystal structure of the γ-phase of 3-chloro cinnamic acid, *Chem. Mat.* **8**, 565-569.

74. Larsen, S. and Marthi, K. (1995). Structures of racemic halogen-substituted 3-hydroxy-3-phenylpropionic acids - relations between spontaneously resolved and racemic compounds, *Acta Cryst.* **B51**, 338-346.

75. Larsen, S., and DeDiego, H.L. (1995) Mandelic acid as a resolving agent, *ACH - Models in Chemistry* **132**, 441-450.

76. Valente, E.J., Miller, C.W., Zubkowski, J., Eggleston, D.S. and Shui, X.Q. (1995) Discrimination in resolving systems. 2. Ephedrine-substituted mandelic acids, *Chirality* **7**, 652-676.

77. Shimoni, L, L. and Glusker, J.P. (1995) Hydrogen-bonding motifs of protein side chains - Descriptions of binding of arginine and amide groups, *Protein Science* **4,** 65-74.

78. Ojala, W.H., Sudbeck, E.A., Lu, L.K., Richardson, T.I., Lovrien, R.E. and Gleason, W.B. (1996) Complexes of lysine, histidine and arginine with sulfonated azo dyes - model systems for understanding the biomolecular recognition of glycosaminoglycans by proteins, *J. Am. Chem. Soc.* **118**, 2131-2142.

79. Kubicki, M., Kindopp, T.W., Capparelli, M.V., and Codding, P.W. (1996) Hydrogen-bond patterns in 1,4-dihydro-2,3-quinoxalinediones - ligands for the glycine modulatory site on the NMDA receptor, *Acta Cryst.* **B52**, 487-499.

80. Fruzinski, A., Karolak-Wojciechowska, J., Mokrosz, M.J., and Bojarski, A.J. (1997) Crystal and molecular structure of N-tert-butyl-3[4-(2-methoxypheny)-1-piperazinyl]-2-phenylpropanamide dihydrochloride, *Polish J. Chem.* **71**, 1611-1617.

CRYSTALLOGRAPHIC DATABASES AND KNOWLEDGE BASES IN MATERIALS DESIGN

FRANK H. ALLEN AND GREGORY P. SHIELDS
Cambridge Crystallographic Data Centre
12 Union Road
Cambridge CB2 1EZ, England

1. Introduction

The development of strategies for the design of novel materials is increasingly a knowledge-based activity and this book highlights a variety of experimental and computational methods that provide structural knowledge about individual compounds or groups of related compounds. However, scientific research and development is an evolutionary process in which the critical analysis and assessment of all existing information is an essential component. In the area of chemical structure, such analyses provide fundamental insights which can lead to the detection of trends and probabilities, and to the formulation of hypotheses and rules. These indications, in their turn, can direct the future course of structural design, synthesis and experiment.

Crystallographic methods, both single-crystal and powder, are now pre-eminent in providing experimental observations of chemical structures. Not only do these techniques provide detailed geometrical pictures of individual structural components, they are also unique in providing detailed information about the non-covalent interactions that are vital to the design of supramolecular systems: a crystal structure is indeed the archetypal supermolecule.

Perhaps already aware of the fundamental importance of their results, crystallographers have a long and distinguished history of self documentation dating back, in printed form, to the early 1930s. Given their early use of emerging computer technology, it is unsurprising that they were amongst the first to turn to the digital computer to solve their information needs. Building on pioneering developments in the electronic representation of chemical data in the 1950s and 1960s, crystallographers began to develop the first truly numerical chemical archives. The central focus of this work was to record the primary results of each crystal structure analysis: cell dimensions, symmetry, atomic coordinates and related data, which were then blended with bibliographic and chemical details to reflect a comprehensive coverage of the original literature.

J.A.K. Howard et al. (eds.),
Implications of Molecular and Materials Structure for New Technologies, 291–302.
© 1999 *Kluwer Academic Publishers. Printed in the Netherlands.*

Today, the results of some 300,000 crystal structures are available in electronic form, representing a huge reservoir of precise structural information that covers the complete chemical spectrum, from metals and alloys to proteins and viruses. Together with continually evolving software systems for database searching, information retrieval, structure visualisation and the analysis of geometrical information, crystallographic database systems now provide facilities for the acquisition of structural knowledge at many different levels of complexity.

Over the past two decades, the processes of data mining and knowledge discovery from the crystallographic databases has proceeded at pace. Some 600 research applications of the Cambridge Structural Database have appeared in the literature and some excellent examples of this type of study are presented elsewhere in this book.

Given the long term value of data mining experiments - and the expertise and time required to undertake the work - it is natural to wonder how these results may themselves be stored in electronic form. These considerations are now leading to the development of knowledge-bases of structural information derived directly from the original databases. Essentially, the atomic coordinates and related information - the raw primary data - are being converted into systematised geometrical descriptions of structure - a form of knowledge that is most readily accessible by potential users. Further, these knowledge bases are being structured not only to be of intrinsic value in themselves, *i.e.* for direct browsing, but also to act as the knowledge resource for intelligent software that is designed to solve specific problems in structural chemistry.

This chapter summaries the history and current status of the crystallographic databases and provides a survey of typical applications in data mining and knowledge discovery, with a particular emphasis on the methods available, pitfalls to avoid, and relationships with computational techniques. The development of crystallographic knowledge bases and prospects for knowledge-based software systems are also discussed.

2. The Crystallographic Databases

Five structural databases, *i.e.* those that contain atomic coordinate data, cover the complete chemical spectrum [1]. They are:

- Protein Data Bank (PDB, 6,500 entries: RCSB, Rutgers University, Piscataway, NJ, USA) [2]

- Nucleic Acids Data Bank (NDB, 731 entries: Rutgers University, Piscataway, NJ, USA) [3]

- Cambridge Structural Database (CSD, 198,000 organics and metallo-organics: CCDC, Cambridge, UK) [4]

- Inorganic Crystal Structure Database (ICSD, 55,000 inorganics and minerals: Fachinformationszentrum, Karlsruhe, Germany) [5]

- Metals Data File (CRYSTMET, 45,000 entries including assigned structures: refer to Toth Information Systems, Ontario, Canada) [6]

2.1. INFORMATION CONTENT

The information content of the crystallographic databases can conveniently be divided into three major components:

- Bibliographic and chemical text: typically compound name(s), molecular formula(e) and literature citation.

- 2D Chemical connectivity representation (CSD): a compact coding of the atom and bond properties of covalently bonded molecular species. Such representations form the basis for substructure search mechanisms and can be displayed. For the PDB, the amino-acid sequence information is a linear representation of connectivity, the underlying 2D chemistry of the units being implied. Connectivity of hetero groups and liganded small molecules must be encoded explicitly in PDB. Connectivity representations are inappropriate for the ICSD and CRYSTMET.

- 3D Crystallographic data: atomic coordinates, cell dimensions, symmetry and, in some databases, the atomic displacement parameters.

2.2. SOFTWARE SYSTEMS

All databases provide software access to the stored information through search, browse and structure display mechanisms. Software for the CSD System also provides extensive facilities for the display and statistical analysis of retrieved geometrical information. The menu-driven graphical software of the CSD System is summarised below [7]:

- QUEST3D is the main search program, permitting searches of all available information fields: (a) 19 text fields, (b) 38 individual numerical fields, (c) element symbols and element counts, (d) full or partial molecular formula, (e) user access to over 100 items of yes/no (bit-encoded) summary information, (f) extensive 2D substructure search capabilities, and (g) full 3D substructure searching at the molecular and extended crystal structure levels. The program will generate output files of user-defined geometrical parameters for any substructure (intramolecular or intermolecular) located in the search process. Other output files allow CSD search results to be rapidly communicated to external modelling software.

- VISTA reads the geometrical table(s) generated by QUEST3D and provides extensive facilities for the graphical representation (histograms, scattergrams, polar plots, *etc.*) and statistical analysis (summary statistics, regression, principal component analysis, *etc.*) of the numerical data.

- PLUTO is used to visualise crystal and molecular structures in a variety of styles, and explore intermolecular networks and packing [8].

3. Research Applications

The Cambridge Structural Database is by far the largest and most widely disseminated of the crystallographic database systems. It is the only database which has established an associated programme of knowledge-base development. In order to present a uniform summary of scientific activity which exemplifies the introductory remarks, the remainder of these notes relate primarily to CSD-based research activities. However, there is a natural synergy between the CSD and the PDB, and some applications of the PDB will also be covered in this chapter.

Because of its widespread use, the CSD System now contains a new, small database component - denoted as DBUSE - which records the literature references of CSD-based research applications together with a short summary of the major results of the work. Another valuable guide to CSD and other database applications is the recent book *Structure Correlation*, edited by Buergi and Dunitz [9].

4. Intramolecular Knowledge Discovery from the CSD

4.1. MEAN MOLECULAR DIMENSIONS

The derivation of simple descriptive statistics for standard geometrical parameters, particularly bond lengths or valence angles, is a relatively straightforward application of the CSD System. These data are provided routinely as part of the standard QUEST3D tabulations, and histograms of individual distributions can be generated rapidly using VISTA. As an aid to structural chemists and modellers, two major compilations of 'standard' bond lengths were produced in the late 1980s, derived for a wide range of organic and metallo-organic bond environments [10,11]. Since this time, software has improved considerably and mean values for environments not covered by these compilations can be simply generated as required in just a few minutes of 'real' time.

4.2. THE METHOD OF STRUCTURE CORRELATION

Each crystallographic experiment provides detailed knowledge of the 3D geometrical structure of a molecule. Within that molecule are contained the 3D substructures of a number of chemical fragments that exist within a specific chemical and crystallographic environment. If we specify a particular chemical substructure of interest as a query fragment to the CSD System, then we will retrieve the geometrical descriptors of tens, hundreds, or even thousands of examples of the fragment taken from as many different environments. Each example, represents a static 3D snapshot of the fragment taken under slightly different conditions. If we use n geometrical parameters to describe each of the i examples, then we obtain a geometric matrix $G(i,n)$ which can be examined for regularities (discrete classifications), and for correlations involving some or all of the geometrical parameters, or between geometrical parameters and other chemical, physical or biological properties. This is the fundamental basis of the *structure correlation* principle [9,12].

4.3. CONFORMATIONAL ANALYSIS

The method of structure correlation can be used to advantage in detecting the conformational preferences of specific substructures from their geometrical attributes, provided we assume that 'environmental' effects are smoothed out over the examples included in the study [13,14]. This assumption has often been found to be viable, especially if the number of observations is large, but in some cases environmental bias, either chemical or crystallographic, can occur.

Mathematically, each of the i substructures retrieved from the CSD can be regarded as a data point or observation in the n-dimensional parameter space. Since conformational analysis involves the numerical analysis of the complete G-matrix (usually, but not always, a matrix of torsion angles), then the lower the dimensionality of the parameter space, the simpler will be the analysis. Often we can describe a conformational problem using just one or two parameters, and simple histograms or scattergrams display the underlying structure of the data. However, and especially in the case of extended chains or ring systems, a much larger number of parameters is needed and suitable methods of multivariate numerical analysis must be employed [15]. Principal component analysis (PCA) [16] and various clustering techniques have been found particularly useful in this context [17].

PCA analyses the total variance (Var) in a multivariate dataset in terms of a new set of uncorrelated, orthogonal variables: the principal components or PC's. The PC's are generated in decreasing order of the percentage of the total variance that is explained by each PC. The hope is that the number of PC's, p, that explains a very large percentage of Var (say, >95%) is such that $p<n$ (the original number of geometric variables). PCA is a method of dimensionality reduction and plots of the data points based on these new (and orthogonal) PC axes can provide valuable visualisations of the complete dataset

[18]. For ring systems, atomic permutational symmetry must be taken into account [19], whence PCA is closely related to Cremer-Pople puckering analysis [20]. PCA maps often reveal discrete clusters of observations that correspond to the preferred conformations of rings or chains that occur in crystal structures [21].

Cluster analysis (CA) is a purely numerical technique that attempts to locate discrete groupings of data points within a multivariate dataset. CA uses the 'distances' or 'dissimilarities' between pairs of points in the n-dimensional space as its basis. A large number of CA algorithms exist, which must be modified to take account of atomic permutational symmetry in many crystallographic applications [22]. CA has been applied to a number of ring systems and, in most cases, has identified global and local energy minima as the crystallographically preferred forms [23]. The case of cyclohepta-1,3-diene illustrates the problem of chemical bias in the dataset: all diene bonds in the CSD entries arose from fusion to aromatic systems and none from purely ethylenic systems [24]. Minimum-energy forms for these two systems are rather different and the clustering process (correctly) identified a conformational preference that corresponded to the bis-fused aromatic energy minimum, an inappropriate model for the parent diene.

Numerical CA also requires considerable attention from the user, both in terms of selection of certain run-time parameters, and in assessing the chemical sensibility of the resulting clusters: it is not a fully automatic process. A recent paper has experimented with conceptual clustering techniques arising from the area of artificial intelligence that is concerned with machine learning [25]. Here, the clustering is carried out in terms of conformational concepts that are familiar to the chemist, *e.g. cis, trans, synperiplanar, antiperiplanar, etc.* The results of a comparison of the numerical and conceptual methods applied to a selection of rings and chains, showed that the fully automatic conceptual method provided more complete clusterings of the data than the numerical method, although the major conformational preferences identified by both methods were equivalent.

A principal result of structure correlation experiments is the existence of a qualitative relationship between the distribution of crystallographic observations and the low-energy features of the conformational hypersurface. Recently, this relationship was studied more systematically for 12 simple systems involving only a single degree of torsional freedom [26]. The analysis compared crystallographically observed conformer distributions with those predicted using high-level *ab initio* calculations. Results showed that: (a) there was close agreement between the condensed phase and in-vacuo data, (b) torsion angles with higher strain energies (>4.5 kJ/mol) were rarely observed in crystal structures, and (c) taken over many structures, conformational distortions due to crystal packing appear to be the exception rather than the rule.

5. Intermolecular Knowledge Discovery from the CSD

In his Nobel Lecture, J.-M. Lehn defined supramolecular chemistry as "the chemistry of the intermolecular bond" [27]: a supermolecule is an assembly of covalently bonded units or ions that is organised according to the diverse weak forces that govern non-covalent interactions. Such a description, of course, is equally applicable to the extended structure of molecular crystals, where the crystalline supermolecule is assembled by molecular self-recognition or through the mediacy of additional components of the system, particularly solvent molecules or counterions. If we are to make advances in the broad area of supramolecular chemistry, which encompasses such topics as materials design, crystal engineering [28] and the the the study of protein - ligand interactions in rational drug design [29], then we must acquire systematic information about non-covalent interactions. In particular, we need to know (a) what are the most important types of non-covalent interactions?, (b) what are their geometrical and directional characteristics? and (c) how strong are the interactions?

Crystal structure analysis is the primary source of experimental data on the geometry and directionality of non-covalent interactions. Historically, the technique has played a major role in the characterisation and understanding of the most ubiquitous of all molecular recognition mechanisms, the hydrogen bond [30]. However, in recent times, there has also been a growing interest in non-covalent interactions that are *not* mediated by hydrogen [31]. Thus, systematic analysis of crystal structure data can provide answers to the first two questions posed above.

Although it is possible to infer qualitative energy relationships from statistical analyses of crystallographic data, as indicated in the previous section, we must turn to computational methods to provide quantitative estimates of the strengths of non-covalent interactions. The intermolecular perturbation theory (IMPT) of Hayes and Stone is an *ab initio* technique that overcomes basis set superposition errors and provides realistic models and good *in vacuo* estimates of interaction energies for small model systems [32]. An important feature of the IMPT method is that it partitions the interaction energy into separate terms which have distinct physical significance. Furthermore, the sum of the significant interaction energy terms yields a total IMPT energy which is free of basis set superposition errors. At first order, these separate terms are (a) E_{es}: the (attractive or repulsive) electrostatic energy that describes the classical Coulombic interaction, and (b) E_{er}: the exchange- repulsion term, the sum of an attractive part due to the exchange of electrons of parallel spin, and a repulsive part as a result of the Pauli exclusion principle which prevents electrons with parallel spins occupying the same region in space. At second order, the IMPT gives (c) E_{pol}: the polarisation or induction energy, (d) E_{ct}: the charge transfer energy, and (e) E_{disp}: the dispersion energy term.

However, the IMPT method is cpu-intensive and is not well suited to the systematic exploration of the full geometrical interaction space. Instead, we use CSD data to limit this exploration to highly populated areas in the geometrical distributions - assumed to be areas of lowest energy in the potential energy hypersurface. This combination of CSD analysis and high-level *ab initio* computations is proving extremely fruitful in the study of hydrogen bonds [33] and a variety of other specific non-covalent interactions [34].

Very recently it has been recognised that crystal engineering and the design of molecular materials depend crucially on the high probabilities of formation of a limited number of strong intermolecular interaction patterns. These motifs have been termed supramolecular synthons by Desiraju [35], and a graph set notation for their description has been proposed by Etter, Bernstein, Davis and others [36]. Despite its successes, much of this activity is based on experiential evidence, hence systematic CSD analysis is now being used to reveal actual probabilities of formation of all common motifs in crystal structures, in the hope that new synthons may be identified for use in materials design [37].

6. Knowledge-Based Libraries of Structural Information

Although the CSD has been used in many diverse research projects, the majority have reported systematic studies of (a) conformational features and (b) intermolecular interactions. This is unsurprising since these are the geometrical attributes that are of most interest to the major users of the database: structural chemists, pharmaceutical researchers and the designers of novel materials. Many of these analyses follow similar paths, in terms of the underlying CSD searches, the geometrical data retrieved, and the ways in which it is displayed and analysed. Some of the searches require a degree of expertise in the use of the database and of its associated software system. Hence, in 1995, the CCDC began to explore the possibility of providing the geometrical results of standard searches in the form of knowledge-based libraries. For a variety of reasons the topic of intermolecular interactions was chosen for prototyping the concept and the resulting library, denoted as IsoStar, was released as part of the CSD System in October 1997 [38]. A library of conformational preferences is currently under development at the CCDC.

These ideas are not necessarily novel. Earlier transformations of the PDB to facilitate specific enquiries are well known [39], while a more comprehensive knowledge-based approach to protein structure determination and molecular modelling (molecular scene analysis) has also been published [40].

6.1. ISOSTAR: A KNOWLEDGE-BASED LIBRARY OF INTERMOLECULAR INTERACTIONS

The previous sections show that the amount of data in the CSD on intermolecular geometries is vast, and CSD-derived information for a number of specific systems is available in the literature at various levels of detail. If not, the CSD must be searched for contacts between the relevant functional groups. To provide structured and direct access to a more comprehensive set of derived information, the IsoStar knowledge-based library of nonbonded interactions has been developed at the CCDC. IsoStar is based on experimental data, not only from the CSD but also from the PDB, and contains some theoretical results calculated using the IMPT method. Version 1.0 of IsoStar (October 1997) contains information on nonbonded interaction formed between 277 common functional groups, referred to as *central groups*, and 28 *contact groups, e.g.* H-bond donors, water, halide ions, *etc.* Information is displayed in the form of scatterplots for each interaction. Version 1.0 contains 6683 scatterplots: 5296 from the CSD and 1387 from the PDB. IsoStar also reports results for 867 theoretical potential-energy minima.

For a given contact between between a central group (A) and a contact group (B), CSD search results were transformed into an easily visualised form by overlaying the A moieties. This results in a 3D distribution (scatterplot) showing the experimental distribution of B around A. The IsoStar software provides a tool that enables the user to quickly inspect the original crystal structures in which specific contacts occur, *via* a hyperlink to original CSD structures. This is helpful in identifying outliers, motifs and biases. Another tool generates contoured surfaces from scatterplots, which show the density distribution of the contact groups. Contouring aids the interpretation of the scatterplot and the analysis of preferred geometries.

The PDB scatterplots in IsoStar only involve interactions between non-covalently bound ligands and proteins, *i.e.* side chain - side chain interactions are excluded. The IsoStar library contains data derived from almost 800 complexes having a resolution better than 2.5Å. Scatterplots derived from PDB data show close similarities to their CSD-based counterparts.

6.2. A LIBRARY OF CONFORMATIONAL PREFERENCES

In essence, the CSD can be regarded as a huge library of individual molecular conformations. However, to be of general value, it is necessary to distil, store and present this knowledge in an ordered manner, in the form of torsional distributions for specific atomic tetrads A-B-C-D. Protein-specific libraries of this type derived from high-resolution PDB structures are commonly used as aids to protein structure determination, refinement and validation [39]. The information can either be stored in external databases, or hard-wired into the program in the form of rules. However, CSD usage has tended to be concentrated on analyses of individual substructures, as noted above, both for their intrinsic interest and to develop novel methods of data analysis.

Recently, Klebe and Mietzner have described the generation of a small library containing 216 torsional distributions derived from the CSD, together with 80 determined from protein - ligand complexes in the PDB [41]. The library was used in a knowledge-based approach for predicting multiple conformer models for putative ligands in the computational modelling of protein - ligand docking. Conformer prediction is accomplished by the computer program MIMUMBA. As part of its programme for the development of knowledge-based libraries from the CSD, the CCDC has now embarked on the generation of a more comprehensive torsional library. Here, information is being ordered hierarchically according to the level of specificity of the chemical substructures for which torsional distributions are available in the library.

7. Prospects for Knowledge-Based Software Systems

This terminology deserves explanation since all computational chemistry programs depend upon various forms of chemical knowledge or parameters which may be read from external files, or encoded as constants or rules within the program.

In the current context, however, we mean software that is written to solve a specific problem in structural chemistry and which accesses one or more large, comprehensive and dynamically updated knowledge bases derived from experimental sources. The MIMUMBA approach to conformer prediction is typical of the concept [41]. Thus, a key feature in the design of CSD-based libraries is to ensure that they can be readily accessed by software of this type. In some cases, it may be possible, even desirable, to amend existing software products so that they can access the developing structural knowledge bases. One case in point is the GOLD program, a genetic algorithm approach to protein-ligand docking, which already uses some experimental information on ligand flexibility and non-covalent interactions derived directly from the CSD [42]. It is likely also that knowledge of the type decribed above will have a role to play in improving algorithms for structure solution from powder diffraction data [43].

8. References

1. Allen, F.H., Bergerhoff, G. and Sievers, R. (1987) *Crystallographic Databases*. International Union of Crystallography, Chester, UK; Allen, F. H. (1998) The development, status and scientific impact of crystallographic databases. *Acta Cryst.*, **A54**, 758-771.
2. Bernstein, F.C., Koetzle, T.F., Williams, G.J.B., Meyer, E.F., Brice, M.D., Rodgers, J.R., Kennard, O., Shimanouchi, T. and Tasumi., M. (1972) The Protein Data Bank. *J. Mol. Biol.*, **112**, 535-547.
3. Berman, H.M., Olson, W.K., Beveridge, D.L., Westbrook, J., Gelbin, A., Demeny, T., Hsieh, S.-H., Srinavasan, A.R. and Schneider, B. (1992) The Nucleic Acid Database: a comprehensive relational database of three-dimensional structures of nucleic acids. *Biophys. J.*, **63**, 751-759.
4. Allen, F.H., Davies, J.E., Galloy, J.J., Johnson, O., Kennard, O., Macrae, C.F., Mitchell, E.M., Mitchell, G.F., Smith, J.M., and Watson, D.G. (1991) The Development of Versions 3 and 4 of the Cambridge Structural Database System, *J. Chem. Inf. Comput. Sci.*, **31**, 187-204.
5. Bergerhoff, G., Hundt., R., Sievers, R. and Brown, I.D. (1983) The Inorganic Crystal Structure Database. *J. Chem. Inf. Comput. Sci.*, **23**, 66-69.

6. Toth Information Systems Inc., 2045 Quincey Avenue, Gloucester, Ontario K1J 6B2, Canada.

7. Kennard, O. and Allen, F.H. (1993) Version 5 of the Cambridge Structural Database System. *Chem Design Automation News*, **8**, pp 1 and 31-37.

8. Motherwell, W.D.S., Shields, G.P. and Allen, F.H. (1999) Visualisation and Characterisation of Non-Covalent Networks in Molecular Crystals: Automated Assignment of Graph Set Descriptors for Asymmetric Molecules, *Acta Cryst.*, **B55**, submitted.

9. Buergi, H.-B. and Dunitz, J.D. (eds.) (1993) *Structure Correlation*, VCH Publishers, Weinheim, Germany.

10. Allen, F.H., Kennard, O., Watson, D.G., Brammer, L., Orpen, A.G. and Taylor, R. (1987) Tables of bond lengths determined by X-ray and neutron diffraction. Part 1. Bond lengths in organic compounds. *J. Chem. Soc., Perkin Trans.* 2, S1-S19.

11. Orpen, A.G., Brammer, L., Allen, F.H., Kennard, O., Watson, D.G. and Taylor, R. (1989). Tables of bond lengths determined by X-ray and neutron diffraction. Part II. Organometallic compounds and coordination complexes of the d- and f-block metals. *J. Chem. Soc. Dalton Trans.*, S1-S83.

12. Auf der Heyde, T.P.E. and Bürgi, H.-B. (1989) *Inorg. Chem.*, **28**, 3960-3939.

13. Dunitz, J.D. (1979) *X-Ray Analysis and the Structure of Organic Molecules*. Cornell University Press, Ithaca, NY, USA.

14. Klebe, G. (1994) Mapping common molecular fragments in crystal structures to explore conformation and configuration space under the conditions of a molecular environment. *J. Mol. Struct.*, **308**, 53-89.

15. Taylor, R. and Allen, F.H. (1993) *Statistical and Numerical Methods of Data Analysis*. Chapter 4 in Buergi, H.-B. and Dunitz, J.D. (1993) *Structure Correlation*. VCH Publishers, Weinheim, Germany.

16. Murray-Rust, P. and Bland, R. (1978) Computer Retrieval and Analysis of Molecular Geometry. II. Variance and its interpretation, *Acta Cryst.*, **B34**, 2527-2533; Auf der Heyde, T. P. E. (1990) *J. Chem. Educ.*, **67**, 461-469.

17. Everitt, B. (1987) *Cluster Analysis*, 2nd. Ed., Wiley, New York; Willett, P. (1987) *Similarity and Clustering in Chemical Information Systems*, Research Studies Press, Wiley, New York.

18. Murray-Rust, P. and Motherwell, S. (1978) Computer Retrieval and Analysis of Molecular Geometry. III. Geometry of the β-1'-aminofuranoside fragment, *Acta Cryst.*, **B34**, 2534-2546; Taylor, R., *J. Mol. Graph.* (1986), **4**, 123-131.

19. Allen, F.H., Doyle, M.J. and Auf der Heyde, T.P.E. (1991) Automated conformational analysis from crystallographic data. 6. Principal component analysis for n-membered carbocyclic rings (n=4,5,6) Symmetry considerations and correlations with ring-puckering parameters. *Acta Cryst.*, **B47**, 412-428.

20. Cremer, D. and Pople, J.A. (1975) A general definition of ring puckering coordinates. *J. Amer. Chem. Soc.*, **97**, 1354-1358.

21. Allen, F.H., Howard, J.A.K. and Pitchford, N.A. (1996) Symmetry-modified conformational mapping and classification of the medium rings. IV. Cyclooctane and related rings. *Acta Cryst.*, **B52**, 882-891.

22. Allen, F.H., Doyle, M.J. and Taylor, R. (1991) Automated conformational analysis from crystallographic data. 1. A symmetry-modified single-linkage clustering algorithm for three-dimensional pattern recognition, *Acta Cryst.*, **B47**, 29-40; 2. Symmetry-modified Jarvis-Patrick and complete-linkage clustering algorithms for three-dimensional pattern recognition, *Acta Cryst.*, **B47**, 41-49.

23. Allen, F.H., Doyle, M.J. and Taylor, R. (1991) Automated conformational analysis from crystallographic data. 3. 3D Pattern recognition within the Cambridge Structural Database System: Implementation and Practical Examples. *Acta Cryst.*, **B47**, 50-61.

24. Allen, F.H., Garner, S.E., Howard, J.A.K. and Pitchford, N.A. (1994) Symmetry-modified conformational mapping and classification of the medium rings from crystallographic data. III. endo-Unsaturated seven-membered rings. *Acta Cryst.*, **B50**, 395-404.

25. Conklin, D., Fortier, S., Glasgow, J.I. and Allen, F.H. (1996) Conformational analysis from crystallographic data using conceptual clustering. *Acta Cryst.*, **B52**, 535-549.

26. Allen, F.H., Harris, S.E. and Taylor, R. (1996) Comparison of conformational preferences in the crystalline state with theoretically predicted gas-phase distributions. *J. Computer-Aided Molec. Design*, **10**, 247-254.

27. Lehn, J.-M. (1988) Supramolecular Chemistry - Scope and Perspectives: Molecules, Supermolecules and Molecular Devices, *Angew. Chem. (Int. Ed. Engl.)*, **27**, 90-112.

28. Desiraju, G.R. (1991) *Crystal Engineering: The Design of Organic Solids*, Academic Press, New York.

29. Klebe, G. (1994) The use of composite crystal field environments in molecular recognition and the *de novo* design of protein ligands. *J. Mol. Biol.*, **237**, 212-235.

302

30. Jeffrey, G.A. and Saenger, W. (1991) *Hydrogen Bonding in Biological Structures*, Springer Verlag, Berlin; Taylor, R. and Kennard, O. (1982) Crystallographic Evidence for the Existence of C-H...O, C-H...N and C-H...Cl Hydrogen Bonds, *J. Am. Chem. Soc.*, **104**, 5063-5070; Desiraju, G.R. (1991) The C-H...O Hydrogen Bond in Crystals: What Is It?, *Acc. Chem. Res.*, **24**, 290-296; Dunitz, J.D. and Taylor, R. (1997) Organic fluorine hardly ever accepts hydrogen bonds. *Eur Chem. J.* **3**, 83-90.

31. Taylor, R., Mullaley, A. and Mullier, G.W. (1990) Use of Crystallographic Data in Searching for Isosteric Replacements: Composite Crystal-Field Environments of Nitro and Carbonyl Groups, *Pestic. Sci.*, **29**, 197-213; Rowland, R.S. and Taylor, R. (1996) Intermolecular nonbonded contact distances in organic crystal structures: comparison with distances expected from van der Waals radii. *J. Phys. Chem.*, **100**, 7384-7391.

32. Hayes, I.C. and Stone, A.J. (1984) Intermolecular Perturbation Theory. *J. Mol. Phys.* , **53**, 83:105.

33. Allen, F.H., Bird, C.M., Rowland, R.S. and Raithby P.R. (1997) Resonance-induced hydrogen bonding at sulphur acceptors in $R^1R^2C=S$ and $R^1CS_2(-)$ systems. *Acta Cryst.*, **B53**, 680-695; Allen, F.H., Lommerse, J.P.M., Hoy, V.J., Howard, J.A.K. and Desiraju, G.R. (1996) The hydrogen bond C-H donor and pi-acceptor characteristics of three-membered rings. *Acta Cryst.*, **B52**, 734-745; Lommerse, J.P.M., Price, S.L. and Taylor, R. (1997) Hydrogen bonding of carbonyl, ether and ester oxygen atoms with alkanol hydroxyl groups. *J. Comp. Chem.*, **18**, 757-780; Nobeli, I., Price, S.L., Lommerse, J.P.M. and Taylor, R. (1997) Hydrogen bond properties of oxygen and nitrogen acceptors in aromatic heterocycles. *J. Comp. Chem.*, **18**, 2060-2074; Nobeli, I., Yeoh, S.L., Price, S.L. and Taylor, R. (1997) On the hydrogen bonding abilities of phenols and anisoles. *Chem. Phys. Lett.*, **280**, 196-202.

34. Lommerse, J.P.M., Stone, A.J., Taylor, R. and Allen, F.H. (1996) The Nature and Geometry of Intermolecular Interactions between Halogens and Oxygen or Nitrogen. *J. Amer. Chem. Soc.*, **118**, 3108-3116; Allen, F.H., Baalham, C.A., Lommerse, J.P.M. and Raithby, P.R. (1998) Carbonyl-carbonyl interactions can be competitive with hydrogen bonds. *Acta Cryst.*, **B54**, 320-329.

35. Desiraju, G.R. (1995) Supramolecular Synthons in Crystal Engineering - A New Organic Synthesis, *Angew. Chem. Int. Ed. Engl.*, **34**, 2311-2327; Desiraju, G. R. (1999) Current challenges in crystal engineering, in *Implications of Materials Structure for New Technologies*, (eds. Allen, F.H. and Howard, J.A.K.), Kluwer, Dordrecht, and references therein.

36. Bernstein, J. and Davis, R. E. (1999) Graph set Analysis of hydrogen bond motifs, in *Implications of Materials Structure for New Technologies*, (eds. Allen, F.H. and Howard, J.A.K.), Kluwer, Dordrecht, and references therein.

37. Allen, F.H., Raithby, P.R., Shields, G.P. and Taylor, R. (1998) Probabilities of formation of bimolecular hydrogen bonded motifs in organic crystal structures: a systematic database analysis. *Chem. Commun.*, 1043-1044; Allen, F.H., Motherwell, W.D.S., Raithby, P.R., Shields, G.P. and Taylor, R. (1999) Systematic analysis of the probabilities of formation of bimolecular hydrogen-bonded motifs in organic crystal structures, *New. J. Chem*, 25-34.

38. Bruno, I.J., Cole, J.C., Lommerse, J.P.M., Rowland, R.S., Taylor R. and Verdonk, M.L. (1997) IsoStar - A library of information about non-bonded interactions. *J. Computer-Aided Mol. Design*, **11**, 525-537.

39. Islam, S.E. and Sternberg, M.J.E. (1989) A relational database of protein structures designed for flexible enquiries about conformation. *Protein Engineering*, **2**, 431-442; Thornton, J.M. and Gardner, S.P. (1989) Protein motifs and database searching. *Trends. Biochem. Sci.*, **14**, 300-304.

40 Allen, F.H., Rowland, R.S., Fortier, S. and Glasgow, J.I. (1990) Knowledge acquisition from crystallographic databases: towards a knowledge-based approach to molecular scene analysis. *Tetrahedron Computer Methodology*, **3**, 757-774.

41. Klebe, G. and Mietzner, T. (1994) *J. Computer-Aided Mol. Design*, **8**, 583-594.

42. Jones, G., Willett, P. and Glen, R.C. (1995) Molecular recognition of receptor sites using a genetic algorithm with a description of desolvation. *J. Mol. Biol.*, **245**, 43-53; Jones, G., Willett, P., Glen, R.C., Leach, A.R. and Taylor, R. (1997) Development and validation of a genetic algorithm for flexible docking. *J. Mol. Biol.*, **267**, 727-748.

43. Harris, K.D.M. and Tremayne, M. (1996) Crystal structure determination from powder diffraction data. *Chem. Mater.*, **8**, 2554-2570; Shankland, K., David, W.I.F., Csoka, T. and McBride, L. (1998) Structure solution from powder diffraction data by the application of a genetic algorithm combined with prior conformational analysis. *Int. J. Pharmaceutics*, **165**, 117-126.

COMPUTATIONAL APPROACHES TO CRYSTAL STRUCTURE AND POLYMORPH PREDICTION

FRANK J.J. LEUSEN[1], STEFFEN WILKE[1], PAUL VERWER[2]
AND GERHARD E. ENGEL[1]
[1]*Molecular Simulations Ltd., 230/250 The Quorum,
Barnwell Road, Cambridge, CB5 8RE, United Kingdom*
[2]*CAOS/CAMM Center, University of Nijmegen,
P.O. Box 9010, 6500 GL Nijmegen, The Netherlands*

Abstract

Detailed knowledge of crystal structure at the atomic level is a prerequisite for rational control of crystallisation processes, polymorphism and solid state properties. Recent advances in computer hardware and software have enabled the prediction of crystal structure and polymorphism for simple organic compounds. Here, we present an overview of crystal structure prediction, and we discuss a number of pitfalls frequently encountered. A concise review of various approaches is given, followed by a description of recent improvements in polymorph prediction methodology developed at Molecular Simulations.

1. Introduction

Organic compounds are often crystallised during their production process. Many pharmaceuticals and agrochemicals are formulated in their solid state, as are nearly all pigments, explosives and non-linear optical materials. For these and other materials, key properties depend on the specific packing of the molecules in the solid state. Examples include the density of explosives, the colour of pigments, the vapour pressure of agrochemicals and the hygroscopic properties of pharmaceuticals. Another key property is crystal shape or morphology, which is also determined - to a large extent - by the internal

303

J.A.K. Howard et al. (eds.),
Implications of Molecular and Materials Structure for New Technologies, 303–314.
© 1999 *Kluwer Academic Publishers. Printed in the Netherlands.*

crystal structure. Rational control or design of solid state properties requires a full understanding of the relationship between structure and property, for which detailed knowledge of crystal structure is an essential starting point [1]. Single crystal X-ray diffraction is the ideal method for solving crystal structures, but it is impossible to grow single crystals of sufficient size and quality for many organic compounds. To complicate matters, compounds may exist in more than one form because the crystallising molecules may find a number of stable arrangements in which to pack together in the solid state. This phenomenon is called polymorphism. Different polymorphic forms of a compound may have a different set of solid state properties. Which polymorph crystallises depends on crystallisation conditions such as concentration, solvent, impurities and temperature [2].

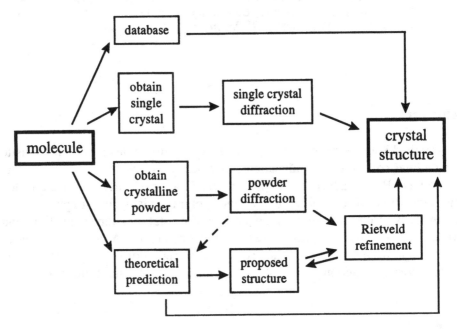

Figure 1. Routes to solve crystal structures

Various approaches to obtain structural information at the atomic level for molecular crystals are shown in Figure 1. Apart from a straightforward database search, computational methods are used routinely to assist the determination of structures from single crystal data [3]. In those cases where single crystals of sufficient quality cannot be grown, computational approaches may be used to solve crystal structures from powder diffraction patterns (which contain much less structural information than single crystal data) [4, 5]. These methods are being applied to increasingly complicated compounds, but rely on high quality powder diffraction data which can be indexed to reveal unit cell and symmetry information of the crystal. If the quality of the powder diffraction pattern is too low to allow indexing, a first-principles approach to crystal structure prediction is required. Even in cases where no powder patterns are available, a first-principles approach may be used to predict possible polymorphs of a given compound. In recent years several such methods have been developed, and some successes have been reported on the polymorph prediction of relatively rigid and non-ionic compounds [6 - 14]. However, accurate and reliable prediction of polymorphic forms requires significant computational resources and expertise, even for these simple compounds, and is thus not yet widely used as a routine tool. In this contribution we review the current state-of-the-art in this field, and outline recent and near future developments that will ultimately lead to a routine tool for polymorph prediction of organic compounds.

2. Theory and Methodology

From a mathematical point of view, polymorph prediction represents a search for the low-lying minima of a high-dimensional potential energy surface representing all possible packing arrangements of molecules in a crystalline environment. The sheer size of the parameter space - three degrees of freedom for each atom in the unit cell and the six parameters describing the shape of the cell - renders polymorph prediction a difficult task. Further complications arise from the small energy differences between various polymorphic forms when compared to the strength of inter-molecular and inter-atomic interactions. Predictions have recently become feasible due to advances in computer technology and the development of more accurate potential energy descriptions. Over the last decade, several computational approaches to predict crystal packing have emerged. However different in detail, all approaches have to address four main points:

- Definition of a reliable and accurate potential energy function
- Reduction of the number of parameters using experimental information and/or chemical intuition
- Location of the low-lying energy minima in the reduced parameter space
- Analysis of the results and verification of the resulting polymorphs against experimental data and/or independent theoretical calculations

In the following each of these points is discussed.

2.1. POTENTIAL ENERGY FUNCTIONS

Nearly all polymorph prediction methods apply molecular mechanics, or force field technology, to calculate the potential lattice energy of crystals. Although an *ab initio* quantum-mechanical approach may be more accurate, in particular for ionic systems, the computational effort involved would be too demanding even for modern computers. A force field is an approximate description of the real lattice energy hyper-surface, and its accuracy depends on many factors, such as the size and composition of the training set used to derive parameters for the force field potentials, and the treatment of electrostatics (see, for instance, refs. [6, 13 - 17]). Bearing in mind that reliable polymorph prediction requires an accuracy better than one kcal/mol (which is the order of magnitude of experimental energy differences between polymorphs), it is apparent that a force field has to be thoroughly tested before being used in a first-principles polymorph prediction procedure. Such a test may be performed against experimentally available crystal structures of similar compounds or against results of high-quality quantum mechanical calculations for a subset of important conformations and/or bonding arrangements of the molecule studied.

In general terms, it is intrinsic to the molecular mechanics approach that the more universal a force field becomes, the less accurate it tends to be. Since polymorph prediction involves the investment of considerable computational resources to evaluate the potential lattice energy of one particular molecule in thousands of crystal packing arrangements, a logical future step would be to derive, specifically for each molecule or class of molecules studied, a non-transferable force field with optimal accuracy [15].

To circumvent the force field accuracy problem, it has been suggested to use so-called scoring functions to evaluate the relative stability of predicted polymorphs [11]. These functions are derived statistically from a set of experimental structures taken from the Cambridge Structural Database (CSD) [18], and can be used to estimate the stability of a certain lattice from the probability that its key structural features are present in other observed crystal structures. This approach has the advantage that the scoring functions measure the frequency of certain structural features in the crystalline environment, whereas traditional force fields are usually derived from molecules in the gas phase. The use of scoring functions, on the other hand, may suppress the chance to predict polymorphs with uncommon or novel structural features. Future software development may include the use of more complex scoring functions that combine a molecular mechanics potential energy function with experimental information (powder pattern; statistical data from the CSD).

Each of the approaches mentioned above have their merits and drawbacks. Other complicating factors, such as the kinetic and thermodynamic stability of polymorphs, the effects of solvent and impurities on crystallisation, and the calculation of atomic charges, will not be discussed here.

2.2. REDUCING COMPLEXITY

The complexity of polymorph prediction grows exponentially with the size of the parameter space. Reduction of the number of degrees of freedom is essential to make first-principles polymorph prediction feasible. There are a number of ways to achieve this. The most common are:

- Constraining the flexibility of the molecules, *i.e.*, rigid body approaches
- Searching subspaces of the phase space in parallel; such subspaces can be defined according to space group symmetry, molecular conformation and/or number of molecules in the unit cell or asymmetric unit
- Exploiting known experimental information, e.g., previously determined crystal structures of similar compounds, powder diffraction data, solid state NMR, and IR spectra

Below, we comment briefly on these ways to reduce complexity.

Molecular flexibility poses a major problem in polymorph prediction. Ideally, the search for low-energy conformations is performed at the same time as the polymorph search, i.e., in the solid state, and all degrees of freedom of the molecule are included. In practice, the flexibility problem is often dealt with first, by performing a full conformational search in the *gas phase*. Each low-energy conformer is then used as input for a separate full polymorph search in which the conformer is treated as a rigid body. This results in a linear relationship between computational effort required and the number of low-energy conformations identified. A complication arises in flexible molecules that are capable of intra-molecular hydrogen bonding, because gas phase conformational analysis will overestimate the importance of intra-molecular hydrogen bonds as compared to the solid state where inter-molecular hydrogen bonds are preferred. A possible solution is the inclusion of a solvent environment, either explicitly or implicitly, during conformational analysis. If experimental IR spectra are available, they may be used to determine whether the molecule has an intra-molecular hydrogen bond in the solid state. In any case, the results of a molecular mechanics conformational analysis should be checked against independent theoretical (other force fields, quantum mechanical calculations) and/or experimental information (solid state conformations of similar molecules, or even the same molecule in a different polymorphic or solvated form) [6].

Although the search for polymorphs can be performed using P1 symmetry and a large number of molecules in the unit cell, the size of the parameter set in such an approach is prohibitively large. In addition, the final result will most likely have certain symmetry elements and only a small number of molecules per asymmetric unit. Indeed, 78% of molecular crystals reported in the CSD are in only 5 out of 230 space groups, and only about 8% of structures have more than one molecule in the asymmetric unit [15, 19, 20]. The occurrence of multiple molecules in the asymmetric unit may be investigated by solid state NMR. The search space can be further reduced for chiral compounds, which crystallise either as optically pure enantiomers (78% in space groups $P2_12_12_1$ and $P2_1$) or as a racemate (usually in space groups $P2_1/c$, P-1 or $C2/c$). For any compound other than the most simple and symmetrical examples such as benzene, the use of space group symmetry and a limited number of molecules in the asymmetric unit is an important step towards the reduction of the number of degrees of freedom in crystal structure prediction.

At this point it should be mentioned that, although rigid body packing calculations under space group symmetry (ideally followed by a lattice energy minimisation with respect to all degrees of freedom) is the current standard, all possible care should be taken to ensure that the search space is not too restricted regarding molecular conformation, space group and number of molecules per asymmetric unit. Restricting a search too much may result in missing the most important polymorph(s).

2.3. LOCATING MINIMA

Distinctions between the various approaches to crystal structure prediction can be made according to the algorithm which is applied to construct three-dimensional crystal structures from molecules, and according to the procedure used to sample the phase space. There are three main approaches to construct the crystal packing:

– Generating clusters of molecules first, which are then used to construct three-dimensional packing possibilities for crystals
– Starting from one-dimensional periodic arrangements and subsequently increasing the dimensionality of packing
– Generating a three-dimensional packing directly

The idea behind the first approach is that strong, relatively short-range interactions between molecules, such as hydrogen bonding, often determine the final crystal packing [21]. If this assumption holds, an effective way to arrive at the final crystal structure is to first find low-energy structural units containing only a small number of molecules and then to derive the three-dimensional crystal from those stable units. A critical point in these approximations is the consequences of boundary effects on the stability of the structural building blocks. Structures which are stable as clusters may become unstable as infinite crystals and *vice-versa*. Examples of methods along these lines are Promet3 [8], FlexCryst [11], and an early study by Williams [22].

Programs implementing the second approach first construct stable one-dimensional structures which then form the building blocks for two-dimensional arrangements which in turn may be used to build up a crystal. An example is the method of Perlstein [23].

The last category of programs is the largest; these methods apply three-dimensional lattice symmetry at all stages. Examples are MPA (Molecular Packing Analysis) [9], CRYSCA (Crystal Structure Calculation) [10], UPACK (Utrecht crystal PACKer) [24], ICE9 [25], MDCP (Molecular Dynamics for Crystal Packing) [26, 27], MOLPAK (MOLecular PAcKing) [28] and the Polymorph Predictor of Molecular Simulations Inc. (MSI) [7, 29].

Apart from the different approaches used to construct crystals from molecules, the various programs can also be distinguished by the methodology applied to locate low-energy minima of the lattice energy hyper-surface. Many programs follow a similar approach: sampling of the phase space, removal of similar structures and finally, energy minimisation. Each of those steps is discussed below.

Basically, three methods are used to sample the phase space. The programs Promet3 [8], FlexCryst [11], UPACK [24], ICE9 [25] and MolPak [28] scan the phase space in a systematic way. Although accurate in principle, systematic searches are practical only if the search can be subdivided into problems with a small number of degrees of freedom. Another frequently applied method is random searching, either using uniform sampling (MPA [9], CRYSCA [10]) or a Monte Carlo simulated annealing approach (MSI's Polymorph Predictor [7, 29]). Random sampling of the phase space, in particular the application of a Monte Carlo technique, is effective and applicable to problems with a larger number of degrees of freedom. Finally, the program MDCP [26, 27] uses crystal dynamics to generate stable structures. Although crystal dynamics represents an elegant approach to simulate transformations between different types of crystal packing, the time scale at which such transformations take place is too long to be simulated with molecular dynamics.

The efficiency of the search may be increased significantly by the removal of similar structures after the initial sampling; the compute intensive task of lattice energy minimisation is then applied to unique structures only. Such methods are often referred to as "clustering" [30]. An optimal clustering algorithm reduces the total set of initial crystal structures to a manageable number of possible structures, without the loss of any distinct crystal structures (as otherwise important polymorphs might be removed from the set of generated crystal structures). Clustering is used by Promet3 [8], FlexCryst [11], UPACK [24, 30] and the MSI Polymorph Predictor [7, 29].

With the exception of FlexCryst [11], the last step in each of the approaches discussed here is lattice energy minimisation of the generated crystal structures. Lattice energy minimisation is a standard procedure and will not be discussed here. It is important to note, however, that minimisation should be performed

310

for all degrees of freedom in the crystal (including molecular flexibility) to allow changes in molecular conformation due to crystal packing effects [6]. Although crystal packing effects on molecular conformation may be small, the effect on lattice energy may well be larger than the energy difference between polymorphs. Rigid body minimisation may therefore be too inaccurate to allow reliable polymorph prediction, in particular for flexible molecules. Minimisation with respect to all degrees of freedom is performed by UPACK [24] and MSI's Polymorph Predictor [7, 29]; all other approaches use rigid body energy minimisation.

2.4. POST-PROCESSING

The result of a polymorph prediction procedure is a number (often several hundred) of possible polymorphs with low energy. Due to force field limitations, thermodynamics, kinetics and restrictions in the phase space, the lowest-energy structures do not necessarily correspond to experimentally found polymorphs, in particular for more complicated compounds. For a high quality prediction, however, the real polymorphs can usually be found among the set of predicted low-energy structures. In such cases, further analysis and evaluation is needed to identify the most promising candidates for observable polymorphs. Such an analysis may include a comparison to available experimental data (e.g., to experimental powder diffraction data). Furthermore, a different force field may be applied to re-evaluate the energetics of predicted polymorphs (i.e., to obtain a "second opinion"). Accurate quantum mechanical calculations on a small selection of proposed structures may also be used to verify the force field accuracy and the stability ranking.

3. Recent Developments

In this section, we report a number of recent improvements to the MSI Polymorph Predictor. These enhancements have brought us closer to the goal of predicting crystal structures of large flexible molecules or systems with more than one molecule in the asymmetric unit. The method consists of four main parts:

– Generation of possible structures using a modified simulated annealing Monte Carlo approach
– Clustering of the Monte Carlo structures to remove similar crystals
– Lattice energy minimisation
– Analysis of results

With the exception of clustering, each of these steps has been scrutinised in the past year. Major speedups have been achieved for both the Monte Carlo search and the lattice energy minimisation steps by the implementation of a variety of algorithmic improvements as outlined below. Moreover, the tedious step of screening resulting structures by comparing simulated with experimental powder patterns has been largely automated. The program now provides a quantitative measure to judge the "closeness" between powder spectra and automatically calculates such a measure for all the structures produced by the polymorph prediction procedure.

3.1. SPEEDUPS TO THE MONTE CARLO SEARCH

During each step in the Monte Carlo search, cell parameters and orientation of the molecules in the unit cell are varied. To ensure reasonably dense packing arrangements of the molecules, this variation is not performed strictly at random. Instead, the program, after an initial random variation of the parameters, employs an "expand – contract – shake" sequence, whereby the parameters are modified gradually until the van der Waals surfaces of some molecules in the resulting structure touch.

This sequence requires a large number of close contacts evaluations to check for the occurrence of molecular overlap. Simply calculating all inter-atomic distances in the unit cell to check for close contacts becomes very time consuming for large structures, as the computational effort scales quadratically with the number of atoms. By implementing a grid based technique, which only calculates inter-atomic distances for atoms within the same or a neighbouring grid cell, and furthermore employing space group symmetry, a speedup of about 2 orders of magnitude was achieved (for a typical example with 80 atoms in the asymmetric unit in space group $P2_1/c$). The computational effort now scales linearly with the size of the system, so speedups are more pronounced for larger system sizes.

3.2. SPEEDUPS TO THE LATTICE ENERGY MINIMISATION

Another time consuming step in a polymorph prediction sequence is the lattice energy minimisation of thousands of crystal structures produced by the Monte Carlo search. By utilising space group symmetry during the force field evaluations of energies and forces, speedup factors of the order of the number of symmetry operators for a given space group were achieved. Moreover, in addition to a full-body minimisation, whereby the co-ordinates of each atom in the asymmetric unit are optimised individually, a rigid body minimisation with user-definable rigid bodies has now been implemented to allow the definition of flexible and rigid parts within a molecule. This results in a considerable performance gain, since the number of minimisation steps required is roughly proportional to the number of degrees of freedom in the system. For a full-body minimisation, the number of degrees of freedom is 3(Number of atoms+2), whereas for a rigid body minimisation, it is 6(Number of rigid bodies+1).

When employing rigid body minimisation, care must be taken to ensure that those parts of the molecule which are combined in a rigid body are chemically sufficiently rigid, i.e., they would not significantly relax even during full-body minimisation. In cases where the accuracy of the rigid body minimisation may not be sufficient, a combination of initial rigid body and subsequent full-body minimisation can still result in a significant speedup.

3.3. AUTOMATIC COMPARISON OF POWDER PATTERNS

The task of manually identifying among a large number of predicted structures the one whose simulated powder pattern most closely matches a given experimental spectrum is often time consuming and prone to misjudgements. Automation of this process is desirable, but requires the availability of a reliable quantitative measure describing the "closeness" between two spectra. Standard measures based on least-squares differences between the spectra are unreliable due to the sharply peaked nature of typical powder spectra, and the great sensitivity of powder spectra to small changes in lattice parameters makes the task even more problematic.

Nonetheless, it was found that improved measures, such as variations of the FOLD measure [31] and a novel Continuous Measure for the Automatic Comparison of Spectra (CMACS) [32], can provide a sufficient degree of accuracy to reliably pick the most closely matching structure out of a large set of structures. These approaches have now been implemented in the Polymorph Predictor, and the program can automatically calculate the match between an experimental powder diffraction pattern and a set of predicted crystal structures.

4. Concluding Remarks

The field of crystal structure prediction via computer simulation has changed rapidly in the last decade due to developments in computer hardware as well as novel simulation approaches. It has advanced from generating trial packings for simple, rigid molecules using experimental unit cell parameters in the 1960's, to the prediction of a set of possible low-energy polymorphs for moderately flexible compounds, based on just the structural formula.

Computer simulation to predict polymorphs extends beyond experimental methods in that it can solve crystal structures in cases where no crystals of sufficient quality can be obtained, and in its ability to produce a whole range of low-energy structures rather than just the structure of the single polymorph that happens to be grown. Thus, it can point to the existence of other crystal packing alternatives which may have specific desired or unwanted properties.

An ideal polymorph prediction approach employs a combination of computational methods including quantum mechanics, molecular mechanics, Monte Carlo methods and clustering techniques, and possibly experimental techniques like IR, X-ray diffraction, solid state NMR and thermodynamic data. Current limits are set in part by the available computational resources (restricting, *e.g.*, the set of space groups to be searched, and the size and flexibility of the molecular system), and in part by the current simulation techniques, that have difficulties handling ionic or strongly polarised systems, or systems containing elements that are not well described by the available force fields.

Future developments will lead to an integrated method that combines experimental knowledge with molecular mechanics based approaches to extend the scope of polymorph prediction to increasingly complex systems. Such a tool may then be used routinely to predict crystal structures for a variety of organic compounds.

Acknowledgements

Many scientists at MSI have contributed to the enhancements described in section 3, in particular N. Austin, P. Bennett, J. Dillen, D. Fincham, O. König, S. Miller, M. Pinches, S. Robertson, J. Turvey and G. Weston.

PV acknowledges support from the Netherlands Foundation for Chemical Research (SON) and the Netherlands Organisation for Scientific Research (NWO), in the framework of the Computational Materials Science Crystallisation project.

314

References

1. G.R. Desiraju (1989) Crystal Engineering: The Design of Organic Solids, Elsevier, Amsterdam.
2. J.D. Dunitz and J. Bernstein (1995) *Acc. Chem. Res.*, **28**, 193.
3. G.M. Sheldrick (1997) SHELXS97. Program for the Solution of Crystal Structures. University of Goettingen, Germany.
4. K.D.M. Harris, M. Tremayne, P. Lightfoot and P.G. Bruce (1994) *J. Am. Chem. Soc.*, **116**, 3543.
5. K. Shankland, W.I.F. David and T. Csoka (1997) *Z. Krist.*, **212**, 550
6. H.R. Karfunkel, Z.J. Wu, A. Burkhard, G. Rihs, D. Sinnreich, H.M. Bürger and J. Stanek (1996) *Acta Cryst.*, **B52**, 555.
7. F.J.J. Leusen (1996) *J. Crystal Growth*, **166**, 900
8. 8. A. Gavezzotti (1996) *Acta Cryst.*, **B52**, 201.
9. D.E. Williams (1996) *Acta Cryst.*, **A52**, 326.
10. M.U. Schmidt and U. Englert (1996) *J. Chem. Soc., Dalton Trans.*, 2077.
11. D.W.M. Hofmann and T. Lengauer (1997) *Acta Cryst.*, **A53**, 225.
12. G.R. Desiraju (1997) *Science*, **278**, 404.
13. R.S. Payne, R.J. Roberts, R.C. Rowe and R. Docherty (1998) *J. Comput. Chem.*, **19**, 1.
14. W.T.M. Mooij, B.P. van Eijck, S.L. Price, P. Verwer and J. Kroon (1998) *J. Comput. Chem.*, **19**, 459.
15. H.R. Karfunkel and F.J.J. Leusen (1992) *Speedup*, **6**, 43
16. H.R. Karfunkel, F.J.J. Leusen and R.J. Gdanitz (1993) *J. Comp.-Aided Mat. Design*, **1**, 177.
17. F.J.J. Leusen, J.H. Noordik and H.R. Karfunkel (1993) *Tetrahedron*, **49**, 5377.
18. F.H. Allen and O. Kennard (1993) *Chem. Design Autom. News*, **8**, 31.
19. W.H. Baur and D. Kassner (1992) *Acta Cryst.*, **B48**, 356.
20. N. Padmaja, S. Ramakumar and M.A. Viswamitra (1990) *Acta Cryst.*, **A46**, 725.
21. M.C. Etter (1990) *Acc. Chem. Res.*, **23**, 120.
22. D.E. Williams (1980) *Acta Cryst.*, **A36**, 715.
23. J. Perlstein, K. Steppe, S. Vaday and E.M.N. Ndip (1996) *J. Am. Chem. Soc.*, **118**, 8433.
24. 24 B.P. van Eijck, W.T.M. Mooij and J. Kroon (1995) *Acta Cryst.*, **B51**, 99.
25. A.M. Chaka, R. Zaniewski, W. Youngs, C. Tessier and G. Klopman (1996) *Acta Cryst.*, **B52**, 165
26. N. Tajima, T. Tanaka, T. Arikawa, T. Sukarai, S. Teramae and T. Hirano (1995) *Bull. Chem. Soc. Jpn.*, **68**, 519.
27. T. Arikawa, N. Tajima, S. Tsuzuki, K. Tanabe and T. Hirano (1995) *J. Mol. Struct. (THEOCHEM)*, **339**, 115.
28. J.R. Holden, Z. Du and H. Ammon (1993) *J. Comput. Chem.*, **14**, 422.
29. Cerius^2 molecular modelling environment, Molecular Simulations Inc., 9685 Scranton Road, San Diego, CA 92121-3752, USA.
30. B.P. van Eijck and J. Kroon (1997) *J. Comput. Chem.*, **18**, 1036.
31. H.R. Karfunkel, B. Rohde, F.J.J. Leusen, R.J. Gdanitz and G. Rihs (1993) *J. Comput. Chem.*, **14**, 1125.
32. G.E. Engel, unpublished results.

DEVELOPING METHODS OF CRYSTAL STRUCTURE AND POLYMORPH PREDICTION

S.L. PRICE
Centre for Theoretical and Computational Chemistry
University College London
20 Gordon Street
London WC1H 0AJ

1. The Varying Aims of Crystal Structure Prediction

The grand goal of being able to predict which crystal structure an organic molecule will adopt when crystallized from different solvents, or varying conditions of temperature and pressure, remains an exciting challenge. However, we are at least discussing the possibility [1] following the recent development of several different methods of searching for possible crystal structures. These methods have shown considerable ability to find the known crystal structure as a favourable structure. This is proving useful for applications such as finding possible crystal structures for starting points [2] for the refinement of crystal structure from powder diffraction data, or rationalizing known, sometimes unusual structures [3]. In this short lecture, I would like to briefly review some of the questions about our understanding of molecular crystal structures, and corresponding assumptions in our computer modelling, that are highlighted by these theoretical studies.

The methods that I am considering are those which predict the complete crystal structure, that is, the cell dimensions, space group and positions and orientations of the molecules within the cell, from just a model for the molecular structure. The aim is a method that could predict the crystal structure and its properties, prior even to the synthesis of the molecule. This would also predict whether the molecule would be polymorphic and so would adopt different crystal forms under different conditions.

2. The Current Assumption

The main assumption behind the computational methods of crystal structure prediction is that the experimental structure will correspond to the global minimum in the static lattice energy. Any competitive local minima are therefore regarded as possible polymorphs. This is just a crude, static (0K) model of the thermodynamics of crystal formation. However, it is a good first criterion, as a structure whose lattice energy is significantly higher than the global minimum certainly will not be observed.

J.A.K. Howard et al. (eds.),
Implications of Molecular and Materials Structure for New Technologies, 315–320.
© 1999 *Kluwer Academic Publishers. Printed in the Netherlands.*

3. The Requirements for Finding the Lowest Energy Crystal Structure

3.1. MOLECULAR MODEL

Ab initio or molecular mechanics calculations are now routinely used to predict the structures of organic molecules, and so there is no difficulty in obtaining a gas phase structure for the molecule to use in its crystal structure prediction. This structure is usually assumed to be rigid. If the molecule cannot be assumed to be rigid, as it is likely to change its conformation from the gas phase structure on crystallization to optimize the intermolecular interactions in the crystal, then the quantity to be minimized is the balance between the intermolecular lattice energy and the intramolecular conformational energy. This is very demanding of the relative accuracy of the inter and intra-molecular force-fields, as well as increasing the dimensionality of the search problem.

Even when the molecule is not obviously flexible, the question arises as to how sensitive the predictions are to the assumed rigid molecular structure. Although the molecular structure found in the experimental crystal structure is often used (with the positions of the hydrogen atoms corrected for the deficiencies in their X-ray determination), this may be biasing the prediction process in favour of the experimental structure. Indeed, since the molecular structure in different polymorphs is never exactly the same, even when the differences are not significant relative to probable experimental error, it is probably more even-handed to use an accurate gas phase structure.

3.2. SIMULATION METHODS

Most methods of crystal structure prediction use static lattice energy minimization, and are therefore predicting the 0K structure. This usually has to be compared with the room temperature structure. We know that the thermal expansion of molecular crystals is very variable [4], and indeed can be so anisotropic that one cell length actually contracts with rising temperature. Although some average thermal expansion effects are implicitly included in model potentials that are empirically fitted to the crystal data, this absorbed effect cannot be accurately transferable. Hence, a discrepancy of a few percent in the lattice parameters between the predicted and observed crystal structure can be crudely attributed to the neglect of thermal effects.

Temperature or dynamical effects are often exploited in the search for the lattice minima, as in the Polymorph Predictor methodology [5], or the heuristic Molecular Dynamics search [6]. However, these are very different from a proper Molecular Dynamics simulation of a finite temperature molecular crystal structure [7].

However, without accurate treatment of temperature effects, we need to know how similar two static crystal structures have to be in order to be considered as being the same structure. Searches for hypothetical structures will often find many minima that are so similar in energy, cell dimensions and packing motif that they would lie on the trajectory of the vibrating, librating molecule in any dynamical model. However, in developing clustering algorithms to pick out the distinct crystal structures from the multitude of minima found, we do not want to miss any structures that would be considered experimentally as distinct polymorphs. Some pairs of polymorphic structures in the Cambridge Structural Database, such as terephthalic acid and indigo, look remarkably similar.

3.3. THE MODEL FOR THE FORCES BETWEEN THE MOLECULES

In order to evaluate the lattice energy of a molecular crystal, it is necessary to assume a model for the forces between the molecules. This model potential must give a minimum in the lattice energy reasonably close to the experimentally known structure, as the best that the search method can do is to find that minimum as the global minimum and therefore the predicted structure. We also want the relative energies of the low energy local minima to be predicted as accurately as possible by the potential. This implies that the potential must be equally accurate in the hypothetical structures as in the known crystal structures of the molecule. We are inevitably limited by the accuracy of the potentials available for organic molecules, but this limitation can be minimized by using either a model potential which is on as firm a theoretical footing as possible, or one which has been tested for its ability to reproduce a wide range of crystal structures of related molecules covering a wide range of plausible relative orientations of the functional groups involved.

The three polymorphic structures of 2-amino-5-nitropyrimidine illustrate [8] the limiting effect of these three assumptions on polymorph prediction. The flat hydrogen-bonded sheet structure of the planar molecule was beautifully reproduced using a model potential [9] comprised of a distributed multipole analysis (DMA) of an MP2 6-31G** wavefunction, along with our usual empirical repulsion-dispersion terms, with negligible differences between using the SCF 6-31G** optimized isolated molecule structure or the molecular structure from the crystal structure of that polymorph. Another ribbon-motif polymorph was also beautifully reproduced by the potential using the molecular structure observed in that polymorph, which had the nitro group rotated by 16°. This slight twist allows the interdigitation of the nitro groups to be stabilized by a C-H...O interaction. The twist is essential to the stability of the polymorph, as when it was modelled by the gas phase molecular structure, the crystal structure minimized to a high symmetry, low density form. The densest polymorph, in which the same hydrogen bonded sheets as the first polymorph were buckled, with a tilt of 33°, gave the most disturbing results. Using the experimental molecular structure (which was essentially the same as in the first polymorph and from the *ab initio* optimization), the structure unbuckled considerably on minimization, and when the *ab initio* structure was used, the sheets unbuckled and slide to the minimum corresponded to that of the first polymorph! Thus the combination of errors in the molecular structure, model potential and use of static minimization meant that the computer modelling could not distinguish between the two sheet structure polymorphs, and gave an unphysical minimum energy structure for the third polymorph. Hence, we could only hope to find the first polymorph by a search for possible crystal structures as dense minima in the lattice energy. This was effectively found. It is salutary to remember that this computer modelling search would have been deemed extremely successful if the other two polymorphs had not been known!

3.4. THE SEARCH METHOD

The search through the vast arrays of possible crystal structures, even when the number of molecules in the unit cell and the space group is specified, is a major task, and it is the advent of these searching methods that has produced the recent interest in predicting crystal structures. A systematic, brute-force, grid search technique, considering all possible variables within a specified space group is possible, and the code which was originally applied to monosaccharides in $P2_12_12_1$ [10], has now been extended to triclinic, monoclinic and orthorhombic space groups [11]. Other, more random, methods rely on simulated annealing to search the multi-dimensional space, as has been described in a previous chapter.

There are other, less computationally demanding methods, which exploit crystallographic experience, to suggest hypothetical structures as starting points for energy minimization. Gavezzotti's method, PROMET [12], seeks promising starting nuclei for likely crystal structures by examining the low energy nuclei formed by small clusters (usually dimers) of the molecules related by the common symmetry elements. This has proved very successful for a range of hydrocarbons, and other specific applications, often used in conjunction with experimental data, but it does depend on the crystallographic intuition of the user. A more automated approach is adopted by MOLPAK [13], which seeks the most dense hypothetical structures within twenty of the commonest coordination types covering the nine space groups most frequently found for organic molecules. The program ICE9 [14] also uses the observation that molecular crystal structures have a high packing coefficient to search for possible hypothetical structures in the 13 space groups that allow close packing of organic molecules.

All these methods have reported promising success in finding the known crystal structure amongst the low energy structures predicted by the search. There have been insufficient publications yet to judge the relative efficiency of the methods. Indeed, since most methods allow the search to be expanded, it may be fairly meaningless to ask which is best. It will be very dependent on the type of molecule - its shape, symmetry and type of intermolecular contacts, as well as whether it adopts a statistically unusual structure.

4. Lessons from Recent Results

Perhaps the most interesting general result to emerge from the various attempts to predict crystal structures is the observation that it is quite common for there to be more low energy minima found than known polymorphs. Gavezzotti's work on hydrocarbons confirms that it is generally possible to construct a large number of crystal structures for a given molecule whose packing energies differ by less than 10% [12]. Around a thousand possible, within 10 kcal/mol of the global minimum, were found for a monosaccharide within one space group [10]. The lack of directionality in the intermolecular bonding, or the molecular flexibility partially accounts for this diversity, and indeed, actually enumerating the number of different structures is problematic because of the difficulties in defining when two structures are effectively the same. However, there are also plenty of examples where structures differing in their hydrogen bonding motif have the same minimized lattice energy, to within a kcal/mol, and therefore are energetically feasible polymorphs. Two separate studies on acetic acid [11, 15], using a range of search methods and model potentials, found many different low energy chain and dimer crystal structures within a small energy range which included the experimental chain structure. Similarly, two different searches [16 & Discussion] for the low energy crystal structures of pyridone, with different potentials, found that the amide dimer motif could pack with a very similar lattice energy to the known catemer structure. Our own work on aromatic molecules with multiple, distinct hydrogen bond donors and acceptors, such as uracil, 6-azauracil and allopurinol [17], found that many possible combinations of donor and acceptor could pack in a crystal structure with a lattice energy that was within a few kcal/mol of the experimental global minimum structure. Thus, even for functional groups which have markedly preferred packing motifs, that can be used for crystal engineering [18, 19], the motifs will often only determine certain aspects of the structure, for example, the formation of chains, but these may pack in a variety of ways with a comparable lattice energy.

Thus, the crystal structure prediction studies are predicting a large propensity for polymorphism, according to the simple thermodynamic model used in the search criterion. However, it would be more accurate to say that the crystal structure prediction methods have shown that the number of crystal structures which are possible on this crude thermodynamic criterion alone, is considerably greater than the number of polymorphs in the Cambridge Structural Database for many organic compounds. How do we interpret these hypothetical crystal structures? Can we refine our theoretical modelling of the molecular crystalline state to predict which ones could be found experimentally?

This will require considerable input from careful experimental studies in collaboration with the theoretical predictions. Firstly, the number of known polymorphs might well increase when experimentalists were looking for them, particularly if only microcrystalline samples were needed. Secondly, there will be kinetic reasons why some structures are unlikely to be found, when other forms can grow more quickly. Indeed some of the hypothetical structures may not nucleate and therefore cannot be formed. There is a clear need to input some kinetic modelling into the computational process. Some pioneering work towards this end has been the molecular dynamics simulations of pyridone [16] and tetrolic acid [20] in CCl_4 solution. These showed that the solute dimers did occasionally break one hydrogen bond, which would be necessary to nucleate the chain catemer motif that is found in the crystal. However, this and the relative lattice energies cannot explain why tetrolic acid has both a dimer and a catemer based polymorph, but only the catemer structure has been observed for pyridone.

Thus the development of a reliable polymorph prediction scheme will both require and help to develop our understanding of crystallization and polymorphism, and be very demanding of our modelling of the intermolecular and intramolecular forces and dynamical effects within crystal structures.

5. References

1. A. Gavezzotti, (1994) Are Crystal-Structures Predictable, *Accounts of Chemical Research*, **27**, 309-314.

2. A. Gavezzotti and G. Filippini, (1996) Computer-Prediction of Organic-Crystal Structures Using Partial X-Ray-Diffraction Data, *Journal of the American Chemical Society*, 118, 7153-7157.

3. D. S. Coombes, G. K. Nagi and S. L. Price, (1997) On the lack of hydrogen bonds in the crystal structure of alloxan, *Chemical Physics Letters*, **265**, 532-537.

4. S. L. Price, (1997) Molecular Crystals, in C. R. A. Catlow, (ed.)*Computer Modelling in Inorganic Crystallography*. San Diego: Academic Press, pp. 269-293.

5. H. R. Karfunkel and R. J. Gdanitz, (1992) Ab initio Prediction of Possible Crystal-Structures For General Organic-Molecules, *Journal of Computational Chemistry*, **13**, 1171-1183.

6. N. Tajima, T. Tanaka, T. Arikawa, T. Sakurai, S. Teramae and T. Hirano, (1995) A Heuristic Molecular-Dynamics Approach For the Prediction of a Molecular-Crystal Structure, *Bulletin of the Chemical Society of Japan*, **68**, 519-527.

7. B. P. van Eijck, L. M. J. Kroon-Batenburg and J. Kroon, (1997) Energy minimisation and Molecular Dynamics calculations for molecular crystals, in A. Gavezzotti, (ed.) *Theoretical Aspects and Computer Modeling of the Molecular Solid State*. Chichester: John Wiley & Sons, pp. 99-146.

8. C. B. Aakeroy, M. Nieuwenhuyzen and S. L. Price, (1998) The three polymorphs of 2-amino-5-nitropyrimidine: Experimental structures and theoretical predictions, *Journal of the American Chemical Society,* **120**, 8986-8993.

9. D. S. Coombes, S. L. Price, D. J. Willock and M. Leslie, (1996) Role of Electrostatic Interactions in Determining the Crystal- Structures of Polar Organic-Molecules - a Distributed Multipole Study, *Journal of Physical Chemistry*, **100**, 7352-7360.

10. B. P. van Eijck, W. T. M. Mooij and J. Kroon, (1995) Attempted Prediction of the Crystal-Structures of 6 Monosaccharides, *Acta Crystallographica Section B-Structural Science*, **51**, 99-103.

11. W. T. M. Mooij, B. P. van Eijck, S. L. Price, P. Verwer and J. Kroon, (1998) Crystal structure predictions for acetic acid, *Journal of Computational Chemistry*, **19**, 459-474.

12. A. Gavezzotti, (1991) Generation of Possible Crystal-Structures From the Molecular-Structure For Low-Polarity Organic-Compounds, *Journal of the American Chemical Society*, **113**, 4622-4629.

13. J. R. Holden, Z. Y. Du and H. L. Ammon, (1993) Prediction of Possible Crystal-Structures For C-Containing, H- Containing, N-Containing, O-Containing and F-Containing Organic- Compounds, *Journal of Computational Chemistry*, **14**, 422-437.

14. A. M. Chaka, R. Zaniewski, W. Youngs, C. Tessier and G. Klopman, (1996) Predicting the Crystal-Structure of Organic Molecular Materials, *Acta Crystallographica Section B-Structural Science*, **52**, 165-183.

15. R. S. Payne, R. J. Roberts, R. C. Rowe and R. Docherty, (1998) The generation of crystal structures of acetic acid and its halogenated analogues, *Journal of Computational Chemistry.*, **19**, 1-20.

16. A. Gavezzotti, (1997) Computer simulations of organic solids and their liquid-state precursors, *Faraday Discussions*, **106**, 63-78.

17. S. L. Price and K. S. Wibley, (1997) Predictions of crystal packings for uracil, 6-azauracil and allopurinol: The interplay between hydrogen bonding and close packing, *Journal of Physical Chemistry A,* **101**, 2198-2206.

18. C. B. Aakeroy, (1997) Crystal Engineering: Strategies and Architectures, *Acta Crystallographica Section B-Structural Science*, **53**, 569-586.

19. G. R. Desiraju, (1989) *Crystal Engineering: The Design of Organic Solids.* Amsterdam: Elsevier, Amsterdam.

20. A. Gavezzotti, G. Filippini, J. Kroon, B. P. van Eijck and P. Klewinghaus, (1997) The crystal polymorphism of tetrolic acid (CH_3C CCOOH): A molecular dynamics study of precursors in solution and a crystal structure generation, *Chemistry-A European Journal*, **3**, 893-899.

CURRENT CHALLENGES IN CRYSTAL ENGINEERING

GAUTAM R. DESIRAJU
School of Chemistry, University of Hyderabad
Hyderabad 500 046, India

Abstract

The subject of crystal engineering continues to attract the interest of organic and physical chemists, crystallographers and materials scientists. This chapter discusses some challenges in the determination, analysis, prediction and design of crystal structures of molecular organic solids with specific properties.

1. Introduction

Crystal engineering or the design of organic solids with specific physical and chemical properties continues to elicit intense interest [1,2]. At the outset, it should be stated that this subject now encompasses a wide variety of research activity ranging from the understanding of crystal packing in organic molecular solids to the design of open network structures based on metal-ligand coordinate bonds, the so-called coordination polymers [3]. On the one hand, one is able to analyse and understand with increasing certainty the vast amounts of accurate crystallographic data currently available from databases [4], while on the other, an appreciation of a phenomenon like polymorphism can lead to immediate benefits in applied areas such as pharmaceutical development [5]. The work of Schmidt on topochemistry [6], is considered by many to represent the formal beginnings of crystal engineering. The seventies could be considered something of a hiatus but interest widened in the eighties to more general studies of intermolecular interactions. With a better appreciation of these interactions, the identification of a crystal as a supermolecule in the nineties appears natural [7,8]. The term 'crystal engineering' is interpreted today in various ways and the scope of the present chapter is subjective.

J.A.K. Howard et al. (eds.),
Implications of Molecular and Materials Structure for New Technologies, 321–339.
© 1999 *Kluwer Academic Publishers. Printed in the Netherlands.*

2. Difficulties in the Prediction of Crystal Structure from Molecular Structure

More than fifty years ago, Pauling and Delbrück stated that in any kind of molecular recognition, and this includes the recognition between identical molecules, it is the dissimilar rather than the similar functionalities that come into closest contact [9]. Steric and electronic complementarity characterise the recognition events that are a prelude to any kind of supramolecular assembly. This fact in itself now appears to be obvious but its consequences in terms of crystal engineering strategies is considerable. Because the supramolecular behaviour of a particular functional group depends on the nature and location of other groups in the molecule, crystal structures of organic molecules need not necessarily be derived in a straightforward way from the functional groups present. Simple relationships between molecular and crystal structure, say as in the series naphthalene, anthracene, tetracene or, benzoic acid, terephthalic acid, trimesic acid are ideal but in practice, only obtained when *interference between orthogonal sets of interactions is minimal.* When and why such minimal interference occurs is still not evident. Accordingly, the crystal structures of many 'simple' organic compounds are hard to understand and predict [10,11].

Sometimes, however, the complementarity between distinct groups in molecular recognition is easily identifiable. Ermer [12] and Hanessian [13] showed that predictable structures can be obtained in systems containing equal stoichiometries of -NH_2 and -OH groups. This predictability arises from the 2:1 and 1:2 hydrogen bond donor:acceptor ratios in these functional groups, leading to tetrahedral configurations at both heteroatoms in the hydrogen bonded network and consequently to variants of the arsenic and wurtzite structures. In particular, 4-aminophenol, **1** (Fig. 1a) is identifiable as an archetype of this family. Yet, how predictable really are these crystal structures? We determined the structures of the isomeric 2- and 3-aminophenols, **2** and **3** with single crystal low temperature neutron diffraction and rather than the expected tetrahedral network, an unusual N-H$\cdots\pi$ hydrogen bond was observed (Figs. 1b,c) [14]. The reason for this unexpected structure is that the packing in all three aminophenols is dominated by the herringbone packing of aromatic rings (Fig. 2) rather than by O-H$\cdots$N and N-H$\cdots$O hydrogen bonding. In 4-aminophenol, the aromatic packing is able to co-exist with the hydrogen bonded wurtzite network, in other words, the interference between these two sets of interactions is minimal. In 2- and 3-aminophenols, however, the same packing of the aromatic rings is incompatible with the tetrahedral network that is replaced by the unconventional hydrogen bond arrangement.

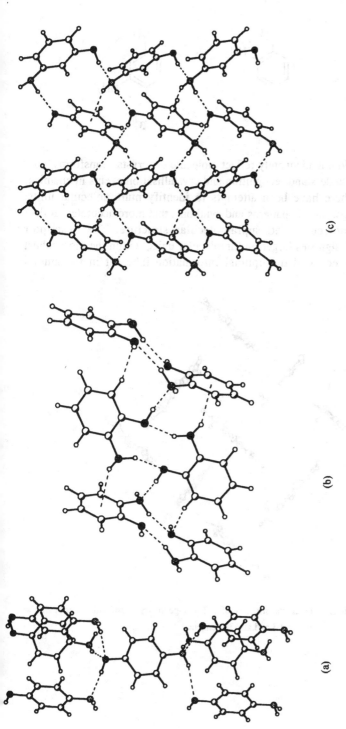

(a)

(b)

(c)

Figure 1. (a) Tetrahedral network formed by N-H···O and O-H···N hydrogen bonds in the crystal structure of **1**; (b) N-H···O, O-H···N, C-H···O and N-H···π interactions in the structure of **2**; (c) Crystal structure of **3**. Notice that synthon **4** and the herringbone arrangement of phenyl rings are common to the three structures.

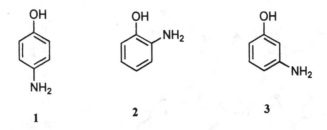

1   2   3

Since functional groups do not adequately reflect molecule → crystal transforms, how then does one attempt to understand recognition and crystallisation events in chemical terms? In this regard, there have been attempts to identify multimolecular units, variously termed couplings, motifs, patterns and synthons, that more accurately reflect the relation between molecular and supramolecular structure. The descriptor 'supramolecular synthon' signifies larger multimolecular units, within the most robust and useful of which is encoded the optimal information inherent in the mutual

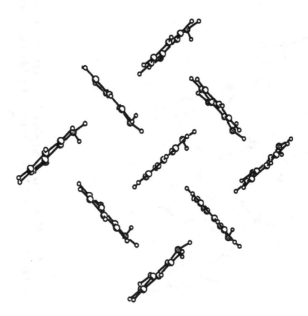

Figure 2. Herringbone interactions in the crystal structure of **2**. The T-geometry of aromatic rings is similar in **1** and **3** and also in benzene.

recognition of molecules to yield solid state supermolecules, that is crystals [15]. Inherent in the term 'synthon' are ideas of size economy and maximum information content. Very large structural units will naturally contain more information but these are accompanied by unnecessary detail, while very small units may lack sufficient and critical information regarding supramolecular structure. For example, the crystal structures of all the three isomeric aminophenols contain the O-H···N and N-H···O synthon **4**. Knowing that the *ortho* and *meta* isomers have a very different structure from the *para* compound, one may then conclude that synthon **4** is too small to discriminate between alternative possibilities. A better synthon is **5**, found in the crystal structures of the *ortho* and *meta* isomers but absent in the structure of the *para* isomer. In effect, the most useful supramolecular synthons combine form with compactness. We note further that, (as is true for synthons in molecules) the number of supramolecular synthons in a supermolecule is very large. However, only some of these are useful. Failure to realise this could lead to a degeneration of the usage of the term 'synthon' to represent complete molecules, an unfortunate development that actually occurred in molecular chemistry, leading even Corey to abandon the term in favour of the alternative 'retron' [16].

3. Weak Intermolecular Interactions and Crystal Packing. What is the Distinction between a Weak Hydrogen Bond and a van der Waals Interaction?

Conventional hydrogen bonding is the pivotal interaction in many crystal structures but a number of weaker and softer interactions have been shown to play a role also in structure stabilisation. This role can vary from unimportant to supportive to one that distorts or even dominates the conventional interactions [17]. Weak hydrogen bonding and polarisation-induced heteroatom interactions are examples of these secondary interactions whose exact nature has still not been fully elucidated [18]. As discussed above, interaction insulation is always desirable and one would like to study the weaker interactions from this viewpoint, even if crystal engineering with weak interactions is not the main aim of the structural chemist.

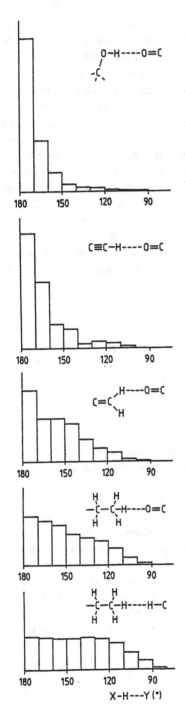

Figure 3. CSD results. Histograms with angular frequencies of X-H···O=C contacts for different donor types, and of C-H···H-C van der Waals contacts (from top to bottom: hydroxyl, ethynyl, vinyl and ethyl donors, van der Waals contacts). The distributions are cone-corrected (weighted by 1/sinθ) and scaled in such a way that they cover the same areas.

In view of the recent debate on the distinction between weak hydrogen bond and a van der Waals interaction [19], it is instructive to consider the angular distributions θ for a variety of X-H···A geometries. When angular data on C-H···O, O-H···O and C-H···H geometries were extracted and cone corrected, the results (Fig. 3) show some interesting features [20]. The histogram for O-H···O hydrogen bonds shows the well-known directional behaviour [mean θ = 154.0(4)°]. The C-H···O geometries involving the acidic alkynyl donors, C≡C-H···O are not very different with the mean θ being 152(2)° and the distribution only slightly broader. For vinyl donors, C=C-H···O, the mean angle θ falls to 143(1)° and the angular distribution widens considerably. For the very weakly polarised CH_3 donor of the ethyl group, the mean angle θ falls further to 137.1(7)° and the angular distribution is correspondingly softened but it still shows directional behaviour. However, the angular distribution for C-H···H-C contacts of methyl groups is almost ideally isotropic in the range 120-180°. This is exactly what is expected for the non-directional van der Waals interaction.

Since C-H···O interactions of the weakest carbon acids are still not isotropic, they should not be classified as mere van der Waals contacts. The observed differences between *any* kind of C-H···O hydrogen bond and the van der Waals interaction is a consequence of the fundamentally different distance and angle fall-off characteristics of these interactions. In summary, a C-H···O hydrogen bond does not become a van der Waals interaction just because the H···O separation exceeds some (arbitrary) threshold.

4. Development of General Supramolecular Synthetic Strategies

Much depends today on individual styles and preferences. The study of a family of compounds provides insight into packing characteristics that may be exploited in the design of new and related structures. The identification of a crystal as a retrosynthetic target follows from the depiction of a crystal structure as a network with the molecules as the nodes and the supramolecular synthons as the node connectors [1,15]. In this context, the organic diamondoid network solids are of relevance. Our interest in the study of such solids originated from the molecular-supramolecular equivalence in the tetraphenylmethane family of structures. It has been shown by us earlier [21] that tetraphenylmethane, 6 and CBr_4 form a 1:1 molecular complex, 7 that is nearly isostructural with tetrakis-(4-bromophenyl)methane, 8 (both *I*). The structural equivalence between 7 and 8 is understood as a switching of molecular and supramolecular synthons. In 7, the Br···phenyl interaction is important while in 8, four Br-atoms form a tetrahedral cluster mediated by Br···Br interactions. In the case of 8 the structure may be described as a triply interpenetrated distorted diamondoid network while there is no interpenetration in 7.

328

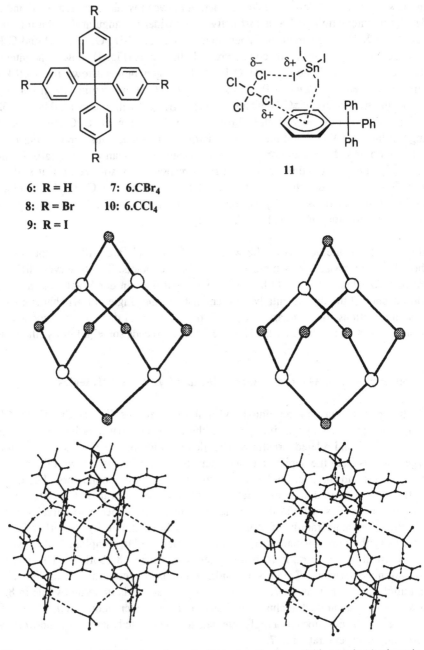

6: R = H   7: 6.CBr₄

8: R = Br   10: 6.CCl₄

9: R = I

Figure 4. Stereoview of the structure of complex **10** to show the diamondoid network: (a) schematic diagram with molecules reduced to spheres; (b) actual structure.

This strategy for producing a diamondoid network has also been extended to the iodo and chloro derivatives. Thus tetrakis-(4-iodophenyl)methane, **9** is isostructural with **8** [22] but crystallisation of **6** with CI_4 did not yield the expected 1:1 complex, possibly in part because of the instability of CI_4. However, when complexation of **6** with SnI_4 was attempted by recrystallisation from CCl_4, the 1:1 complex **10**, was unexpectedly produced. The crystal structure of complex **10** (Fig. 4) is of the diamondoid type and is isostructural to complex **7**. The packing is characterised by a long Cl···phenyl centroid contact of 3.79Å (corresponding Br···phenyl centroid distance in **7** is 3.68Å). Crystallisation of pure **6** from CCl_4, did not yield complex **10** suggesting a definite role for SnI_4 in the formation of **10**. The role of SnI_4 in promoting the inclusion of CCl_4 is not clear but it could be involved in the mutual polarisation of molecules as shown in **11**. What is clear is that CCl_4 forms a complex with tetraphenylmethane with more difficulty than does CBr_4. It is also lost more easily; complex **7** is stable under ambient conditions for about a month but the CCl_4 is lost from the crystals of complex **10** in less than five minutes. All this suggests that the smaller CCl_4 molecule is held more loosely in the tetraphenylmethane cavity than is CBr_4 and that SnI_4 has a specific function in the crystallisation of complex **10**.

5. Obtaining Structures from Limited Crystallographic Information

With the tremendous increase in the range of chemical systems being studied, an all-too common occurrence is the failure to obtain diffraction quality single crystals. Clearly one would not like to abandon a research project for such a reason. Indeed it could be argued that freeing oneself from the burden of growing a single crystal of a material is almost a worthwhile goal in itself. The growth and development of methods of structure determination from X-ray diffraction patterns of polycrystalline samples is surely of importance for the future of crystal engineering [23-25]. Among the difficulties encountered commonly in the use of such methods are the overlap of non-equivalent reflections in powder patterns, the indexing of patterns of low crystal symmetry and the dependence of the pattern on inhomogeneities in microcrystal size and orientation. With regard to the latter, layered materials are especially problematic. Synchrotron X-ray and high resolution neutron diffraction sources are very useful but are not commonly available and conventional X-rays are therefore used also. Structures of purely organic compounds are difficult to solve because of the absence of strong scatterers, but a recent report of the structure solution of the drug chlorothiazide is noteworthy in that all the 17 non-hydrogen atoms are clearly visible in the E-map generated in the direct methods solution despite the non-centrosymmetric triclinic symmetry [26]. These computational methods have been shown to be extremely powerful in obtaining crystal structures of microcrystalline materials if partial crystallographic information is available.

6. Functionalised Solids

Crystal engineering targeted at solids exhibiting large optical non-linearities has attracted much interest [27,28]. The traditional design of NLO active materials has concentrated exclusively on dipolar molecules and in effect ignored the wealth of possibilities that could arise from two- and three-dimensional self-assembly. Diversified investigations of more isotropic molecules, such as 2,4,6-triamino-1,3,5-trinitrobenzene, with attached octupolar and multipolar non-linearities have been proposed [29] but these investigations have been confined to the molecular rather than extended to the supramolecular or crystalline level.

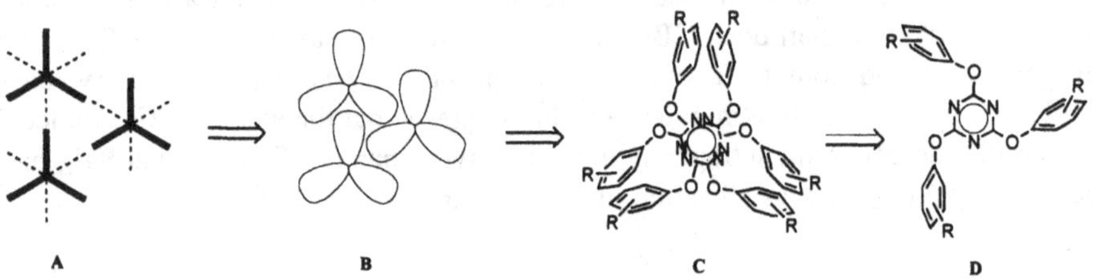

Figure 5. Retrosynthesis of a trigonal octupolar network. A. Trigonal network; B. Recognition of trigonal species; C. Stacked molecular diads of triazines; D. 2,4,6-Triaryloxy-1,3,5-triazine.

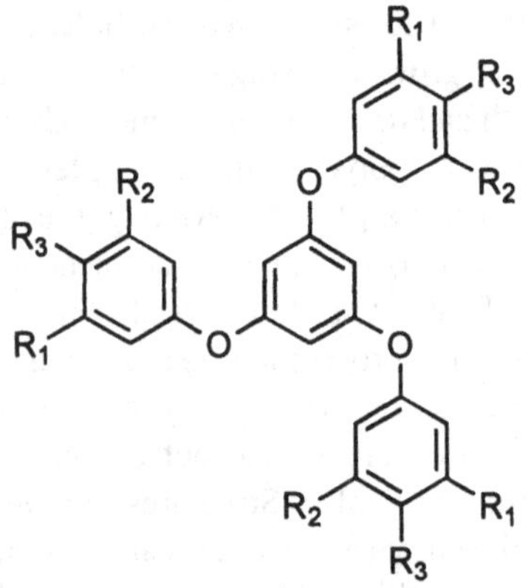

12: $R_1 = R_2 = R_3 = H$
13: $R_1 = Cl, R_2 = R_3 = H$
14: $R_1 = Br, R_2 = R_3 = H$
15: $R_1 = Me, R_2 = R_3 = H$
16: $R_1 = R_2 = Cl, R_3 = H$
17: $R_1 = R_2 = Me, R_3 = H$

A typical symmetry pattern that leads to crystalline octupolar non-linearity is the trigonal network **A** constituted with trigonal molecules (Fig. 5). Retrosynthetic analysis applied to **A** leads one to think in terms of synthons such as **B**. Crystalline triaryloxy substituted 1,3,5-triazines form stacked molecular diads called *Piedfort Units*

(PUs) and such PUs, which in themselves are supramolecular species, **C** may be derived retrosynthetically from **B**. The ultimate molecular starting point of this supramolecular synthesis is a series of substituted 2,4,6-triphenoxy-1,3,5-triazines, **D**. To summarise, the retrosynthetic analysis **A** $\Rightarrow$ **B** $\Rightarrow$ **C** $\Rightarrow$ **D** has to ensure that the three-fold molecular symmetry in **D** is faithfully transmitted at each level of the supramolecular hierarchy. Implied here is that in each level from **D** to **A**, only those combinations of interactions that are consistent with high symmetry are optimised while numerous others that would be expected to lower the crystal symmetry are suppressed. The crystal structures of triazines **12-17** show that all of them form quasi-trigonal or trigonal networks that are two-dimensionally non-centrosymmetric [30]. Molecular non-linearities have been measured by Harmonic Light Scattering experiments. In particular, triazine **12** adopts a non-centrosymmetric crystal structure. It also has a measurable SHG powder signal of around 0.10 x urea. Such octupolar uniaxial structures are of significance from the viewpoint of crystal optics and are seen to optimise phase matching requirements. Their structurally built-in polarisation independence is a major asset in contrast to the situation in the more traditional one-dimensional dipolar structures [31].

7. Understanding Polymorphic Structures

Originally an enigma, then a curiosity, polymorphism [32] today may be central to our understanding of crystallisation mechanisms. Crystallisation is inherently a very efficient process and Nature seems to have developed ways and means to eliminate alternative pathways. Given the complementary nature of molecular recognition, it would appear reasonable that the events during crystallisation should be quite specific to the molecule in question and that the possibility of obtaining two or more forms (especially in the same crystallisation batch) is unlikely. Many simple compounds, say naphthalene, benzoic acid or 1,4-benzoquinone, occur only in a single crystalline form under ambient pressure conditions, despite the fact that these substances have been studied for decades, giving lie to McCrone's oft-quoted dictum that the number polymorphs for a given organic compound is proportional to the time and money spent in looking for them [33]. At the same time, one notes that polymorphic forms are obtained with ease for say, perylene, terephthalic acid or (4'-methoxyphenyl)-1,4-benzoquinone. Clearly, polymorphism is a complex phenomenon. Its occurrence is more frequent in compounds that are conformationally flexible and also contain groups that are able to form strong hydrogen bonds, like -OH, -NH$_2$, -CO$_2$H, and -CONH$_2$. The fact that these groups are commonly found in drug molecules makes this phenomenon of outstanding significance in pharmaceutical development [34].

18   19

19a   19b   19c

Polymorphism may also be expected when the molecular structure contains multiple occurrences of the same functionality in non-equivalent sites. Take pyrazine carboxamide, **18**. With its four or so polymorphic modifications, it has often been invoked to justify the argument that polymorphism is an all-pervasive and capricious phenomenon, playing Devil's advocate as it were to the entire subject of crystal engineering. Yet an inspection of the molecule shows why polymorphism could be expected or even inevitable. So, McCrone is correct in some cases! An explanation in terms of supramolecular synthons is appropriate. The C-H··N dimer synthon, **19** which is an important motif can be formed in many distinct ways, three of which (**19a, b** and **c**) are actually found in the observed polymorphs, as the dimer or as the open helical variants which are virtually isoenergetic. Recent computational approaches to polymorph prediction including modelling of solvent effects are very promising. With Polymorph Predictor [35], the problem of predicting stable polymorphic modifications in any given space group may be considered essentially solved for molecules with less than three or so torsional degrees of freedom and with a single molecule in the asymmetric unit [36].

20   21   22

In general, it might be said that the occurrence of polymorphic forms under widely different crystallisation conditions need not be surprising. However, when the phenomenon occurs under similar conditions, or more exceptionally, in the same crystallisation batch, it is worthy of a more detailed study [37]. The extent of differences between polymorphs may themselves vary. While there are polymorphs in which the same synthons occur in slightly different ways, as say the pyrazine carboxamide **18**, there are others which contain completely different supramolecular synthons. A case in point is furnished by 4,4-diphenyl cyclohexadienone, **20**. This material was investigated in the expectation of obtaining simple benzoquinone-type C-H···O motifs. Reality proves to be far more complex and two polymorphic modifications were obtained from EtOAc-hexane. The structure of the monoclinic form ($P2_1$, Z = 2) is shown in Fig. 6 and consists of a linear C-H···O pattern. Curiously, however, the acidic quinonoid C-H groups are not involved in the C-H···O network.

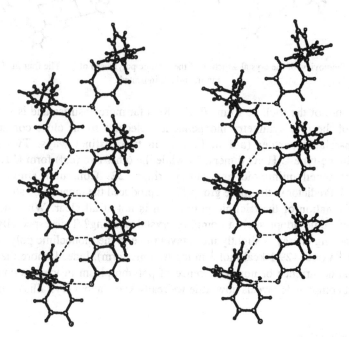

Figure 6. Stereoview of the crystal structure of the monoclinic polymorph of **20**. Notice the C-H···O hydrogen bonded chains.

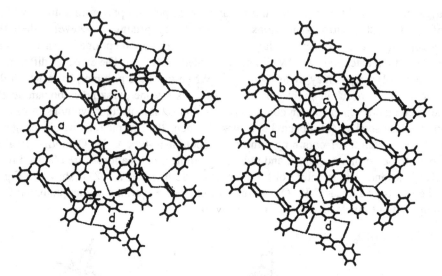

Figure 7. Stereoview of the crystal structure of the triclinic polymorph of **20**. The four inversion-related motifs a-d are indicated.

The structure of the triclinic form (P, $Z = 8$) is far more abstruse and is shown in Fig. 7. Each of the four symmetry independent molecules forms close contacts with its inversion-related partner (a-d in Fig. 7) in two distinct ways. Two (a,b) form benzoquinone-type C-H···O dimers, **21** while the other two (c,d) form C-H···O dimers with overlapped quinonoid and phenyl rings, **22**. Computer simulation using Polymorph Predictor in the space group $P2_1$ reproduced the monoclinic structure readily but the formation of the triclinic polymorph is not at all clear. Why should such a simple molecule adopt such a complex crystal packing? This especially when the crystal packing energy is slightly more favourable in the monoclinic polymorph (-33.5 kcal mol^{-1} versus -29.1 kcal mol^{-1} in the triclinic form). Indeed, more than any other factor, the occurrence or non-occurrence of polymorphism in various compounds or groups of compounds reveals how little we really know about crystallisation itself.

8. Crystallisation

Crystal engineering uses experimental crystal structures as its starting point. The subject as it stands today is based on the model that when patterns of interactions repeat in crystal structures, the robust synthons derived from them can be used in subsequent crystal design efforts [1,4]. One can also attempt to modify a known and stable packing arrangement to produce new and also presumably stable arrangements using well-accepted chemical arguments and strategies. All of this, however, rests on the premise that the experimental crystal structure is thermodynamically favoured or that if it is kinetically favoured, the kinetic effects are comparable in the given family under study. Generally, questions relating to the events during crystallisation are avoided or ignored. The reason for this is simple. Crystallisation is far too complex a phenomenon for proper experimental or theoretical study at the present time. The fate of molecules in solution up to, during and after nucleation is largely unknown and is therefore subject to speculation. However, one already senses that it is probably only through an understanding of the crystallisation process that many of the difficult issues in crystal engineering can be appreciated, even comprehended.

23

Advances in instrumentation may create opportunities to study the complex phenomenon of crystallisation. The crystal structure of quinoxaline, **23** (Fig. 8a) is intriguing. At the outset it should be stated that the material has a low melting point of 30 °C and that the data were collected with a CCD at 120 K. The structure is in the space group $P2_12_12_1$ and there are five molecules in the asymmetric unit [37]. Inspection of Fig. 8a shows a C-H⋯N catemer and a general packing arrangement that are in themselves unsurprising. However, the environments around each of each of the five symmetry independent molecules are nearly the same. It is not at all difficult to conceive of a simpler structure in the same space group but with just one symmetry-independent molecule. Such a structure was indeed simulated using the Polymorph Predictor and is shown in Fig. 8b. The experimental ($Z' = 5$) and simulated ($Z' = 1$) structures have calculated X-ray powder patterns that are virtually identical.

In this light, what if any is the significance of the observed crystal structure of **23**? The occurrence of five symmetry independent molecules in the asymmetric unit is very unusual and the CSD contains only 17 other such cases. Molecular Dynamics [35] studies on the experimental crystal structure of **23** show that the preferred displacements about the atomic positions are such that they tend to remove the minor differences between the orientations of the five symmetry-independent molecules. All this hints that the experimental structure is kinetically locked in because of the method of crystal growth and that the system represents a case of arrested crystallisation. It is a

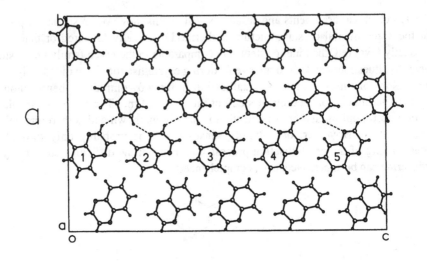

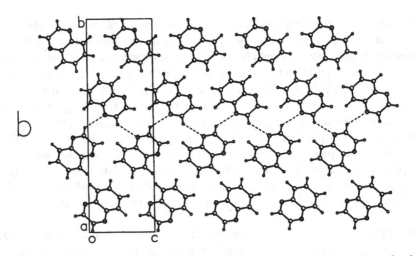

Figure 8. Crystal structure of quinoxaline, **23** showing the unit cell and the C-H···N hydrogen bonded catemer: (a) Experimental structure with Z' = 5. The symmetry independent molecules are marked; (b) Structure determined with the Polymorph Predictor and having Z' = 1. Notice the similarity of this structure to the experimental one.

matter of debate if such a crystal will transform to the simpler structure with $Z' = 1$ when suitably annealed. Obtaining information regarding the behaviour of molecules during crystallisation is as difficult as studying the transition state of a reaction [38] and experimental approaches to this end will depend in part on fortuitous crystallographic observations such as the one described above. For example, is it possible to suggest that during the crystallisation of the $Z' = 1$ structure, the C-H···N catemer synthons are formed initially while the precise adjustments of the molecular tilting come later?

9. Conclusions and Future Objectives

The design of organic solids with potential applications as new and useful materials is an evolving subject. However, this new subject of crystal engineering contains features additional to those characteristic of molecule-based organic chemistry. Ability to plan and design crystal structures will depend very largely on thinking supramolecularly and in viewing a crystal structure as a blend of a very large number of interactions of various types and strengths. In this complex mélange, one must try to identify interactions or sets of interactions, that is supramolecular synthons, that will carry through across an entire family of structures. In this way, and only under such a condition, one may return to the tenets of organic chemistry. This is operationally convenient because crystals are made from molecules and molecules can only be made using the methods and techniques of organic chemistry. The need to minimise interaction interference is therefore paramount and this introduces modularity, that is orthogonality into the crystal design process with all its implications for three-dimensional structure control. The control of polymorphism is not a distinct endeavour from the control of crystal packing in that the need to understand the nature of weak interactions is acutely felt in both activities. The challenging nature of many of these tasks should ensure fairly lively activity in this area during the coming years.

10. Acknowledgements

Financial assistance from the Department of Science and Technology, Government of India and collaborations with Molecular Simulations Inc. (Cambridge and San Diego) are gratefully acknowledged. I would like to thank Dr. Ashwini Nangia for helpful discussions and for his assistance in the preparation of this chapter.

338

11. References

1. Desiraju, G.R. (1997) Designer crystals: intermolecular interactions, network structures and supramolecular synthons, *Chem. Commun.* 1475-1482.
2. Desiraju, G.R. (1997) Crystal engineering: solid state supramolecular synthesis, *Curr. Opin. Solid State Mater. Sci.* **2** 451-454.
3. Losier, P. and Zaworotko, M.J. (1996) A noninterpenetrated molecular ladder with hydrophobic cavities, *Angew. Chem. Int. Ed. Engl.* **35** 2779-2782.
4. Allen, F.H., Kennard, O. and Taylor, R. (1983) Systematic analysis of structural data as a research technique in organic chemistry, *Acc. Chem. Res.* **16** 146-153.
5. Desiraju, G.R. (1997) Crystal gazing: structure prediction and polymorphism, *Science* **278** 404-405.
6. Ginsburg, D. (1976) in G.M.J. Schmidt (Ed.), *Solid State Photochemistry*, Verlag Chemie, Weinheim.
7. Dunitz, J.D. (1996) Thoughts on crystals as supermolecules, in G.R. Desiraju (ed.), *The Crystal as a Supramolecular Entity*, Wiley, Chichester, pp. 1-30.
8. Lehn, J.-M. (1994) Perspectives in supramolecular chemistry – From the lock-and-key image to the information paradigm, in J.-P. Behr (ed.), *The Lock-and-Key Principle. The State of the Art – 100 Years on*, Wiley, Chichester, pp. 307-317.
9. Pauling, L. and Delbrück, M. (1940) Nature of the intermolecular forces operative in biological processes, *Science* **92** 77-79.
10. Desiraju, G.R. (1989) *Crystal Engineering. The Design of Organic Solids*, Elsevier, Amsterdam.
11. Gavezzotti, A. (1994) Are crystal structures predictable? *Acc. Chem. Res.* **27** 309-314.
12. Ermer, O. and Eling, A. (1994) Molecular recognition among alcohols and amines: super-tetrahedral crystal structures of linear diphenol-diamine complexes and aminophenols, *J. Chem. Soc., Perkin Trans.* 2 925-944.
13. Hanessian, S., Simard, M. and Roelens, S. (1995) Molecular recognition and self-assembly by non-amidic hydrogen bonding: an exceptional assembler of neutral and charged supramolecular structures, *J. Am. Chem. Soc.* **117** 7630-7645.
14. Allen, F.H., Hoy, V.J., Howard, J.A.K., Thalladi, V.R., Desiraju, G.R., Wilson, C.C. and McIntyre G.J. (1997) Crystal engineering and correspondence between molecular and crystal structures. Are 2- and 3-aminophenols anomalous? *J. Am. Chem. Soc.* **119** 3477-3480.
15. Desiraju, G.R. (1995) Supramolecular synthons in crystal engineering — A new organic synthesis, *Angew. Chem. Int. Ed. Engl.* **34** 2311-2327.
16. Corey, E.J., (1988) Retrosynthetic thinking — Essentials and examples, *Chem. Soc. Rev.* **17** 111-133.
17. Desiraju, G.R. (1996) The C-H...O hydrogen bond: structural implications and supramolecular design, *Acc. Chem. Res.* **29** 441-449.
18. Steiner, T. (1996) C-H···O hydrogen bonding in crystals, *Cryst. Rev.* **6** 1-57.
19. Cotton, F.A., Daniels, L.M., Jordan, G.T. and Murillo, C.A (1997) The crystal packing of bis(2,2'-dipyridylamiso)cobalt(II), $Co(dpa)_2$, is stabilised by C-H···N bonds: are there any real precedents? *Chem. Commun.* 1673-1674.
20. Steiner, T. and Desiraju, G.R. (1998) Distinction between the weak hydrogen bond and the van der Waals interaction, *Chem. Commun.* 891-892.

21. Reddy, D.S., Craig, D.C. and Desiraju, G.R. (1996) Supramolecular synthons in crystal engineering. 4. Structure simplification and synthon interchangeability in some organic diamondoid solids, *J. Am. Chem. Soc.* **118** 4090-4093.

22. Thaimattam, R., Reddy, D. S., Xue, F., Mak, T. C. W., Nangia, A. and Desiraju, G. R. (1998) Molecular networks in the crystal structures of tetrakis-(4-iodophenyl)methane and (4-iodophenyl) triphenylmethane, *New J. Chem.*, 143-148.

23. Poojary, D.M., and Clearfield, A. (1997) Application of X-ray powder diffraction techniques to the solution of unknown crystal structures, *Acc. Chem. Res.* **30** 414-422.

24. Gavezzotti, A. and Filippini, G. (1996) Computer prediction of organic crystal structures using partial X-ray diffraction data, *J. Am. Chem. Soc.* **118** 7153-7157.

25. Shankland, K., David, W.I.F. and Csoka, T. (1997) Crystal structure determination from powder diffraction data by the application of a genetic algorithm, *Z. Krist.* **212** 550-552.

26. Shankland, K., David, W.I.F. and Sivia, D.S. (1997) Routine ab initio structure determination of chlorothiazide by X-ray powder diffraction using optimised data collection and analysis strategies, *J. Mater. Chem.* **7** 569-572.

27. Chemla, D.S. and Zyss, J. (eds.) (1987) *Nonlinear Optical Properties of Organic molecules and Crystals*, Academic, Boston.

28. Zyss, J. (ed.) (1994) *Molecular Nonlinear Optics: Materials, Physics and Devices*, Academic, Boston.

29. Zyss, J. and Ledoux, I. (1994) Nonlinear optics in multipolar media: Theory and experiment, *Chem. Rev.* **94** 77-105.

30. Thalladi, V.R., Brasselet, S., Weiss, H.-C., Bläser, D., Katz, A.K., Carrell, H.L., Boese, R., Zyss, J., Nangia, A. and Desiraju, G.R. (1998) Crystal engineering of some 2,4,6-triaryloxy-1,3,5,-triazines: Octupolar non-linear materials, *J. Am. Chem. Soc.* **120**, 2563-2577.

31. Zyss, J., Brasselet, S., Thalladi, V.R. and Desiraju, G.R. (1998) Octupolar versus dipolar crystalline structures for nonlinear optics: A dual crystalline and propogative engineering approach, *J. Chem. Phys.* **109**, 658-669.

32. Threlfall, T.L. (1995) Analysis of organic polymorphs — A review, *Analyst* **120** 2435-2460.

33. McCrone, W.C. (1965) Polymorphism, in D. Fox, M.M. Labes, A. Weissberger (eds.), *Physics and Chemistry of the Organic Solid State*, Vol. 2, Interscience, New York, p. 725.

34. DeCamp, W.H. (1996) Regulatory considerations in crystallisation processes for bulk pharmaceutical chemicals – A reviewer's perspective, in A.S. Myerson, D.A. Green and P. Meenan (eds.), *Crystal Growth of Organic Materials*, ACS Proceedings Series, American Chemical Society, Washington, DC, pp. 66-71.

35. Cerius[2] suite of modules from Molecular Simulations, 9685 Scranton Road, San Diego, CA 92121-3752,USA and 240/250 The Quorum, Barnwell Road, Cambridge CB5 8RE UK.

36. Leusen, F.J.J. (1996) Ab initio prediction of polymorphs, *J. Cryst. Growth* **166** 900-903.

37. Anthony, A., Desiraju, G.R., Jetti, R.K.R., Kuduva, S.S., Madhavi, N.N.L., Nangia, A., Thaimattam, R. and Thalladi, V.R. (1998) Crystal engineering: Some further strategies, *Materials Science Bulletin,* **SS**, 1-18.

38. Seebach, D. (1988) Structure and reactvity of lithium enolates. From pinacolones to selective C-alkylations of peptides. Difficulties and opportunities afforded by complex strucfures, *Angew. Chem. Int. Ed. Engl.* **27** 1624-1654.

A

I